TEACHER'S PLANNING GUIDE

Project-Based Inquiry Science™

LIVING TOGETHER

IT's ABOUT TIME®

HERFF JONES EDUCATION DIVISION

D1430575

IT's ABOUT TIME®
HERFF JONES EDUCATION DIVISION

84 Business Park Drive, Armonk, NY 10504
Phone (914) 273-2233 Fax (914) 273-2227
www.its-about-time.com

Program Components

Student Edition	Durable Equipment Kit
Teacher's Planning Guide	Consumable Equipment Kit
Teacher's Resources Guide	Multimedia

— Far Mill River Water Sampling Investigations DVD
— NetLogo Software CD
— Macroinvertebrate Classification Card Set
— Food Web Card Set

Printed and bound in the United States of America.

ISBN-13: 978-1-58591-608-5
1 2 3 4 5 VH 12 11 10 09 08

This project was supported, in part, by the **National Science Foundation** under grant nos. 0137807, 0527341, 0639978.
Opinions expressed are those of the authors and not necessarily those of the National Science Foundation.

PBIS Principal Investigators

Janet L. Kolodner is a Regents' Professor in the School of Interactive Computing in the Georgia Institute of Technology's College of Computing. Since 1978, her research has focused on learning from experience, both in computers and in people. She pioneered the Artificial Intelligence method called *case-based reasoning*, providing a way for computers to solve new problems based on their past experiences. Her book, *Case-Based Reasoning*, synthesizes work across the case-based reasoning research community from its inception to 1993.

Since 1994, Dr. Kolodner has focused on the applications and implications of case-based reasoning for education. In her approach to science education, called Learning by Design™ (LBD), students learn science while pursuing design challenges. Dr. Kolodner has investigated how to create a culture of collaboration and rigorous science talk in classrooms, how to use a project challenge to promote focus on science content, and how students learn and develop when classrooms function as learning communities. Currently, Dr. Kolodner is investigating how to help young people come to think of themselves as scientific reasoners. Dr. Kolodner's research results have been widely published, including in *Cognitive Science, Design Studies,* and the *Journal of the Learning Sciences.*

Dr. Kolodner was founding Director of Georgia Tech's EduTech Institute, served as coordinator of Georgia Tech's Cognitive Science program for many years, and is founding Editor in Chief of the *Journal of the Learning Sciences.* She is a founder of the International Society for the Learning Sciences, and she served as its first Executive Officer. She is a fellow of the American Association of Artificial Intelligence.

Joseph S. Krajcik is a Professor of Science Education and Associate Dean for Research in the School of Education at the University of Michigan. He works with teachers in science classrooms to bring about sustained change by creating classroom environments in which students find solutions to important intellectual questions that subsume essential curriculum standards and use learning technologies as productivity tools. He seeks to discover what students learn in such environments, as well as to explore and find solutions to challenges that teachers face in enacting such complex instruction.

Dr. Krajcik has authored and co-authored over 100 manuscripts and makes frequent presentations at international, national, and regional conferences that focus on his research, as well as presentations that translate research findings into classroom practice. He is a fellow of the American Association for the Advancement of Science and served as president of the National Association for Research in Science Teaching. Dr. Krajcik co-directs the Center for Highly Interactive Classrooms, Curriculum and Computing in Education at the University of Michigan and is a co-principal investigator in the Center for Curriculum Materials in Science and The National Center for Learning and Teaching Nanoscale Science and Engineering. In 2002, Dr. Krajcik was honored to receive a Guest Professorship from Beijing Normal University in Beijing, China. In winter 2005, he was the Weston Visiting Professor of Science Education at the Weizmann Institute of Science in Rehovot, Israel.

Daniel C. Edelson is Vice President for Education and Children's Programs at the National Geographic Society. Previously, he was the director of the Geographic Data in Education (GEODE) Initiative at Northwestern University, where he led the development of Planetary Forecaster and Earth Systems and Processes. Since 1992, Dr. Edelson has directed a series of projects exploring the use of technology as a catalyst for reform in science education and has led the development of a number of software environments for education. These include My World GIS, a geographic information system for inquiry-based learning, and WorldWatcher, a data visualization and analysis system for gridded geographic data. Dr. Edelson is the author of the high school environmental science text, *Investigations in Environmental Science: A Case-Based Approach to the Study of Environmental Systems*. His research has been widely published, including in the *Journal of the Learning Sciences,* the *Journal of Research on Science Teaching*, *Science Educator*, and *Science Teacher*.

Brian J. Reiser is a Professor of Learning Sciences in the School of Education and Social Policy at Northwestern University. Professor Reiser served as chair of Northwestern's Learning Sciences Ph.D. program from 1993, shortly after its inception, until 2001. His research focuses on the design and enactment of learning environments that support students' inquiry in science, including both science curriculum materials and scaffolded software tools. His research investigates the design of learning environments that scaffold scientific practices, including investigation, argumentation, and explanation; design principles for technology-infused curricula that engage students in inquiry projects; and the teaching practices that support student inquiry. Professor Reiser also directed BGuILE (Biology Guided Inquiry Learning Environments) to develop software tools for supporting middle school and high school students in analyzing data and constructing explanations with biological data. Reiser is a co-principal investigator in the NSF Center for Curriculum Materials in Science. He served as a member of the NRC panel authoring the report Taking Science to School.

Mary L. Starr is a Research Specialist in Science Education in the School of Education at the University of Michigan. She collaborates with teachers and students in elementary and middle school science classrooms around the United States who are implementing *Project-Based Inquiry Science*. Before joining the PBIS team, Dr. Starr created professional learning experiences in science, math, and technology, designed to assist teachers in successfully changing their classroom practices to promote student learning from coherent inquiry experiences. She has developed instructional materials in several STEM areas, including nanoscale science education, has presented at national and regional teacher education and educational research meetings, and has served in a leadership role in the Michigan Science Education Leadership Association. Dr. Starr has authored articles and book chapters, and has worked to improve elementary science teacher preparation through teaching science courses for pre-service teachers and acting as a consultant in elementary science teacher preparation. As part of the PBIS team, Dr. Starr has played a lead role in making units cohere as a curriculum, in developing the framework for PBIS Teachers Planning Guides, and in developing teacher professional development experiences and materials.

Acknowledgements

Three research teams contributed to the development of *Project-Based Inquiry Science* (PBIS): a team at the Georgia Institute of Technology headed by Janet L. Kolodner, a team at Northwestern University headed by Daniel Edelson and Brian Reiser, and a team at the University of Michigan headed by Joseph Krajcik and Ron Marx. Each of the PBIS units was originally developed by one of these teams and then later revised and edited to be a part of the full three-year middle-school curriculum that became PBIS.

PBIS has its roots in two educational approaches, Project-Based Science and Learning by Design™. Project-Based Science suggests that students should learn science through engaging in the same kinds of inquiry practices scientists use, in the context of scientific problems relevant to their lives and using tools authentic to science. Project-Based Science was originally conceived in the hi-ce Center at the University of Michigan, with funding from the National Science Foundation. Learning by Design™ derives from Problem-Based Learning and suggests sequencing, social practices, and reflective activities for promoting learning. It engages students in design practices, including the use of iteration and deliberate reflection. LBD was conceived at the Georgia Institute of Technology, with funding from the National Science Foundation, DARPA, and the McDonnell Foundation.

The development of the integrated PBIS curriculum was supported by the National Science Foundation under grants no. 0137807, 0527341, and 0639978. Any opinions, findings and conclusions, or recommendations expressed in this material are those of the authors and do not necessarily reflect the views of the National Science Foundation.

PBIS Team

Principal Investigator
Janet L. Kolodner

Co-Principal Investigators
Daniel C. Edelson
Joseph S. Krajcik
Brian J. Reiser

NSF Program Officer
Gerhard Salinger

Curriculum Developers
Michael T. Ryan
Mary L. Starr

Teacher's Planning Guide Developers
Rebecca M. Schneider
Mary L. Starr

Literacy Specialist
LeeAnn M. Sutherland

NSF Program Reviewer
Arthur Eisenkraft

Project Coordinator
Juliana Lancaster

External Evaluators
The Learning Partnership
Steven M. McGee
Jennifer Witers

Field-Test Teachers

National Field Test
Tamica Andrew
Leslie Baker
Jeanne Bayer
Gretchen Bryant
Boris Consuegra
Daun D'Aversa
Candi DiMauro
Kristie L. Divinski
Donna M. Dowd
Jason Fiorito
Lara Fish
Christine Gleason
Christine Hallerman
Terri L. Hart-Parker
Jennifer Hunn
Rhonda K. Hunter
Jessica Jones
Dawn Kuppersmith
Anthony F. Lawrence
Ann Novak
Rise Orsini
Tracy E. Parham
Cheryl Sgro-Ellis
Debra Tenenbaum
Sarah B. Topper
Becky Watts
Debra A. Williams
Ingrid M. Woolfolk
Ping-Jade Yang

New York City Field Test
*Several sequences of PBIS
units have been field-tested
in New York City under the
leadership of Whitney Lukens,
Staff Developer for Region 9,
and Greg Borman, Science
Instructional Specialist,
New York City Department of
Education*

6th Grade
Norman Agard
Tazinmudin Ali
Heather
   Guthartz Aniba
Asher Arzonane
Asli Aydin
Shareese Blakely
John J. Blaylock
Joshua Blum
Tsedey Bogale

Filomena Borrero
Zachary Brachio
Thelma Brown
Alicia Browne-Jones
Scott Bullis
Maximo Cabral
Lionel Callender
Matthew Carpenter
Ana Maria Castro
Diane Castro
Anne Chan
Ligia Chiorean
Boris Consuegra
Careen Halton Cooper
Cinnamon Czarnecki
Kristin Decker
Nancy Dejean
Gina DiCicco
Donna Dowd
Lizanne Espina
Joan Ferrato
Matt Finnerty
Jacqueline Flicker
Helen Fludd
Leigh Summers Frey
Helene Friedman-Hager
Diana Gering
Matthew Giles
Lucy Gill
Steven Gladden
Greg Grambo
Carrie Grodin-Vehling
Stephan Joanides
Kathryn Kadei
Paraskevi Karangunis
Cynthia Kerns
Martine Lalanne
Erin Lalor
Jennifer Lerman
Sara Lugert
Whitney Lukens
Dana Martorella
Christine Mazurek
Janine McGeown
Chevelle McKeever
Kevin Meyer
Jennifer Miller
Nicholas Miller
Diana Neligan
Caitlin Van Ness
Marlyn Orque
Eloisa Gelo Ortiz
Gina Papadopoulos
Tim Perez
Albertha Petrochilos
Christopher Poli

Kristina Rodriguez
Nadiesta Sanchez
Annette Schavez
Hilary Sedgwitch
Elissa Seto
Laura Shectman
Audrey Shmuel
Katherine Silva
Ragini Singhal
C. Nicole Smith
Gitangali Sohit
Justin Stein
Thomas Tapia
Eilish Walsh-Lennon
Lisa Wong
Brian Yanek
Cesar Yarleque
David Zaretsky
Colleen Zarinsky

7th Grade
Mayra Amaro
Emmanuel Anastasiou
Cheryl Barnhill
Bryce Cahn
Ligia Chiorean
Ben Colella
Boris Consuegra
Careen Halton Cooper
Elizabeth Derse
Urmilla Dhanraj
Gina DiCicco
Lydia Doubleday
Lizanne Espina
Matt Finnerty
Steven Gladden
Stephanie Goldberg
Nicholas Graham
Robert Hunter
Charlene Joseph
Ketlynne Joseph
Kimberly Kavazanjian
Christine Kennedy
Bakwah Kotung
Lisa Kraker
Anthony Lett
Herb Lippe
Jennifer Lopez
Jill Mastromarino
Kerry McKie
Christie Morgado
Patrick O'Connor
Agnes Ochiagha
Tim Perez
Nadia Piltser

Chris Poli
Carmelo Ruiz
Kim Sanders
Leslie Schiavone
Ileana Solla
Jacqueline Taylor
Purvi Vora
Ester Wiltz
Carla Yuille
Marcy Sexauer Zacchea
Lidan Zhou

8th Grade
Emmanuel Anastasio
Jennifer Applebaum
Marsha Armstrong
Jenine Barunas
Vito Cipolla
Kathy Critharis
Patrecia Davis
Alison Earle
Lizanne Espina
Matt Finnerty
Ursula Fokine
Kirsis Genao
Steven Gladden
Stephanie Goldberg
Peter Gooding
Matthew Herschfeld
Mike Horowitz
Charlene Jenkins
Ruben Jimenez
Ketlynne Joseph
Kimberly Kavazanjian
Lisa Kraker
Dora Kravitz
Anthony Lett
Emilie Lubis
George McCarthy
David Mckinney
Michael McMahon
Paul Melhado
Jen Miller
Christie Morgado
Ms. Oporto
Maria Jenny Pineda
Anastasia Plaunova
Carmelo Ruiz
Riza Sanchez
Kim Sanders
Maureen Stefanides
Dave Thompson
Matthew Ulmann
Maria Verosa
Tony Yaskulski

Living Together

Living Together was developed by the PBIS development team based on two previously-developed units: *Water Quality*, developed at University of Michigan, and *Can a Bug Save a Farm?* developed at University of Illinois, Urbana-Champaign. *Water Quality* was developed as part of the work of the Center for Learning Technologies in Urban Schools and as a joint project of University of Michigan's Center for Highly Interactive Computing in Education and the Detroit Public Schools Urban Systemic Initiative. *Can a Bug Save a Farm?* was developed as part of the PBIS project.

Living Together

Lead Developers:
Michael T. Ryan
Mary L. Starr

Other Developers:
Francesca Casella

Contributing field-test teachers
Asher Arzonane
Matthew Carpenter
Anne Chan
Lizanne Espina
Enrique Garcia
Steven Gladden
Dani Horowitz
Stephan Joanides
Sunny Kam
Crystal Marsh
Tim Perez
Christopher Poli
Nadiesta Sanchez
Caitlin Van Ness
Cesar Yarleque
Renee Zalewitz

Water Quality

Project Directors:
Joseph Krajcik
Ron Marx

Lead Developers:
Jonathon Singer
Margaret Roy

Other Developers:
Karen Amati
Steven Best
Elena S. Takaki
Valerie Talsma
Rebecca M. Schneider

Detroit Schools Liaison
Deborah Peek-Brown

Can a Bug Save a Farm?

Project Director:
Brian J. Reiser

Lead Developers:
Barbara Hug
M. Elizabeth Gonzalez

Other Developers:
Issam Abi-El-Mona
Jiehae Lee
Heidi Leuszler
Faith Sharp

Pilot teachers:
Bonnie McArthur
Lara Fish

The development of *Living Together* was supported by the National Science Foundation under grants no. 0137807, 0527341, and 0639978. The development of *Water Quality* was supported the Center for Learning Technology in Urban School funded by the National Science Foundation under grant nos. 0830 310 A605. We are indebted to teachers from the Detroit Public Schools for their feedback on the unit. The development of *Can a Bug Save a Farm?* was supported by the National Science Foundation under grant no. 0137807. We are grateful for the recommendations of Whitney Lukens of the NYC Public Schools during development of this unit. Any opinions, findings, and conclusions or recommendations expressed in this material are those of the authors and do not necessarily reflect the views of the National Science Foundation.

Living Together Teacher's Planning Guide

Learning Set 2

How Do You Determine the Quality of Water in a Community?

Water quality, plant growth, effects of nitrates and phosphates, dissolved oxygen, pH, pH scale, acidity, temperature, turbulence, independent and dependent variables, indicators, data collection, explanation, use of evidence.

Project-Based Inquiry Science

Learning Set 3

How Can Changes in Water Quality Affect the Living Things in an Ecosystem?

Ecosystems, ecology, habitat, diversity, abundance, macroinvertebrates, species, classification, dichotomous key, photosynthesis, cell respiration, food chains, food webs, populations, biomes, modeling and simulation, explanation, use of evidence.

Welcome to Project-Based Inquiry Science!

Welcome to Project-Based Inquiry Science (PBIS): A Middle-School Science Curriculum!

This year, your students will be learning the way scientists learn, exploring interesting questions and challenges, reading about what other scientists have discovered, investigating, experimenting, gathering evidence, and forming explanations. They will learn to collaborate with others to find answers and to share their learning in a variety of ways. In the process, they will come to see science in a whole new, exciting way that will motivate them throughout their educational experiences and beyond.

What is PBIS?

In project-based inquiry learning, students investigate scientific content and learn science practices in the context of attempting to address challenges in or answer questions about the world around them. Early activities introducing students to a challenge help them to generate issues that need to be investigated, making inquiry a student-driven endeavor. Students investigate as scientists would, through observations, designing and running experiments, designing, building, and running models, reading written material, and so on, as appropriate. Throughout each project, students might make use of technology and computer tools that support their efforts in observation, experimentation, modeling, analysis, and reflection. Teachers support and guide the student inquiries by framing the guiding challenge or question, presenting crucial lessons, managing the sequencing of activities, and eliciting and steering discussion and collaboration among the students. At the completion of a project, students publicly exhibit what they have learned along with their solutions to the specific challenge. Personal reflection to help students learn from the experience is embedded in student activities, as are opportunities for assessment.

The curriculum will provide three years of piloted project-based inquiry materials for middle-school science. Individual curriculum units have been defined that cover the scope of the national content and process standards for the middle-school grades. Each Unit focuses on helping students acquire qualitative understanding of targeted science principles and move toward quantitative understanding, is infused with technology, and provides a foundation in reasoning skills, science content, and science process that will ready them for more advanced science. The curriculum as a whole introduces students to a wide range of investigative approaches in science (e.g., experimentation, modeling) and is designed to help them develop scientific reasoning skills that span those investigative approaches.

Technology can be used in project-based inquiry to make available to students some of the same kinds of tools and aids used by scientists in the field. These range from pencil-and-paper tools for organized data recording, collection, and management to software tools for analysis, simulation, modeling, and other tasks. Such infusion provides a platform for providing prompts, hints, examples, and other kinds of aids to students as they are engaging in scientific reasoning. The learning technologies and tools that are integrated into the curriculum offer essential scaffolding to students as they are developing their scientific reasoning skills, and are seamlessly infused into the overall completion of project activities and investigations.

Standards-Based Development

Development of each curriculum Unit begins by identifying the specific relevant national standards to be addressed. Each Unit has been designed to cover a specific portion of the national standards. This phase of development also includes an analysis of curriculum requirements across multiple states. Our intent is to deliver a product that will provide coverage of the content deemed essential on the widest practical scope and that will be easily adaptable to the needs of teachers across the country.

Once the appropriate standards have been identified, the development team works to define specific learning goals built from those standards, and takes into account conceptions and misunderstandings common among middle-school students. An orienting design challenge or driving question for investigation is chosen that motivates achieving those learning goals, and the team then sequences activities and the presentation of specific concepts so that students can construct an accurate understanding of the subject matter.

Inquiry-Based Design

The individual curriculum Units present two types of projects: engineering-design challenges and driving-question investigations. Design-challenge Units begin by presenting students with a scenario and problem and challenging them to design a device or plan that will solve the problem. Driving-question investigations begin by presenting students with a complex question with real-world implications. Students are challenged to develop answers to the questions. The scenario and problem in the design Units and the driving question in the investigation Units are carefully selected to lead the students into investigation of specific science concepts, and the solution processes are carefully structured to require use of specific scientific reasoning skills.

Pedagogical Rationale

Research shows that individual project-based learning units promote excitement and deep learning of the targeted concepts. However, achieving deep, flexible, transferable learning of cross-disciplinary content (e.g., the notion of a model, time scale, variable, experiment) and science practice requires a learning environment that consistently, persistently, and pervasively encourages the use of such content and practices over an extended period of time. By developing project-based inquiry materials that cover the spectrum of middle-school science content in a coherent framework, we provide this extended exposure to the type of learning environment most likely to produce competent scientific thinkers who are well grounded in their understanding of both basic science concepts and the standards and practices of science in general.

Evidence of Effectiveness

There is compelling evidence showing that a project-based inquiry approach meets this goal. Working at Georgia Tech, the University of Michigan, and Northwestern University, we have developed, piloted, and/or field-tested many individual project-based units. Our evaluation evidence shows that these materials engage students well and are manageable by teachers, and that students learn both content and process skills. In every summative evaluation, student performance on post-tests improved significantly from pretest performance (Krajcik, et al., 2000; Holbrook, et al., 2001; Gray et. al. 2001). For example, in the second year in a project-based classroom in Detroit, the average student at post-test scored at about the 95th percentile of the pre-test distribution. Further, we have repeatedly documented significant gains in content knowledge relative to other inquiry-based (but not project-based) instructional methods. In one set of results, performance by a project-based class

in Atlanta doubled on the content test while the matched comparison class (with an excellent teacher) experienced only a 20% gain (significance p < .001). Other comparisons have shown more modest differences, but project-based students consistently perform better than their comparisons. Most exciting about the Atlanta results is that results from performance assessments show that, within comparable student populations, project-based students score higher on all categories of problem-solving and analysis and are more sophisticated at science practice and managing a collaborative scientific investigation. Indeed, the performance of average-ability project-based students is often statistically indistinguishable from or better than performance of comparison honors students learning in an inquiry-oriented but not project-based classroom. The Chicago group also has documented significant change in process skills in project-based classrooms. Students become more effective in constructing and critiquing scientific arguments (Sandoval, 1998) and in constructing scientific explanations using discipline-specific knowledge, such as evolutionary explanations for animal behavior (Smith & Reiser, 1998).

Researchers at Northwestern have also investigated the change in classroom practices that are elicited by project-based units. Analyses of the artifacts students produce indicate that students are engaging in ambitious learning practices, requiring weighing and synthesizing many results from complex analyses of data, and constructing scientific arguments that require synthesizing results from multiple complex analyses of data (Edelson et al, 1998; Reiser et al, 2001). Students are engaged in planning, performing, monitoring and revising their investigations, and reporting on their investigation processes as well as their results (Loh et al. 1998). In general, the classrooms engaging in project-based activities reveal substantial moves toward a scientific discourse community in which students focus on arguing from evidence, critiquing ideas, and conjecturing, rather than simply reporting on what they have read or been told (Tabak & Reiser, 1997).

Introducing PBIS

What Do Scientists Do?
1) Scientists...address big challenges and big questions.

Students will find many different kinds of *Big Challenges* and *Questions* in PBIS Units. Some ask them to think about why something is a certain way. Some ask them to think about what causes something to change. Some challenge them to design a solution to a problem. Most are about things that can and do happen in the real world.

Understand the Big Challenge or Question

As students get started with each Unit, they will do activities that help them understand the *Big Question* or *Challenge* for that Unit. They will think about what they already know that might help them, and they will identify some of the new things they will need to learn.

Project Board

The *Project Board* helps you and your students keep track of their learning. For each challenge or question, they will use a *Project Board* to keep track of what they know, what they need to learn, and what they are learning. As they learn and gather evidence, they will record that on the *Project Board*. After they have answered each small question or challenge, they will return to the *Project Board* to record how what they have learned helps them answer the *Big Question* or *Challenge*.

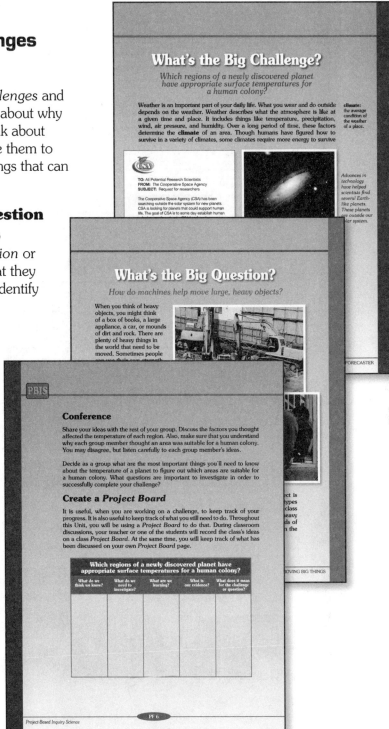

Learning Set 1

How Do Flowing Water and Land Interact in a Community?

The big question for this unit is *How does water quality affect the ecology of a community?* So far you have considered what you already know about what water quality is. Now you may be wondering where the water you use comes from. If you live in a city or town, the water you use may come from a river. You would want to know the quality of the water you are using. To do so, it is important to know how the water gets into the river. You also need to know what happens to the water as the river flows across the land.

You may have seen rivers or other water bodies near your home, your school, or in your city. Think about the river closest to where you live. Consider from where the water in the river comes. If you have traveled along the river, think about what the land around the river looks like. Try to figure out what human activities occur in the area. Speculate as to whether these activities affect the quality of water in the river.

To answer the big question, you need to break it down into smaller questions. In this *Learning Set*, you will investigate two smaller questions. As you will discover, these questions are very closely related and very hard to separate. The smaller questions are *How does water affect the land as it moves through the community?* and *How does land use affect water*

Address the Big Challenge

How Do Scientists Work Together to Solve Problems?

You began this unit with the question, *how do scientists work together to solve problems?* You did several small challenges. As you worked on those challenges you learned about how scientists solve problems. You will now watch a video about real-life designers. You will see what the people in the video are doing that is like what you have been doing. Then you will think about all the different things you have been doing during this unit. Lastly, you will write about what you have learned about doing science and being a scientist.

Watch

IDEO Video

The video you will watch follows a group of designers at IDEO. IDEO is an innovation and design firm. In the video, they face the challenge of designing and building a new kind of shopping cart. These designers are doing many of the same things that you did. They also use other practices that you did not use. As you watch the video, record the interesting things you see.

After watching the video, answer the questions on the next page. You might want to look at them before you watch the video. Answering these questions should help you answer the big question of this unit: *How do scientists work together to solve problems?*

100

Learning Sets

Each Unit is composed of a group of *Learning Sets*, one for each of the smaller questions that needs to be answered to address the *Big Question* or *Challenge*. In each *Learning Set*, students will investigate and read to find answers to the *Learning Set's* question. They will also have a chance to share the results of their investigations with their classmates and work together to make sense of what they are learning. As students come to understand answers to the questions on the *Project Board*, you will record those answers and the evidence they collected. At the end of each *Learning Set*, they will apply their knowledge to the *Big Question* or *Challenge*.

Answer the Big Question/ Address the Big Challenge

At the end of each Unit, students will put everything they have learned together to tackle the *Big Question* or *Challenge*.

2) Scientists...address smaller questions and challenges.

What Students Do in a Learning Set

Understanding the Question or Challenge

At the start of each *Learning Set*, students will usually do activities that will help them understand the *Learning Set's* question or challenge and recognize what they already know that can help them answer the question or achieve the challenge. Usually, they will visit the *Project Board* after these activities and record on it the even smaller questions that they need to investigate to answer a *Learning Set's* question.

Investigate/Explore

There are many different kinds of investigations students might do to find answers to questions. In the *Learning Sets,* they might

- design and run experiments;
- design and run simulations;
- design and build models;
- examine large sets of data.

Don't worry if your students haven't done these things before. The text will provide them with lots of help in designing their investigations and in analyzing thier data.

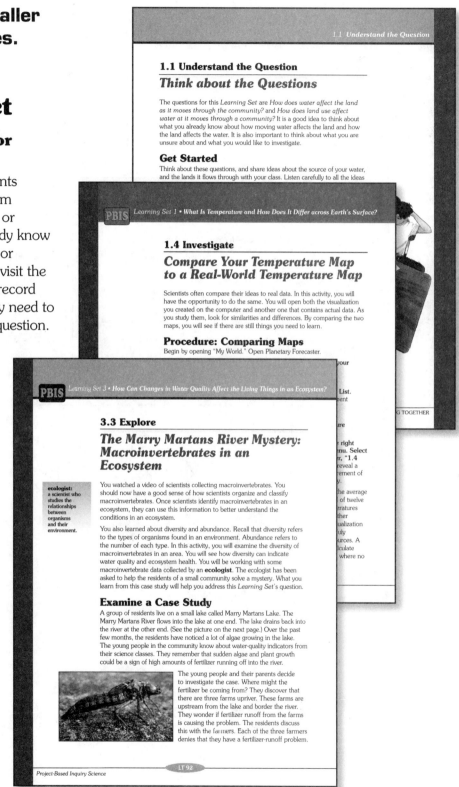

1.1 Understand the Question
1.1 Understand the Question

Think about the Questions

The questions for this *Learning Set* are *How does water affect the land as it moves through the community?* and *How does land use affect water at it moves through a community?* It is a good idea to think about what you already know about how moving water affects the land and how the land affects the water. It is also important to think about what you are unsure about and what you would like to investigate.

Get Started

Think about these questions, and share ideas about the source of your water, and the lands it flows through with your class. Listen carefully to all the ideas

PBIS *Learning Set 1 • What Is Temperature and How Does It Differ across Earth's Surface?*

1.4 Investigate

Compare Your Temperature Map to a Real-World Temperature Map

Scientists often compare their ideas to real data. In this activity, you will have the opportunity to do the same. You will open both the visualization you created on the computer and another one that contains actual data. As you study them, look for similarities and differences. By comparing the two maps, you will see if there are still things you need to learn.

Procedure: Comparing Maps

Begin by opening "My World." Open Planetary Forecaster.

PBIS *Learning Set 3 • How Can Changes in Water Quality Affect the Living Things in an Ecosystem?*

3.3 Explore

The Marry Martans River Mystery: Macroinvertebrates in an Ecosystem

ecologist:
a scientist who studies the relationships between organisms and their environment.

You watched a video of scientists collecting macroinvertebrates. You should now have a good sense of how scientists organize and classify macroinvertebrates. Once scientists identify macroinvertebrates in an ecosystem, they can use this information to better understand the conditions in an ecosystem.

You also learned about diversity and abundance. Recall that diversity refers to the types of organisms found in an environment. Abundance refers to the number of each type. In this activity, you will examine the diversity of macroinvertebrates in an area. You will see how diversity can indicate water quality and ecosystem health. You will be working with some macroinvertebrate data collected by an **ecologist**. The ecologist has been asked to help the residents of a small community solve a mystery. What you learn from this case study will help you address this *Learning Set*'s question.

Examine a Case Study

A group of residents live on a small lake called Marry Martans Lake. The Marry Martans River flows into the lake at one end. The lake drains back into the river at the other end. (See the picture on the next page.) Over the past few months, the residents have noticed a lot of algae growing in the lake. The young people in the community know about water-quality indicators from their science classes. They remember that sudden algae and plant growth could be a sign of high amounts of fertilizer running off into the river.

The young people and their parents decide to investigate the case. Where might the fertilizer be coming from? They discover that there are three farms upriver. These farms are upstream from the lake and border the river. They wonder if fertilizer runoff from the farms is causing the problem. The residents discuss this with the farmers. Each of the three farmers denies that they have a fertilizer-runoff problem.

Project-Based Inquiry Science

LT 92

The following text appears within the illustrated pages shown in the image:

5.3 Read

What is Different between Lower Elevations and Higher Elevations?

analogy:
the similarity
between things
that are different.

fluid: a
substance that
is able to flow
(takes the shape
of its container).

In the previous investigation, you noticed that the temperature decreased as elevation increased. Mountain climbers also notice this difference in temperature. It gets very cold as they reach the top of a high mountain. What is different about lower elevations and higher elevations that causes the temperature to be lower at high elevations?

The Atmosphere is an Ocean of Air

Learning Set 5 • How Does Elevation Affect Surface Temperature?

Project-Based In...

Learning Set 1 • The Book-Support Challenge

Plan Your Book-Support Design

The first time you built a book support, it was for the purpose of understanding the design challenge. You built it quickly and without a lot of planning. During this second attempt, you are aiming to design and build a book support that really works. Consider what you learned from your first attempt. You might also get ideas by thinking about other products that are similar to a book support. Consider the positives and negatives of each idea. Discuss them with your group members. This will make your design better.

Build and Test Your Design

Now you will iteratively build and test a working book support. Keep records of each **iteration**.

iteration:
a repetition
that attempts
to improve on
a process o...
product.

Iteration

When people design things, they usually call the thing a product. Often, designers do not create the best or most successful product the first...

Project-Based In...

1.2 Design

1.2 Design

A Better Book-Support Design

You have already taken some time to explore the materials you will be using. You have built and tested your first book support. You will soon build a better version of the book support. In this science class, you will have many chances to re-engineer solutions to problems or challenges. This time when you build the book support, you will need to record what you are doing. You will also need to communicate your results to others in the class. Before you start, read about the importance of recording your scientific work. You will then have ten minutes to build a working book support.

Materials
· 100 note cards
· 50 paper clips
· 50 rubber bands
· ruler

7

DIVING INTO SCIENCE

Project-Based Inquiry Science

Read

Like scientists, students will also read about the science they are investigating. They will read a little bit before they investigate, but most of the reading they do will be to help them understand what they have experienced or seen in an investigation. Each time they read, the text will include *Stop and Think* questions after the reading. These questions will help students gauge how well they understand what they have read. Usually, the class will discuss the answers to *Stop and Think* questions before going on so that everybody has a chance to make sense of the reading.

Design and Build

When the *Big Challenge* for a Unit asks them to design something, the challenge in a *Learning Set* might also ask them to design something and make it work. Often students will design a part of the thing they will design and build for the *Big Challenge*. When a *Learning Set* challenges students to design and build something, they will do several things:

- identify what questions they need to answer to be successful

- investigate to find answers to those questions

- use those answers to plan a good design solution

- build and test their design

Because designs don't always work the way one wants them to, students will usually do a design challenge more than once. Each time through, they will test their design. If their design doesn't work as well as they would like, they will determine why it is not working and identify other things they need to investigate to make it work better. Then, they will learn those things and try again.

Explain and Recommend

A big part of what scientists do is explain, or try to make sense of why things happen the way they do. An explanation describes why something is the way it is or behaves the way it does. An explanation is a statement one makes built from claims (what you think you know), evidence (from an investigation) that supports the claim, and science knowledge. As they learn, scientists get better at explaining. You will see that students get better, too, as they work through the *Learning Sets*.

A recommendation is a special kind of claim—one where you advise somebody about what to do. Students will make recommendations and support them with evidence, science knowledge, and explanations.

3.5 Explain

Create an Explanation

After scientists get results from an investigation, they try to make a claim. They base their claim on what their evidence shows. They also use what they already know to make their claim. They explain why their claim is valid. The purpose of a science explanation is to help others understand the following:

* what was learned from a set of investigations
* why the scientists reached this conclusion

Later, other scientists will use these explanations to help them explain other phenomena. The explanations will also help them predict what will happen in other situations.

You will do the same thing now. Your claim will be the trend you found in your experiment. You will use data you collected and science knowledge you have read to create a good explanation. This will help you decide whether your claim is valid. You will be reporting the results of the investigation to your classmates. With a good explanation that matches your claim, you can convince them that your claim is valid.

Because your understanding of the science of forces is not complete, you may not be able to fully explain your results. But you will use what you have read to come up with your best explanation. Scientists finding out about new things do the same thing. When they only partly understand something, it is impossible for them to form a "perfect" explanation. They do the best they can based on what they understand. As they learn more, they make their explanations better. This is what you will do now and what you will be doing throughout PBIS. You will explain your results the best you can based

---

4.3 Explain and Recommend

Explanations and Recommendations about Parachutes

As you did after your whirligig experiments, you will spend some time now explaining your results. You will also try to come up with recommendations. Remember that explanations include your claims, the evidence for your claims, and the science you know that can help you understand the claim. A recommendation is a statement about what someone should do. The best recommendations also have evidence, science, and an explanation associated with them. In the *Whirligig Challenge*, you created explanations and recommendations separately from each other. This time you will work on both at the same time.

Create and Share Your Recommendation and Explanation

Work with your group. Use the hints on the *Create Your Explanation* pages to make your first attempt at explaining your results. You'll read about parachute science later. After that, you will probably want to revise your explanations. Right now, use the science you learned during the *Whirligig Challenge* for your first attempt.

Write your recommendation. It should be about designing a slow-falling parachute. Remember that it should be written so that it will help someone else. They should be able to apply what you have learned about the effects of your variable. If you are having trouble, review the example in *Learning Set 3*.

LIVING TOGETHER

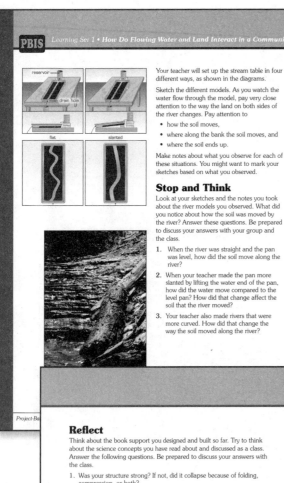

Your teacher will set up the stream table in four different ways, as shown in the diagrams.

Sketch the different models. As you watch the water flow through the model, pay very close attention to the way the land on both sides of the river changes. Pay attention to

- how the soil moves,
- where along the bank the soil moves, and
- where the soil ends up.

Make notes about what you observe for each of these situations. You might want to mark your sketches based on what you observed.

Stop and Think

Look at your sketches and the notes you took about the river models you observed. What did you notice about how the soil was moved by the river? Answer these questions. Be prepared to discuss your answers with your group and the class.

1. When the river was straight and the pan was level, how did the soil move along the river?

2. When your teacher made the pan more slanted by lifting the water end of the pan, how did the water move compared to the level pan? How did that change affect the soil that the river moved?

3. Your teacher also made rivers that were more curved. How did that change the way the soil moved along the river?

1.3 Read

Reflect

Think about the book support you designed and built so far. Try to think about the science concepts you have read about and discussed as a class. Answer the following questions. Be prepared to discuss your answers with the class.

1. Was your structure strong? If not, did it collapse because of folding, compression, or both?

2. How could you make the structure stronger to resist folding or compression?

3. Was your book support stable? That is, did it provide support so that the book did not tip over? Did it provide this support well? Draw a picture of your book support showing the center of mass of the book and the places in your book support that resist the load of your book.

4. How could you make your book support more stable?

5. How successful were the book supports that used columns in their design?

6. How could you make your book support work more effectively by including columns into the design?

7. Explain how the pull on the book could better be resisted by the use of columns in your design. Be sure to discuss both the strength and the stability of the columns in your design. You might find it easier to draw a sketch and label it to explain how the columns do this.

8. Think about some of the structures that supported the book well. What designs and building decisions were used?

You are going to get another chance to design a book support. You will use the same materials. Think about how your group could design your next book support to better meet the challenge. Consider what you now know about the science that explains how structures support objects.

---

3) Scientists...reflect in many different ways.

PBIS provides guidance to help students think about what they are doing and to recognize what they are learning. Doing this often as they are working will help students be successful student scientists.

Tools for Making Sense

Stop and Think

Stop and Think sections help students make sense of what they have been doing in the section they are working on. *Stop and Think* sections include a set of questions to help students understand what they have just read or done. Sometimes the questions will remind them of something they need to pay more attention to. Sometimes they will help students connect what they have just read to things they already know. When there is a *Stop and Think* in the text, students will work individually or with a partner to answer the questions, and then the whole class will discuss the answers.

Reflect

Reflect sections help students connect what they have just done with other things they have read or done earlier in the Unit (or in another Unit). When there is a *Reflect* in the text, students will work individually or with a partner or small group to answer the questions. Then, the whole class will discuss the answers. You may want to ask students to answer *Reflect* questions for homework.

Analyze Your Data

Whenever students have to analyze data, the text will provide hints about how to do that and what to look for.

Mess About

"Messing about" is a term that comes from design. It means exploring the materials to be used for designing or building something or examining something that works like what is to be designed. Messing about helps students discover new ideas—and it can be a lot of fun. The text will usually give them ideas about things to notice as they are messing about.

What's the Point?

At the end of each *Learning Set*, students will find a summary, called *What's the Point?*, of the important information from the *Learning Set*. These summaries can help students remember how what they did and learned is connected to the *Big Question* or *Challenge* they are working on.

PBIS | *Learning Set 3 • How Does a Planet's Tilt Affect Surface Temperatures?*

Analyze Your Data

Calculate the temperature range for each location using a table like the one shown.

	Temperature Ranges				
Location	High Temperature	Month	Low Temperature	Month	Yearly Temperature Change (high-low)
Greenland (81°N 36°W) *polar*					
Helsinki, Finland (60°N 24°E) *mid latitude north*					
Atlanta, USA (33°N 84°W) *mid latitude north*					
Quito, Ecuador (0° 78°W) *tropic*					
Darwin, Australia (14°S 131°E) *tropic*					

PBIS | *Learning Set 3 • The Whirligig Challenge*

Messing About: an exploratory activity that gives you a chance to become familiar with the materials you will be using or the function of the product you will be designing.

structure: the way the parts of an item are put together. (This is a different definition of structure than the one you saw while making your book support.)

mechanism: the way the

Mess About with the Whirligig

To help you think about how to achieve your challenge, you will begin by **messing about** with the whirligig. You will use the basic whirligig that now appears on the back of the cereal boxes.

You will get a template (pattern) of a whirligig. It will look like the one shown below. The whirligig has several parts: blades, paper clips, and a stem. If you call them by those names when you talk about the whirligig, everyone will know what you are talking about.

Cut out the template. To form the whirligig, fold the cutout template. Attach two paper clips to the stem.

blade

blades

PBIS | *Learning Set 4 • The Parachute Challenge*

What's the Point?

Through *messing about*, you became familiar with the way parachutes work. You developed a *feel* for the materials you will use later. You were also able to identify some of the variables that might affect how slowly a parachute will fall. This allowed you to do two things:

- Identify the criteria and constraints of the challenge (what you need to accomplish and the limitations).
- Identify questions you need to investigate to be able to design the best parachute.

In your class discussions around the *Project Board* you made a list of factors that would be appropriate to investigate. Different groups came up with different ideas of what affects a parachute's fall. It was only by collaborating (working together) as a class that you were able to record a full set of questions about how the parachute might work.

Project-Based Inquiry Science

4) Scientists...collaborate.

Scientists never do all their work alone. They work with other scientists (collaborate) and share their knowledge. PBIS helps students by giving them lots of opportunities for sharing their findings, ideas, and discoveries with others (the way scientists do). Students will work together in small groups to investigate, design, explain, and do other science activities. Sometimes they will work in pairs to figure out things together. They will also have lots of opportunities to share their findings with the rest of their classmates and make sense together of what they are learning.

Investigation Expo

In an *Investigation Expo*, small groups report to the class about an investigation they've done. For each *Investigation Expo*, students will make a poster detailing what they were trying to learn from their investigation, what they did, their data, and their interpretation of the data. The text gives them hints about what to present and what to look for in other groups' presentations. *Investigation Expos* are always followed by discussions about the investigations and about how to do science well. You may want to ask students to write a lab report following an investigation.

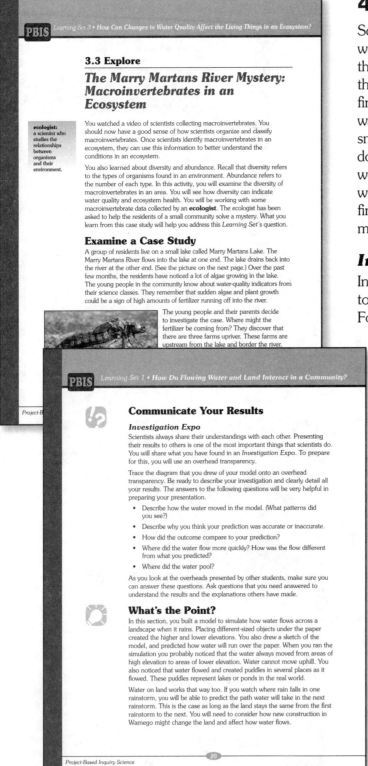

Plan Briefing/Solution Briefing/ Idea Briefing

Briefings are presentations of work in progress. They give students a chance to get advice from their classmates that can help them move forward. During a *Plan Briefing*, students present their plans to the class. They might be plans for an experiment for solving a problem or achieving a challenge. During a *Solution Briefing*, students present their solutions in progress and ask the class to help them make their solutions better. During an *Idea Briefing*, students present their ideas, including their evidence in support of their plans, solutions, or ideas. Often, they will prepare posters to help them make their presentation. Briefings are almost always followed by discussions of their investigations and how they will move forward.

Solution Showcase

Solution Showcases usually happen near the end of a Unit. During a *Solution Showcase*, students show their classmates their finished products—either their answer to a question or solution to a challenge. Students will also tell the class why they think it is a good answer or solution, what evidence and science they used to get to their solution, and what they tried along the way before getting to their answers or solutions. Sometimes a *Solution Showcase* is followed by a competition. It is almost always followed by a discussion comparing and contrasting the different answers and solutions groups have come up with. You may want to ask students to write a report or paper following a *Solution Showcase*.

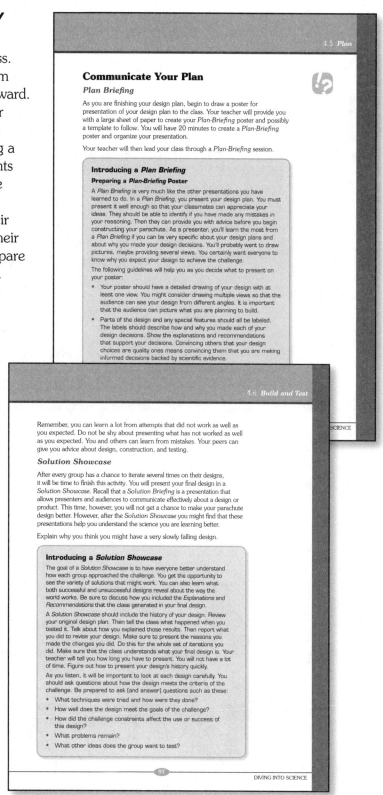

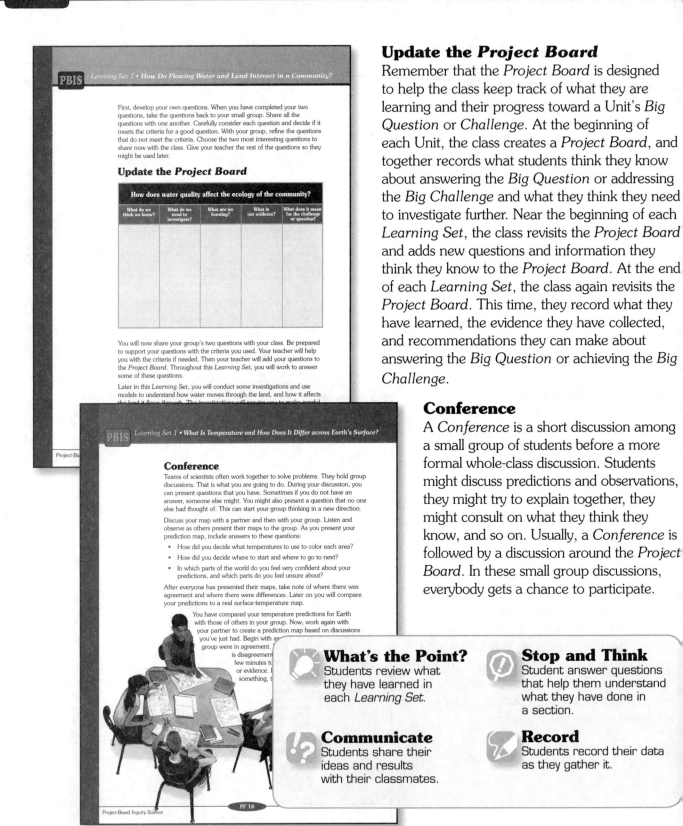

PBIS Learning Set 1 • How Do Flowing Water and Land Interact in a Community?

First, develop your own questions. When you have completed your two questions, take the questions back to your small group. Share all the questions with one another. Carefully consider each question and decide if it meets the criteria for a good question. With your group, refine the questions that do not meet the criteria. Choose the two most interesting questions to share now with the class. Give your teacher the rest of the questions so they might be used later.

Update the *Project Board*

How does water quality affect the ecology of the community?

What do we think we know?	What do we need to investigate?	What are we learning?	What is our evidence?	What does it mean for the challenge or question?

You will now share your group's two questions with your class. Be prepared to support your questions with the criteria you used. Your teacher will help you with the criteria if needed. Then your teacher will add your questions to the *Project Board*. Throughout this *Learning Set*, you will work to answer some of these questions.

Later in this *Learning Set*, you will conduct some investigations and use models to understand how water moves through the land, and how it affects the land it flows through. The investigations will require you to make careful

Project-Ba

PBIS Learning Set 1 • What Is Temperature and How Does It Differ across Earth's Surface?

Conference

Teams of scientists often work together to solve problems. They hold group discussions. That is what you are going to do. During your discussion, you can present questions that you have. Sometimes if you do not have an answer, someone else might. You might also present a question that no one else had thought of. This can start your group thinking in a new direction.

Discuss your map with a partner and then with your group. Listen and observe as others present their maps to the group. As you present your prediction map, include answers to these questions:

• How did you decide what temperatures to use to color each area?

• How did you decide where to start and where to go to next?

• In which parts of the world do you feel very confident about your predictions, and which parts do you feel unsure about?

After everyone has presented their maps, take note of where there was agreement and where there were differences. Later on you will compare your predictions to a real surface-temperature map.

You have compared your temperature predictions for Earth with those of others in your group. Now, work again with your partner to create a prediction map based on discussions you've just had. Begin with a[...] group were in agreement. [...] is disagreement[...] few minutes to[...] or evidence. [...] something, t[...]

Project-Based Inquiry Science   PF 18

Update the *Project Board*

Remember that the *Project Board* is designed to help the class keep track of what they are learning and their progress toward a Unit's *Big Question* or *Challenge*. At the beginning of each Unit, the class creates a *Project Board*, and together records what students think they know about answering the *Big Question* or addressing the *Big Challenge* and what they think they need to investigate further. Near the beginning of each *Learning Set*, the class revisits the *Project Board* and adds new questions and information they think they know to the *Project Board*. At the end of each *Learning Set*, the class again revisits the *Project Board*. This time, they record what they have learned, the evidence they have collected, and recommendations they can make about answering the *Big Question* or achieving the *Big Challenge*.

Conference

A *Conference* is a short discussion among a small group of students before a more formal whole-class discussion. Students might discuss predictions and observations, they might try to explain together, they might consult on what they think they know, and so on. Usually, a *Conference* is followed by a discussion around the *Project Board*. In these small group discussions, everybody gets a chance to participate.

What's the Point?
Students review what they have learned in each *Learning Set*.

Stop and Think
Student answer questions that help them understand what they have done in a section.

Communicate
Students share their ideas and results with their classmates.

Record
Students record their data as they gather it.

NOTES

NOTES

NOTES

NOTES

PBIS

LIVING TOGETHER

As a student scientist, you will...

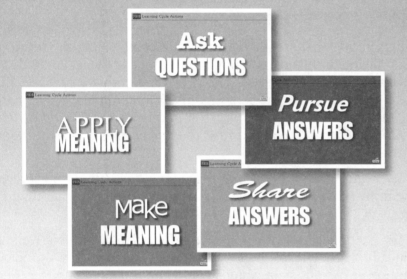

Ask QUESTIONS

Pursue ANSWERS

APPLY MEANING

Make MEANING

Share ANSWERS

UNIT OVERVIEW

In this Unit, Living Together, *students learn about the importance of water quality for all living things. Using a fictional community, students learn how everyday use of water in homes, local businesses, farming, industry, and recreation, affects the quality of water that living things depend upon. As a result, students conclude that the quality of water can change the land and affect organisms that depend on the water.*

Students engage in the practices of science and social practices of the classroom throughout the Unit. They work together as scientists, sharing their developing knowledge through presentations and Project Board *discussions.*

Content

The goal of this Unit is for students to learn about ecology and the relationships between the nonliving and living components in an ecosystem. To do this, they first focus on water, how water moves, and how different uses of water affect its quality and the quality of the lives of organisms in communities served by a river. The *Big Question* that students focus on in this Unit is: *How does water quality affect the ecology of a community?* To answer that question, students follow water through a watershed.

In *Learning Set 1* and *Learning Set 2*, students investigate the nonliving parts of an ecosystem. In *Learning Set 3*, they look more closely at living things in the ecosystem.

In *Learning Set 1*, students test how the shape of land affects the movement of water in a watershed. They learn, by investigation, that water flows predictably from higher to lower elevations in watersheds. Then they look at the variety of ways that land use affects water quality, including point and non-point sources that pollute water as they move through a community. Students build a watershed and consider what activities along the Rouge River in Michigan have affected the quality of the water in that watershed.

In *Learning Set 2*, students brainstorm and test indicators that scientists look at when determining water quality. Students learn how to test for pH, dissolved oxygen, temperature, turbidity, and fecal coliform bacteria in water. They learn that different organisms are sensitive to changes in pH.

In *Learning Set 3*, students study how changes in water quality affect living things by observing changes in actual organism populations. They practice classifying common aquatic macroinvertebrates. They use this knowledge to examine a case study that illustrates the effects of pollution on aquatic life. Students begin to look at food chains by analyzing a typical meal and tracing it back to its sources. Later, students predict how the presence or absence of

LOOKING AHEAD

Save Jar 3 and Jar 4 for use in a demonstration in Section 2.3.

28 class periods

*A class period is considered to be one 40 to 50 minute class.

certain water quality indicators affects the numbers of members in a food chain. Students learn that stable ecosystems remain in balance but that this balance can change with a sudden explosion in the population of one member or the decrease of other members. These changes can occur due to changes in the non-living factors in the ecosystem.

To conclude the Unit, students use evidence they have collected to predict how water quality changes might affect a real-life situation involving the life and economic future of two communities along the Crystal River.

Investigations

In *Learning Set 1*, students work in groups to design and build a model watershed that provides them with clues as to how water flows through a watershed. Their understanding of watersheds is expanded when they use raised relief maps of the state of Michigan to track watersheds to the Great Lakes. Experimenting with stream tables, students model and predict how different kinds of land use contribute to erosion and deposition in the land. Observing erosion and deposition of soil provides evidence that the way land is used can contribute to changes in water quality.

In *Learning Set 2*, students learn ways to test for a variety of water-quality indicators. Using duckweed, they begin by designing an investigation to test how various concentrations of common fertilizer components affect plant growth in water. This test is run over a period of five to ten days during which time, students track data each day. In a subsequent investigation, students test five different substances to learn about pH range. Students learn that many aquatic organisms are limited in where they can survive by the pH of the environment. A teacher-led demonstration using probe ware allows students to consider how high and low temperature of water in a stream and the speed of a stream affect the amount of dissolved oxygen the stream water can hold. This in term affects organisms that live there.

During *Learning Set 3*, students use a dichotomous key to classify aquatic macroinvertebrates. This provides a base for understanding that if populations in a community begin to change in some way, the ecology of the area also is changing. Students further investigate the effects turbidity on water quality. By studying the effects of light on photosynthesis in aquatic plants (Elodea) in an area, scientists determine whether enough oxygen and sugar are being produced to support animal life there. Students investigate how many individuals from different groups in a food chain interact with each other. Model populations are tracked with the use of a computer program, NetLogo® to determine how change in one group affects other populations. Students work in groups to find connections in feeding relationships that culminate in food webs.

Nature of Science

In this Unit, students are made aware that science has everyday importance. When the fictional town of Wamego on the Crystal River faces economic problems, students compile evidence on how the proposed solution might impact the town's land and water resources. Students engage in behaviors and habits of scientists by working in groups to design investigations to test a set of water quality indicators, both living and nonliving. Finally, they act as expert witnesses by presenting evidence based on experimentation to the town council. This process reflects science as a human endeavor and illustrates the role that scientific evidence can play in societal decisions.

Artifacts

Overall, the class uses a P*roject Board* to keep track of design plans, drawings of models, and accumulated evidence from investigations that support how students should answer the *Big Question*. In *Learning Set 1*, students begin their *Project Board* after being introduced to the Unit challenge. They draw a design and model a watershed. They build a stream table incorporating a specific kind of land use (*residential, commercial, industrial, or agricultural*), keeping track of the effects of water flow that results from specific types of land use. They also create and present a poster of their land-use model indicating where erosion, deposition, and runoff occur. Students use a *Create Your Explanation* page to organize claims and evidence for their poster presentations.

In *Learning Set 2*, students update the *Project Board* with new thoughts about how to check water quality. In groups, students use a *Plant Growth Experiment Planning* Page to design an experiment to test the effect of various concentrations of fertilizers on growth of duckweed. This plan is then presented to the class. Data over 5 to 10 days are recorded on a *Plant-Growth Data and Observations* page. Students record results on a data chart while they test for pH as a water quality indicator.

In *Learning Set 3*, students update their *Project Board*. They classify aquatic organisms using a dichotomous key and record the names of the organisms as they are identified. An *Elodea Investigation* page is used to record data from an experiment in which students test the effect of turbidity on photosynthesis. Students prepare a poster in which they share the analysis of a meal to its sources. They draw a food chain and record factors that affect populations on a *Model Population Prediction and Observations* page. Students then use NetLogo® to simulate how populations of grass, mice, and coyotes change and how this affects an ecosystem. Teams utilize *Food-Chain Records* pages and posters to depict food webs.

Students use a *Create Your Explanation* page to think about and justify their recommendations for the town of Wamego. The Unit culminates with a presentation to the town council. Final recommendations from each group are in the form of a poster, a PowerPoint® presentation, or a skit.

Targeted Concepts, Skills, and Nature of Science	Section
Scientists often work together and then share their findings. Sharing findings makes new information available and helps scientists refine their ideas and build on others' ideas. When another person's or group's idea is used, credit needs to be given.	All Learning Sets
Criteria and constraints are important in design.	LS 1, LS 2
Scientists must keep clear, accurate, and descriptive records of what they do so that they can share their work with others and consider what they did, why they did it, and what they want to do next.	All Learning Sets
In a fair test, only the manipulated (independent) variable and the responding (dependent) variable change. All other variables are held constant.	LS 2
Scientists make claims (conclusions) based on evidence obtained (trends in data) from reliable investigations.	All Learning Sets
Explanations are claims supported by evidence, accepted ideas, and facts.	All Learning Sets
Scientists use models to simulate processes that happen too fast, too slow, on a scale that cannot be observed directly (either too small or too large), or that are too dangerous.	LS 1,
Scientists often break down big questions into smaller questions that they can investigate.	All Learning Sets
Predicting, observing, and explaining are important investigative skills.	LS 1
Students read raised relief maps and know the difference between three-dimensional and two-dimensional maps.	LS 1
An ecosystem is a complex community of interdependent organisms and the environment they share.	All Big Question sections related to the Unit challenge
Water is an essential substance in an ecosystem.	LS 1
Water and land interact with each other.	All Learning Sets

Targeted Concepts, Skills, and Nature of Science	Section
Water in a watershed travels predictably, from higher to lower elevations.	LS 1
Watersheds define the flow of water from an area of land into a river system and the flow of rivers systems into lakes and oceans. Watersheds are nested.	LS 1, LS 2, LS 3
Land structures and the materials that make up land, can change the quality of water moving in the ecosystem.	LS 1
Water flow transports and redistributes materials in a stream.	LS 1, LS 2
pH is a measure of how acidic or basic a substance is and is an indicator of water quality.	LS 2
Mixing less acidic (alkaline) solutions with acidic solutions changes the overall pH (acidity) of a solution.	LS 2
Some aquatic organisms are very sensitive to pH.	LS 2
Aquatic organisms use dissolved oxygen for respiration.	LS 2, LS 3
Dissolved oxygen is an indicator of water quality.	LS 2
The amount of dissolved oxygen in water increases as temperature decreases and as turbulence increases.	LS 2, LS 3
Scientists classify organisms based on criteria including physical appearance and feeding relationships and relationships to the environment.	LS 2, LS 3
All living things need energy to survive.	LS 3
Living things that produce their own food, such as plants, are producers.	LS 3
Living things that consume other organisms are consumers, which include herbivores, carnivores, and omnivores.	LS 3
Predators kill other organisms (prey) for food.	LS 3
Scientists use tools such as the dichotomous key to classify and identify different organisms.	LS 3

Targeted Concepts, Skills, and Nature of Science	Section
Food chains and webs help show who eats what in an ecosystem.	*LS 3*
Living organisms are made of cells, get energy from the environment, grow and develop; most reproduce and respond to changes in their environment.	*LS 3*
Plants absorb energy from sunlight using chlorophyll in their cells; they produce food by photosynthesis.	*LS 3*
Scientists can determine water quality using biotic indicators.	*LS 2, LS 3*
Plant growth can affect water quality.	All Learning Sets
Biotic and abiotic components interact in an ecosystem.	LS 2, LS 3
By following how water flows in and over land in an ecosystem, ecologists can learn how water is affected by organisms and by the land in the ecosystem.	LS 1, LS 3
Human activity can affect the ecology of a community. Humans use rivers for residential, commercial, industrial, and agricultural purposes. These activities affect water quality along a river.	All Big Question sections related to the Unit challenge

NOTES

....

....

....

....

....

Unit Material List

Quantities for groups of 4-6 students.		
Unit Durable Group Items	**Section**	**Quantity**
Assorted building blocks, pkg. of 7	1.2, 1.6	1
Stream table apparatus	1.2, 1.3, 1.6	7
Stream table, 27 qt	1.2, 1.3, 1.6	7
2" X 2" black laminated squares, pkg. of 40	1.6	1
2" X 11" black laminated strips, pkg. of 40	1.6	1
Wooden slats	1.6	1
120 Volt light fixture with shield	2.2, 3.4	1
Large ringstand	2.2, 3.4	1
Burette clamp with coated jaw	2.2, 3.4	1
Plastic beaker, 400 mL	2.2, 2.3	14
Graduated cylinder, 10 mL	2.2, 2.3	1
Plastic test tube rack	2.3	1
Large test tube	2.3	5
Rubber stopper, with no hole, size #2	2.3	1
Beaker, 250 mL	2.3	2
Plastic shoe box	3.4	2
Macroinvertebrate Classification Card set, laminated	3.2	1

Quantities for 5 classes of 8 groups.		
Unit Durables Classroom Items	**Section**	**Quantity**
Project Board, laminated	ABQ, 1.1, 1.2, 1.8, 2.1, 3.1, 3.4, 3.7	5
Project Board transparency	ABQ, 1.1, 1.2, 1.8, 2.1, 3.1, 3.4, 3.7	1
Graduated cylinder, 250 mL	ABQ; 2.2	1
Modeling clay, 1 lb	1.2	1
Raised relief map of Michigan	1.4, 1.5	2
Transparency of two-dimensional relief map of Michigan	1.4	1
Box of interlocking building blocks	1.6	1
Beaker, 600 mL	2.4	4
Funnel set	2.3	1
Dominoes game set	3.5	1
Far Mill River sampling DVD	3.2	1
NetLogo software CD	3.6	1
Food Web Cards, set of 74, laminated, with instructions	3.7	1

Quantities for groups of 4-6 students.		
Unit Consumable Group Items	**Section**	**Quantity**
60 Watt frosted bulb	3.4	1
Transparency erasable marker set	1.2, 1.4, 1.5, 3.7	1
Landscape materials for stream table, 5 lb	1.6	2
Spanish moss	1.6	1
China marker, black	2.2, 2.3	1

Quantities for 5 classes of 8 groups.		
Unit Consumables Classroom Items	**Section**	**Quantity**
pH paper strips, pkg. of 100	ABQ	1
Vinegar, 64 oz	ABQ, 2.3	2
Food coloring, red	ABQ	1
Food coloring, blue	ABQ, 1.6	1
Baking soda	ABQ, 2.3, 3.4	5
Masking tape, 1 roll	1.2, 2.2, 3.7	1
Rubber gloves, pkg. of 100	ABQ, 2.2	1
Restickable easel pad	ABQ, 1.6, 2.2, 3.5, 3.6, 3.7	4
Colored markers, set of 8	ABQ, 1.6, 2.2, 3.5, 3.6, 3.7	6
Grow lamp to use in heat shield for duckweed growth	2.2	2
Duckweed plants, 2 oz	2.2	1
Liquid houseplant fertilizer (N:P:K =8:7:6), 8oz	2.2	1
Plastic wrap	2.2	1
Plastic pipettes, 3 mL, pkg. of 10	2.3	15
Coffee filter	2.3	20
Rubbing alcohol	2.3	3
Plastic spoon	ABQ, 2.3, 2.4, 3.4	8
Elodea-10" sprig, pkg. of 10	3.4	6
Giant water bug	3.2	1
Tongue depressors, pkg. of 100	1.6, 3.7	1

Additional Items Needed Not Supplied	Section	Quantity
Clear plastic jar with lid, 32 oz	ABQ, 2.3	5
Butcher paper	1.2	1 roll
Access to river water for five water samples	ABQ, 2.4	2.5 L
Coffee grounds	ABQ	36 oz
Cocoa powder	ABQ	36 oz
Overhead transparency	1.2, 1.4	2 per group
Spray bottle	1.2, 1.3, 1.6	2 per group
Kitchen sieve	1.6	1 per classroom
Roll of paper towels	1.2, 1.3, 1.6, 2.3, 2.4	1 per classroom
Access to water	ABQ, 1.2, 1.3, 1.6	1 per classroom
Map of Michigan	1.4	1 per classroom
Books to stack stream table on	1.6	3 per group
Distilled water	2.2, 2.3, 2.4	4 gal per class taught
Paperclips to move duckweed to beakers	2.2	2 per group
Rubber band to hold plastic wrap over beaker	2.2	4 per classroom
Goggles	2.3, 2.4	1 per student
Rubberized apron	2.3, 2.4	1 per student
Head of red cabbage, diced	2.3	1 per 3 sections
Knife to cut cabbage	2.3	1
Colorless, flat soft drink 2 L	2.3	1 per classroom
Clear window cleaner	2.3	1 per classroom

Additional Items Needed Not Supplied	Section	Quantity
Method for heating and cooling water to produce pH indicator and conduct dissolved oxygen water quality testing	2.3, 2.4	1 per classroom
LCD projector, or TV monitor	2.4, 3.2, 3.6	1 per classroom
Large map of local, county, or state area (Optional)	1.1, 1.3	1 per classroom
Projection of local body of water (Optional)	1.1	1 per classroom
Kitchen sponge (Optional)	1.3	1 per classroom
Small bucket to collect water runoff from stream table	1.6	1 per group
Large pot with lid	2.3	1 per classroom
Pitcher, 2 qt	2.4, 3.6	1 per classroom
Laptop, or probeware display	2.4	1 per classroom
Temperature probe	2.4	1 per classroom
Dissolved oxygen probe	ABQ	1 per classroom
Plastic sandwich bags	3.4	2 per group
Photosynthesis probe	3.6	1 per classroom
Blank transparency map of the U.S.A.	3.7	1 per group
Blank U.S. biomes map	3.7	1 per student
Map of U.S. biomes	3.7	1 per classroom

UNIT INTRODUCTION

What's the Big Question?

How does water quality affect the ecology of a community?

◄ *2 class periods*

A class period is considered to be one 40 to 50 minute class.

Overview

The goal of this Unit is for students to learn about ecology and the relationships within an ecosystem. Students are introduced to the *Big Question* of *Living Together*: *How does water quality affect the ecology of a community?* They are also introduced to the challenge, which is to determine how the towns of Wamego and St. George, located on the Crystal River, might survive ecologically, culturally, and economically if a new business were to move to Wamego. To begin to answer this challenge, the class considers five jars of different fluids. They think about the quality of each sample and discuss what it would be like to live in or drink the sample water. They work in groups and consider why it might be important to know the quality of water and think about factors in the environment that affect water quality. Students then think about what they might need to know to answer the *Big Question*. They create a *Project Board* for the Unit, which they will update as they gather evidence and information to answer the *Big Question*.

Targeted Concepts, Skills, and Nature of Science	Performance Expectations
Scientists often work together and then share their findings. Sharing findings makes new information available and helps scientists refine their ideas and build on other's ideas. When another person's or groups' idea is used, credit needs to be given.	Students should be able to cite instances during the *Learning Set* when they have behaved much as scientists would when sharing information with each other while discussing a problem.
Scientists often break down big questions into smaller questions that they can investigate.	Throughout the Unit, you should hear students discuss parts of *Big Question* in the context of the Unit challenge.

Targeted Concepts, Skills, and Nature of Science	Performance Expectations
Water is an essential substance in an ecosystem.	You should be able to hear students relate that water is essential to plant and animal life and that changes in water quality can affect plant and animal life.
Water and land interact with each other.	Students should begin to think about how water and land interact as they study the examples depicted in the photographs of the Rouge River.
An *ecosystem* is a complex community of interdependent organisms and the environment they share.	Students should be able to iterate that an ecosystem contains living organisms whose populations interact with each other and that when some nonliving factors change, these populations change in number and character.
Humans use rivers for residential, commercial, industrial, and agricultural purposes. These activities affect water quality along a river.	Students should be able to conclude from studying the photographs of areas along the Rouge River that human activity might be responsible for what is depicted and that water quality might be affected by these activities.

Materials

1 per class	Five jars labeled Jar 1, Jar 2, Jar 3, Jar 4, and Jar 5 with lids
per classroom	Source of tap water
per classroom	Coffee grounds
1 per classroom	Container of cocoa powder
1 per classroom	Red food coloring
1 per classroom	Blue food coloring
1 per classroom	250-mL graduated cylinder
1 per classroom	Vinegar
1 per classroom	Box of baking soda

Materials	
per classroom	200 mL pond or river water
1 pair per teacher	disposable gloves

What is Water Quality?

Set up a classroom demonstration consisting of five small jars, each with 200 mL of tap water, and each with a lid. Label each jar with a number and treat each as a different water sample by adding the following items. Allow time before the demonstration to set these up. You may have to experiment to get colors or pH readings exactly as they need to be.

- **Jar 1:** Add coffee grounds and cocoa powder to 200 mL water until the water looks very dirty.

- **Jar 2:** Add a few drops of red and blue food coloring to 200 mL water. The water will turn purple but should be transparent.

- **Jar 3:** Add about 40 mL of vinegar to the water until it has a pH of approximately 5. Test the solution using pH paper while adjusting the pH.

- **Jar 4:** Add about ½ tablespoon of baking soda to the water so that the pH is about 8-9. Test the solution using pH paper while adjusting the pH. Jar 4 must test differently from Jar 3.

- **Jar 5:** Wearing disposable gloves, collect water from a local pond or river. It is all right to have a small amount of soil in the water. If you cannot get water from a local source, add a small amount of soil to tap water. Thoroughly wipe the outside of the jar and do not open it for students. When it is time to dispose of the water, dispose of it at its collection site if possible.

PBIS

Homework Options

Reflection

- **Nature of Science:** Describe two ways that meeting in your group to solve the *Big Question* is similar to the way you think scientists solve problems. *(Students' answers might include that their group meeting was similar to scientists working together because they share ideas and may develop new ideas because of what is said by someone else in the group.)*

- **Science Process:** Summarize one idea from the *Project Board* that interested you. *(Student choices will vary but should come from the information listed on the* Project Board.*)*

Preparation for 1.1

⚠️ Check with your local environmental agency to make sure it is legal for you to collect water from local ponds.

- **Science Content:** How do you think you would model what happens to rain after it hits the ground? *(Accept all reasonable responses. Students may suggest using colored water to track the movement of rain.)*

- **Science Process:** What are two questions you might ask about how water and land interact with each other? *(Accept all reasonable responses. Students might ask, "How does water affect the land as rain falls?" and "Does rain change after it falls on the ground?")*

NOTES

....

....

....

....

....

....

....

....

◀ *2 class periods**

What's the Big Question?

How does water quality affect the ecology of a community?

Water is very important in your life. You drink it, you wash with it, you use it to cook, and you use it for play and exercise. You also know that plants and animals depend on water to stay alive.

If you need water, you can turn on a tap. Towns and cities in the United States have municipal water systems in place. That is where most people get their water. To make sure the quality of water you use is good, it is important to know where it comes from.

In this Unit, you are going to investigate how water use affects water quality. You will then look at how water quality affects the plants, animals, and humans in a community. **Ecology** is the study of how plants and animals, including humans, interact with one another and the physical environment.

ecology: the study of the relationships between organisms and their environment.

Look at the *Big Question* for this Unit: *How does water quality affect the ecology of a community?* This is a very big question. To answer the question, you will need to break it down into smaller questions you can answer. You probably already have some smaller questions you might want to ask. You will have a chance to ask those questions when you start working on your class *Project Board.*

Welcome to *Living Together.*
This is a great opportunity for you to work
as a student scientist.

LT 3

LIVING TOGETHER

What's the Big Question?

How does water quality affect the ecology of a community?

10 min.

Introduce the Big Question *to the class.*

META NOTES

Consider planning a field trip to a local river, pond, or lake for water to test in *Learning Set 2.*

○ Engage

During this introduction, you want to get students thinking about the importance of water. Begin by asking students to think about how they use water every day. (*Possible responses include taking a shower, washing clothes, drinking, cooking, and washing hands.*) Create a class list of their responses.

*A class period is considered to be one 40 to 50 minute class.

"Where do you think water for these activities comes from? Have you ever wondered about where water from the tap comes from? Let's make a list of places where you see water in the community. (*Probable responses are rain, snow, rivers, streams, lakes, ponds, water in the ground.*)

Now that you have thought about how you use water every day and listed some sources of water, think about this: Would you use water straight from the river to drink? Would you swim in water near the paint (*or some other local example*) factory down by the river? What are some reasons why you would or would not use this water? (*Students might say the water does not look clean or that it has an odor.*)"

☐ Assess

Turn students' attention to the *Big Question* of the Unit. Ask them to describe in their own words what they think the question means so that you can determine their understanding of the big picture. Listen to how they use some of the terms; in particular, water quality, ecology, and community.

"*Ecology* is the study of plants and animals and how they interact with their physical environment or surroundings. What do you think makes up the physical environment? Think about what you interact with each day. (*Responses should include water, land, and air.*)"

NOTES

....

....

....

....

....

PBIS

Think about the Big Question

Before you start, it is a good idea to think about what you might already know about the *Big Question*. You will do two activities. They will help you think about how you use water in your daily life. You will also need to think about what is important about the quality of that water.

Get Started: What Is Water Quality?

Your teacher will show you five jars. Each jar contains a different water sample. Observe the water in the jars. Record your observations. Then decide whether or not you would use the water in the jars to fish, swim, boat, or drink. Describe how you arrived at these decisions.

LT 4

Project-Based Inquiry Science

Think about the Big Question
20 min.

Get Started: What is Water Quality?

Students work on their observational skills and begin to articulate their ideas of what they consider to be important factors in water quality.

○ Engage

Ask the class what they think water quality means. Give students a few minutes to suggest ideas. If students have difficulty verbalizing a description, ask them to think about ads they might have seen or heard where the word "quality" was used.

TEACHER TALK

"If you hear about a product that has quality, does that make you want to have it or avoid it? What characteristics do you think a quality product has? *(List responses.)* Now if that is true for a breakfast cereal, or a car, or a piece of clothing, what characteristics do you want in the water that you drink or swim in?"

△ Guide

Tell students that you are going to provide samples of water for them to observe in their groups. Inform them that they should record what they observe about each sample.

TEACHER TALK

META NOTES

As an alternate approach, set up the five jars as a demonstration to reduce the number of glass jars and the potential for glass breakage in the classroom.

"I'm going to put out five jars of water. Each has a number to make it easier to refer to when you write notes about it. I want you to look at each sample and think of how you might describe it. Make a guess as to where you think the sample came from? What is your reason? Describe how might you use each sample? Can you really decide on the quality just by looking at it? What else would you like to know about it?"

☐ Assess

META NOTES

Caution students not to open any of the jars. Ask them to observe and describe the water in each jar. For each jar, have students discuss what the water might be used for.

On a table, place five water samples marked Jar 1 through 5 respectively. Make sure that each group has time to look at the jars. Listen to the discussions in each group. Are students thinking aloud about where each sample might have come from? Listen to hear what students might say is different about each sample. If necessary, draw their attention to one particular sample.

△ Guide

Listen to find out if students are discussing reasons for using or not using each sample. If you do not hear that as you walk around the classroom, then ask a question of each group. Possible questions might be: "Which sample would you keep a goldfish in? Which sample would you swim in? Would you use the water in Jar 4 to cook in? Which sample would help if you needed a drink of water?" Remind them to come up with reasons for their responses.

TEACHER TALK

"What is there about Jar 3 that makes you think that a person might be able to drink the water? What is your reason? What do you see that makes it different from the water in the other jars?"

Stop and Think

You have made some decisions about the quality of different water samples.

1. What is meant by quality? What is water quality?

2. How did you determine water quality for the bottles? Was this an adequate method?

3. How else could you measure water quality?

4. If you were walking along a river, lake, or stream, how could you determine the quality of the water?

5. You probably judged the quality of the water from a human's point of view. What if you were a water plant instead of a human? How would you judge the quality of water in each jar?

6. What if you were a fish? Which jar would have good water quality for a fish?

Get Started: What Affects Water Quality?

As you were looking at the samples of water in the jars, you may have wondered where the samples came from and what could make them look so different.

Look at the photos on the next two pages. They are taken in different locations along a river. The river runs through many different types of landscapes and areas. Examine the photos carefully. Think about what the quality of the water in the river might be at each location.

Try to match each water sample in the jar to one of the locations in the photos. Write down which photo you are matching to each jar and why you are making that match.

LT 5

LIVING TOGETHER

Stop and Think

15 min.

Discuss students' observations of the water samples using the Stop and Think *questions.*

META NOTES

Determining the quality of the five samples by looking at them is a qualitative measure. In *Learning Set 2*, students will measure pH, temperature, dissolved oxygen, and turbidity—all quantitative tests of water quality.

◯ Get Going

As groups complete their observations of the water samples, turn their attention to the *Stop and Think* questions. Tell them they have a few minutes to discuss possible responses within their groups and then they will discuss them as a whole.

Stimulate the discussion by focusing on questions 2 and 3. Students have just observed the jars with their eyes. Ask, "What else might you need to know or observe to detect if a fish or plant could live in the water or if you could drink the water?" Explain that later in the Unit, students will have the opportunity to test different characteristics of water to help them determine a more complete picture of the quality of these samples.

△ Guide and Assess

Answers to *Stop and Think* questions will vary but during the discussion, listen to responses and guide the discussion as needed to bring out the responses listed below.

1. Students should be able to say that the term quality means the characteristics of something, generally the good or more positive characteristics of something. Water quality refers to how good or useful the water is, generally in terms of human health.

2. Most students will say that they determined the water quality in the five jars by sight. Students should be able to say that sight alone is inadequate for determining water quality.

3. Students' answers will vary because they probably have not experienced using quantitative methods but some may refer to some way to measure types and amounts of chemicals in the samples of water. Accept other reasonable responses.

4. Most students will respond that they would use their senses to determine the quality of the river water. Most will probably say sight and smell.

5. Even though plants cannot "judge" situations, they can respond positively or negatively. Water quality would be good if the plant received the nutrients it needed to stay alive. Water quality would be poor if the water does not supply the plant with what it needs and it dies.

6. Students may guess that the jars with the cleanest or clearest water would support a fish and keep it alive, but in fact, students cannot make this judgment at all because they do not know the composition of the water samples.

META NOTES

It is all right, if at this point, students do not know what factors scientists use to determine water quality.

5. You........ the quality of........ a human........ view. ...... you were a water plant in........ a human? How ...... you judge the quality of water in each jar?

6. What if you were a fish? Which jar would have good water quality for a fish?

Get Started: What Affects Water Quality?

As you were looking at the samples of water in the jars, you may have wondered where the samples came from and what could make them look so different.

Look at the photos on the next two pages. They are taken in different locations along a river. The river runs through many different types of landscapes and areas. Examine the photos carefully. Think about what the quality of the water in the river might be at each location.

Try to match each water sample in the jar to one of the locations in the photos. Write down which photo you are matching to each jar and why you are making that match.

LT 5

LIVING TOGETHER

Get Started: What Affects Water Quality?

10 min.

Students are introduced to the idea that land use around a river can affect river water. Students begin to make this connection and articulate their ideas about how and why this happens.

○ Engage

Divide the class into their groups or teams. Tell students that they are going to look at the water samples in the jars again and match them with the photographs on pages 6 and 7 in the student edition. Remind students to prepare for a class discussion by having reasons for matching a particular sample with the scene in the photograph.

TEACHER TALK

"Picture yourself in the setting shown in each photograph. Think of the kinds of things might you observe about the water. Remember to use your senses. How does the water look to you? Does it have a pleasant or unpleasant odor? What color is the water? Does the air smell fresh? Do you hear birds or see any fish swimming in the water? Which of the five jars looks like water from that setting? Make sure that you have a reason for the match to share with the class. Once you have made a decision, record your reason."

Here the river runs through a farming community.

The river runs through a golf course. Notice the fairways and sand bunkers. In the distance there are several homes.

Here the river winds through a shipping yard. Notice the docked barges. In the right front there is a plant that produces paint. The paint is then shipped by barge, train, and truck from the plant.

Here the river widens and moves very slowly. In fact, the river enters into a lake at one end, and then it exits the lake through a small stream at the other end. This picture was taken from the yard of a small cottage on the lake. The dock belongs to the cottage owner.

Left: *The river runs past several housing communities and wildlife habitats.*

Below: *At one point, the river is very wide. It is often used for boating. There is even a powerboat race every Fourth of July.*

Above left and right: *This is a small drainage ditch near a highway. This ditch drains into the sewer pipe, and eventually the water flows into a larger part of the river.*

Left: *At the end of the 130-km (80-mile) river, it flows into an even larger river. At this point, there is a large manufacturing plant that makes cars.*

LT 7

LIVING TOGETHER

Conference

15 min.

Working in groups, students may find that there is more than one way to look at evidence.

Conference

Share your decisions with the rest of your group. Discuss why you made the matches you did. For some of the jars, you may agree with your group members on the matches you made. For some, you will have disagreements. It is important for each member of your group to discuss why they made their choices. See if you can come to an agreement. Make sure to clearly discuss the reasons for the matches each of you made, as you are trying to see if you agree.

This activity may have reminded you of some things you think you know or don't know about water quality. Jot down notes during the discussion so you will remember what was said when you share again with the class.

Decide as a group what are the most important things to know about water quality and what affects it. What questions are important to investigate to answer the *Big Question*?

Wamego is a farming community.

Your Challenge
Wamego Needs Help!

To answer the *Big Question*, you will need to respond to a challenge. A small town requires help in making a decision that will affect its future. Wamego (hwah-MEE-goh) is the small town that needs your advice. It has a population of about 1800. It is located on the banks of the Crystal River. This town has always been a farming community. Most of the farmers grow corn and soybeans. These are the best crops to grow in this area. Nearly 95% of the residents are employed by Wamego's farming businesses. The local economy depends on farming. The other businesses in town all depend on the farmers and their employees (workers). These businesses include a grocery store, gas stations, a movie theater, and several restaurants.

△ Guide

After students have matched the jars and photographs, list the nine photographs and ask students to say which jar matches the first photograph for use during a class discussion. As students talk about each photograph, they may quickly realize that they do not all agree on each match. Let students know that it is all right to disagree. Explain that scientists often disagree when they begin to discuss a topic, but that each person presents a reason to support their viewpoint.

Your Challenge
Wamego Needs Help!

To answer the *Big Question*, you will need to respond to a challenge. A small town requires help in making a decision that will affect its future. Wamego (hwah-MEE-goh) is the small town that needs your advice. It has a population of about 1800. It is located on the banks of the Crystal River. This town has always been a farming community. Most of the farmers grow corn and soybeans. These are the best crops to grow in this area. Nearly 95% of the residents are employed by Wamego's farming businesses. The local economy depends on farming. The other businesses in town all depend on the farmers and their employees (workers). These businesses include a grocery store, gas stations, a movie theater, and several restaurants.

Wamego is a farming community.

LT 8

Project-Based Inquiry Science

Your Challenge
10 min.

Telling the story of Wamego engages students in the challenge.

○ Engage

Introduce the challenge by first describing the story of Wamego and its concerns about its economic and ecological future.

TEACHER TALK

"Wamego is a small community on the Crystal River. The cold and clean river water supports an abundance of trout. In fact, each year Wamego has a Trout Festival. Most of Wamego's business is farming, but the Festival supplies a lot of income for the town.

In the past five years, farming hasn't been so profitable. People have moved away. FabCo, a company that manufactures cloth is interested in building a new factory in Wamego. FabCo will benefit because it needs the river water and the cost of living in the town is low. The town will benefit because FabCo will provide new jobs, new homes, and increased taxes. Some members of the community however, are concerned that the new factory would change the land and the water quality of the river."

META NOTES

The story provides the motivation for students to begin to think about how they will solve the challenge.

The Crystal River is also important to Wamego. The river is a source of water for the crops. The river is also known as a good trout-fishing river. Trout need very clean, cold water to thrive. Crystal River suits their needs. Every summer Wamego has a Trout Festival. Many people who enjoy fishing travel to the area. The festival celebrates trout fishing and preservation.

The festival also educates people about what trout need to thrive. The goal of the education effort is to keep the number of trout at a healthy level. In that way, people can enjoy fishing there for many years to come. This festival is fun for many residents and tourists. It is also another income source for the residents of Wamego.

Lately, the farming business has not been good. Crop prices have dropped. The farmers are not making very much money. There is not enough to pay their workers or to support themselves. Some of the farmers have gone bankrupt. As a result, Wamego has lost 15% of its population during the last five years.

LT 9

LIVING TOGETHER

Present the challenge to students. Inform them that first they will design and build a model that shows how water flows through land before it gets to a river. Then, they will investigate a variety of ways that land can affect water and water can affect land. Finally, they will learn about the organisms that live in the river water and how changes to the water can affect these organisms. Tell them that the evidence that they collect will enable them to answer the *Big Question* and advise the town on what might happen ecologically if FabCo develops a new factory.

PBIS

The town council is very concerned. They know farming will always be a part of life in Wamego. They do, however, worry about the town losing too many people. They do not want to get so small that there will be very few businesses and residents in Wamego.

FabCo Wants to Move In

A mid-sized manufacturing company called FabCo has contacted the town council. FabCo manufactures cloth. The cloth is sold to companies that make clothes. FabCo is looking for a new location to build its company headquarters and manufacturing plant. FabCo is very interested in relocating to Wamego for several reasons.

- Wamego has a fairly large river and a train line running through the town. This, along with roads, would provide transportation routes for their products.

- The cost of living in the town is low. Their employees would like that.

- The river provides a natural resource (water). Water is important to the production of their cloth.

If FabCo is allowed to move to Wamego, the town could benefit as well. It would mean the following benefits:

- About 15,000 new residents would relocate to Wamego. This would require the building of many new homes, roads, and parks. A new school would need to be built. New businesses offering services to the company and the new residents would be needed. This means more buildings, parking lots, and roads would appear in Wamego.

- FabCo would offer many new jobs to Wamego's residents.

- The town would have money from taxes collected from FabCo and the new residents. This extra money could be used to improve life in Wamego in many ways, including a new hospital.

- The town would not have to depend on farming alone.

LT 10

TEACHER TALK

"Your challenge is to investigate to find out what changes might occur to the land and the water around Wamego if FabCo moves in. You will use the evidence you collect from models, investigations, and the science knowledge you read throughout the Unit to advise the town council about how to save their town, especially how to maintain the river's water quality. As you prepare to talk with the town council, you will answer the *Big Question* and learn how important a healthy environment can be."

Sounds Great! So, What's the Problem?

Many of the residents, including some town council members, are concerned. They worry that FabCo could mean problems for their community. Currently, the land is used for agriculture. If FabCo comes to town, the use of the land will change. The land will be needed for residential, commercial, and industrial purposes. Some people, including the organizers of the Trout Festival, wonder if this will change the river and the wildlife of Wamego.

Wamego residents are not the only ones concerned. Ten miles downstream is the town of St. George. It is also located along the Crystal River.

St. George is an even smaller town than Wamego. It is a resort town. People travel from all over to vacation in St. George. They use the river for recreation. There is fishing, swimming, boating, hiking, and camping in the area. There are several hotels and bed & breakfasts that provide accommodations for tourists. The Crystal River's water quality is very important to St. George's economy and residents. The residents of St. George are worried that the changes in Wamego might affect their lives.

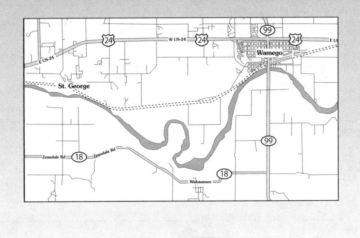

LT 11

LIVING TOGETHER

△ Guide

At this point you may want to allow class groups to read the more detailed description of the Wamego story in their textbook and then conference for about five minutes so that they can begin to sort through the parts of the story on their own. Listen for whether students grasp the scope of the town's problem. What kinds of questions do you hear them asking other students within their own group? Watch faces to see if anyone displays confusion over what is going on. At the end of that time, let the class discuss or almost retell the story to insure that you know they understand the various parts. If necessary, list events as they recount the story. You may want to post a map of Wamego and the river.

Setting up a *Project Board* might point out where and how a very large problem might need to be broken down into more manageable parts. If students have trouble getting the big picture, ask them to verbalize what they do understand. Treat each student's version as a piece in a puzzle. Then, ask students how to draw the pieces together. If necessary, write each part on a separate piece of poster paper, then let students put them in order. Transition the whole class into focusing on the use of a *Project Board* as a place to brainstorm ideas and use as an organizational tool.

TEACHER TALK

"As you begin to think about the challenge that the town faces, you need to think about ways to organize your thoughts so that you can investigate their problem logically. A *Project Board* is a way to keep track of ideas and progress. Let's begin by recording what you think you already know about water quality and ecology."

NOTES

Create a
Project Board

15 min.

A Project Board *can reflect progress in what students learn about a challenge.*

As you answer the *Big Question*, you will also take on the challenge of giving advice to the town council of Wamego. What should they take into account in deciding whether or not to let FabCo move in? What will be the ecological advantages of FabCo building its plant in Wamego? What ecological problems might the project cause?

What ecological problems do you think might arise if FabCo moves in? What do you need to know more about to give the Wamego town council advice? Share your ideas with your group. Discuss the reasons for your ideas. Make lists of what you think might happen and what you think you need to investigate. You will share these with the class when you create a *Project Board*.

Create a *Project Board*

It is useful, when you are working on a design project or trying to answer a hard question or solve a hard problem, to keep track of your progress. You also want to keep track of what you know and what you still need to do. Throughout this Unit, you will be using a *Project Board* to do that. During classroom discussions, your teacher or one of the students will record
the class's ideas on a class *Project Board*. At the same time, you will keep track of what has been discussed on your own *Project Board* page.

Recall that a *Project Board* has space for answering five guiding questions:

- What do we think we know?

- What do we need to investigate?

- What are we learning?

- What is our evidence?

- What does it mean for the challenge or question?

To get started on this *Project Board*, you need to identify and record the important science question you need to answer: *How does water quality affect the ecology of a community?* You also need to record your challenge: *What advice should we give Wamego?*

LT 12

△ Guide

At this point, the class will be supplying start-up information for the first two columns of the *Project Board* (*What do we think we know?* and, *What do we need to investigate?*). Emphasize that more can be added to these two columns throughout the Unit.

How does water quality affect the ecology of a community? What advice should we give Wamego?				
What do we think we know?	What do we need to investigate?	What are we learning?	What is our evidence?	What does it mean for the challenge or question?

What do we think we know?

In this column of the *Project Board*, you'll record what you think you know about water quality and ecology. Discuss and post the things you think you and your classmates know about water quality and ecology. Have you studied these concepts before? What knowledge did you gain from your studies? Even if it is a small fact or idea, talk about it. Discuss any factors that you think might affect water quality, the ecology of a community, and the ecology of Wamego.

What do we need to investigate?

In this column, you will record the things you need to know about to answer the question and address the challenge. During your group conference, you may have found that you and others in your group disagreed about some ideas. You may not know how else to measure water quality. You and your group may not have agreed on where a particular water sample may have been taken. This second column is designed to help you keep track of things that are debatable or unknown and need to be investigated.

Later in this Unit, you will return to the *Project Board*. For now, work with your classmates and follow your teacher's instructions as you begin filling in the first two columns.

LT 13

Distribute the student *Project Board* pages at this time. Remind students to keep a personal copy of the class *Project Board* to refer to as needed. Draw students' attention to the first column of the class *Project Board* to have students volunteer information to be recorded.

How does water quality affect the ecology of a community? What advice can you give Wamego?

What do you think you know?	What do you need to investigate?	What are you learning?	What is your evidence?	What does it mean for the challenge or question?
1. We know that water quality needs to be such that it can support life. 2. We know the layout of Wamego and the river.	1. How might water quality change if the land use is different from what it is now? 2. How is water quality measured?			

Assessment Options

Targeted Concepts, Skills, and Nature of Science	How do I know if students got it?
Scientists often work together and then share their findings. Sharing findings makes new information available and helps scientists refine their ideas and build on others' ideas. When another person's or group's idea is used, credit needs to be given.	**ASK:** In what way do you play the role of scientists during conferencing or when you work in small groups? **LISTEN:** You should hear students relate that during conferencing or when working in small groups, they act like scientists because they share ideas, even if the ideas are different from one another.
Scientists often break down big questions into smaller questions that they can investigate.	**ASK:** When did you begin to think about the many facets that make up water quality? **LISTEN:** Answers will vary. Some students may verbalize the necessity of thinking about needing a variety of ways to study water quality when they look at the five different water samples or when they look at the nine different settings pictured along the Rouge River.

Targeted Concepts, Skills, and Nature of Science	How do I know if students got it?
Water is an essential substance in an ecosystem.	**ASK:** In what ways is water important to you and everything around you? **LISTEN:** You should be able to hear students say that plant and animal life are dependent on water to stay alive; that changes in water can affect all living things.
Water and land interact with each other.	**ASK:** How do you know that water and land interact with each other? **LISTEN:** Students should be able to say that water and land interact as they study the examples depicted in the photographs of the Rouge River. Some students may know from personal experience something about erosion or changes in the quality of the water that they drink.
An *ecosystem* is a complex community of interdependent organisms and the environment they share.	**ASK:** How is an ecosystem a complex community of interdependent organisms? **LISTEN:** By the end of the Unit, students should be able to iterate that an ecosystem contains living organisms whose populations interact with each other and that when some nonliving factors change, these populations can change in number and character.
Humans use rivers for residential, commercial, industrial, and agricultural purposes. These activities affect water quality along a river.	**ASK:** How does the use of rivers by humans for residential, commercial, industrial, and agricultural purposes affect the quality of a river? **LISTEN:** By the end of the Unit, students should be able to conclude from their investigations throughout the Unit and through studying information about areas along the Rouge River watershed that water quality is affected by human activities.

Teacher Reflection Questions

- The Unit challenge was introduced. It is designed to motivate students to learn the concepts presented throughout this Unit. Could you tell if students were motivated by the challenge story? What can you do to maintain their motivation throughout the Unit?

- Were students able to successfully match the various jars with photographs?

- Were most students able to follow the story of Wamego and its problems? Are there too many factors for most students to follow and account for? Should the story be modified to reduce the amount of detail?

NOTES

....

....

....

....

....

....

....

....

....

....

....

LEARNING SET 1 INTRODUCTION

Learning Set 1

How Do Flowing Water and Land Interact in a Community?

Students think about and begin to study the effects that water and land have on each other. Using models, they observe the normal flow of water in a watershed. Using stream tables, they experiment with how land use can affect the quality of water resources.

Overview

Students are probably familiar with various kinds of bodies of water, but few students understand what a watershed is. In *Learning Set 1*, students focus on what a watershed is and investigate how water drains from land to form bodies of water such as rivers, streams, lakes, and ponds. Students consider various land uses and the effects these uses have on water in a watershed. Students begin by building a physical model of a watershed. They observe that the movement of water is dependent on differences in height of various parts of their model. Students then map the movement of water within the Rouge River watershed and through major watersheds in the state of Michigan to the Atlantic Ocean. Students also will build a physical model of a river using a stream table. With this, they explore how materials move within a river and infer the relationships between land use and its effect on water quality.

> **LOOKING AHEAD**
>
> Throughout this *Learning Set*, students will investigate watersheds to help them get closer to answering the *Big Question*.

Targeted Concepts, Skills, and Nature of Science	Section
Scientists often work together and then share their findings. Sharing findings makes new information available and helps scientists refine their ideas and build on others' ideas. When another person's or group's idea is used, credit needs to be given.	1.1
Scientists often break down big questions into smaller questions they then investigate.	1.1
Scientists use models to simulate processes that happen too fast, too slow, on a scale that cannot be observed directly (either too small or too large), or that are too dangerous.	1.2, 1.3, 1.4, 1.6, 1.7

Targeted Concepts, Skills, and Nature of Science	Section
A watershed is the land area from which water drains into a particular stream, river, or lake.	1.1, 1.2, 1.5
Criteria and constraints are important in design.	1.1
Water in a watershed travels predictably, from higher to lower elevations.	1.2, 1.3
Land structures and the materials that make up land, can change the quality of water moving in the ecosystem.	1.5, 1.6. 1.7
Predicting, observing, and explaining are important investigative skills.	1.2, 1.3
Water and land interact with each other.	1.4, 1.5, 1.6, 1.7
Water movement shapes the land, carving out rivers and streams.	1.3, 1.6
Human activity can affect the ecology of a community. Humans use rivers for residential, commercial, industrial, and agricultural purposes. These activities affect water quality along a river.	1.5, 1.6, 1.7, 1.8
Nested watersheds are smaller watersheds that are part of larger watersheds.	1.4, 1.5
Water flow transports and redistributes materials in a stream.	1.6
Scientists must keep clear, accurate, and descriptive records of what they do so that they can share their work with others and consider what they did, why they did it, and what they want to do next.	1.6
Pollutants to a watershed may occur from point or non-point sources of human activity.	1.7, 1.8

Students' Initial Conceptions and Capabilities

- Students are aware of the atmospheric portions of the water cycle, but few, at lower middle grades have an idea of groundwater as a part of the hydrologic cycle. (Ben-zvi-Assarf, 2005.)

- Many students, through late middle school, have misconceptions about groundwater. They have little idea of the scale of places in which groundwater is found. Many think that it is generally held in various structures underground such as one finds above ground (lakes, ponds) rather than rock pores. (Dickerson, 2005.)

Understanding for Teachers

Water is essential to ecology. Water is responsible for carrying sediments of various sizes, nutrients, and pollutants that help or harm ecosystems. Water moves across land structures of all sizes, and through land, eventually forming rivers, streams, lakes, and ponds. Some of this water evaporates along the way. Some of it moves via rivers, to larger bodies of water, such as oceans. The structures, uses, and materials that water encounters along the way affect its quality.

The goal of this Unit is to help students understand that as water moves across and through land, it changes the land. The water might play a role in changing the shape of the land as it flows from higher to lower elevations. It seeps into the ground and rocks and supplies plants with water to grow. The way the land is used can also change the water. Water picks up and dissolves some materials. This may change the quality of the water, making it useful for some things and useless for others.

Watersheds

In this *Learning Set*, students develop an understanding of a local watershed. They set up physical models of a watershed. In the process, they observe the forces that form and shape a river. They will also observe influences that change the quality of the water as it flows through a watershed.

A watershed is an area of land that drains into a river system. Any water entering the watershed will travel from higher to lower elevations due to gravity. Streams and rivers form as the water moves downward along the slope of the land. The shape or topography of the land determines if the water pools into structures such as lakes or channels into rivers. So long as there are sufficient differences in elevation, water continues to move downward as rivers join lower rivers, eventually moving toward an ocean.

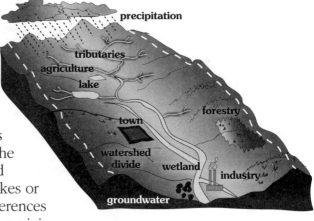

LIVING TOGETHER

The Rouge River Watershed
(Rouge River Wet Weather Demonstration Project, 2001)

The Rouge River watershed in southeastern Michigan, drains about 438 square miles of land in 48 municipalities in three counties. The river itself has a total of 126 river miles and four branches, the Main, Upper, Middle, and Lower. The Rouge River empties into the Detroit River, which connects Lake St. Claire with Lake Erie. The entire Rouge River watershed affects the water and land of the lower 20 miles of the Detroit River and a part of Lake Erie.

The land that drains or seeps water into a small stream is its watershed. When this stream empties into a larger stream, it is "nested" with the larger watershed. It may also be referred to as a sub-watershed of the larger system. The Rouge River watershed has many smaller watersheds nested within it. The Rouge River watershed empties into the Detroit River and is therefore, a sub-watershed of the Detroit River watershed. The Detroit River is nested within the Lake Erie watershed. Lake Erie is part of the Great Lakes/Laurentian watershed because it ultimately discharges into the Atlantic Ocean through the St. Lawrence River.

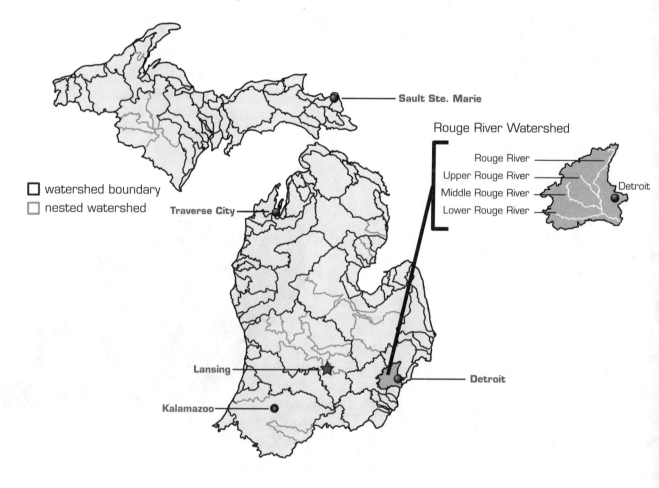

Recreational Uses of the Rouge River Watershed

The Rouge River watershed presents a variety of opportunities for public enjoyment. More than 50 miles of the Rouge River flows through public parkland, making it one of the most publicly accessible rivers in the state. There are more than 400 lakes, impoundments, and ponds in the watershed. These aquatic features present recreational opportunities for more than 1.5 million people who live within the watershed area.

Sources of Water Quality Impairment in Rivers, Lakes, and Estuaries			
Rank	Causes in rivers	Causes in lakes	Causes in estuaries
1	Agricultural runoff	Agricultural runoff	Municipal point sources
2	Changes in streams and changes in habitats	Changes in streams and changes in habitats	Urban runoff
3	Municipal point sources	Urban runoff	Atmospheric deposition

Source: National Water Quality Inventory: 1998, Office of Wetlands, Oceans, and Watersheds

Affect of Urbanization on the Watershed

In this *Learning Set*, students explore the concept of a watershed and its functions. Through physical models, they learn about the nature of watersheds and the forces that shape rivers. Through investigations, they learn about the influence that changing land use has on processes that occur in streams. Students discover that urbanization alters a watershed, ultimately leading to the degradation of water quality in its rivers.

While the Rouge River watershed has many opportunities for recreation, it also is the most densely populated land area in the state. Only 25 percent of the watershed is undeveloped. More than 50 percent of the remaining land area is "urbanized," mainly in the southeast of the watershed. Public enjoyment of the Rouge River watershed is therefore limited by the effects of urbanization. Urbanization restricts the movement of water into groundwater supplies because in urban areas, most of the vegetation has been removed and replaced by surfaces such as roads, parking lots, and sidewalks. All of these impervious surfaces promote runoff. Groundwater is not recharged but runoff increases the mass of water delivered over the surface directly to rivers. Urbanization disrupts the natural water cycle because there is less evaporation from water that would naturally stand in pools. There is also less transpiration because of the loss of vegetation.

The Rouge River suffers from untreated sewage, unstable river flows, and the naturally slow flow of the river because of the nearly level slope of the land. Thick mats of sewage blanket the streambed. Slow-moving streams mean less-dissolved oxygen for organisms. In turn, water temperatures rise and curtail or cause a change in the variety of organisms that normally thrive there. Oxygen levels fluctuate, causing contaminated sediments. This reduces the number of organisms that might inhabit the watershed if it were supplied with a normal amount of dissolved oxygen. In addition, stream banks are undercut by massive log and debris jams.

NOTES

....

....

....

....

....

....

....

....

....

....

....

....

....

LEARNING SET 1 IMPLEMENTATION

Learning Set 1

How Do Flowing Water and Land Interact in a Community?

The *Big Question* for this Unit is *How does water quality affect the ecology of a community?* So far you have considered what you already know about water quality. Now you may be wondering where the water you use comes from. If you live in a city or town, the water you use may come from a river. You would want to know the quality of the water you are using. To do so, it is important to know how the water gets into the river. You also need to know what happens to the water as the river flows across the land.

You may have seen rivers or other water bodies near your home, your school, or in your city. Think about the river closest to where you live. Consider where the water in the river comes from. If you have traveled along the river, think about what the land around the river looks like. Try to figure out what human activities occur in the area. Speculate as to whether these activities affect the quality of water in the river.

To answer the *Big Question*, you need to break it down into smaller questions. In this *Learning Set*, you will investigate two smaller questions. As you will discover, these questions are very closely related and very hard to separate. The smaller questions are *How does water affect the land as it moves through the community?* and *How does land use affect water as it moves through a community?*

Project-Based Inquiry Science

Learning Set 1

How Do Flowing Water and Land Interact in a Community?
20 min.

Introduce students to the questions of this Learning Set: How does water affect the land as it moves through the community? *and* How does land use affect water as it moves through a community?

META NOTES

The questions for the *Learning Set* provide a context for students to probe and suggest solutions to the bigger problem faced by the town of Wamego. Students will collaborate to investigate and put together science-based advice for the town to consider as it undergoes the significant changes described in the *Big Question*.

○ **Engage**

To get students thinking about the importance and complexity of where they live, display a large map of your town or county that labels towns and depicts residential streets, major roads, creeks, rivers, schools, playgrounds, shopping areas, parks, and ponds or lakes of the area. Have students classify these features in two lists as land and water areas.

△ Guide

Divide the class into groups of three or four students. Provide each group with a copy of the state map. Assign each group a quadrant of the map to determine where most cities and towns are located on the map. Have them look at the relationship between where people live and water resources. Listen for students to observe that most towns and cities are located near a body of water, even if it is a small stream. Lead students to observe that most large cities will almost always be located on a large river or near an ocean, lake, or gulf.

To get students thinking about how land and water in a community can affect each other, pose some questions about this relationship.

TEACHER TALK

"Imagine that you were a town planner in pioneer times. Your job was to study the state to find a location for a new town that would provide jobs, homes, businesses, and places for community activities. At the same time, you had to maintain the natural beauty and natural balance of the area you selected. What kind of place would you select? How could you plan to maintain the natural balance of the land and the water in the area? What questions might you ask before the town was developed?"

NOTES

....

....

....

....

....

....

....

1.1 Understand the Question

Think about the Questions

◀ *1 class period*

*A class period is considered to be one 40 to 50 minute class.

Overview

Students think about how moving water affects land and how land affects the movement of water. Students begin to connect what they already know and think they know to their local water system. A Unit *Project Board* is begun as students indicate what they think they know about water quality and what they would like to investigate about water quality. All of this prepares students for building a model watershed in *Section 1.2*.

Targeted Concepts, Skills, and Nature of Science	Performance Expectations
Water and land interact with each other.	Students should be able to conclude that water and land in a community interact and impact each other in positive and negative ways.
Criteria and constraints are important in design.	Students should be able to explain how the questions they develop for their *Project Board* meet the criteria listed on page 15 in the student edition.
Scientists often work together and then share their findings. Sharing findings makes new information available and helps scientists refine their ideas and build on others' ideas. When another person's or group's idea is used, credit needs to be given.	Students should give reasons for the questions that they thought of individually. As a group, they might refine and prioritize which questions to write on the *Project Board*.

Materials	
1 per group	Map of local, county, or state area
1 per class	Projection of a local body of water or an industry on water
1 per class	Class *Project Board*

Activity Setup and Preparation

Obtain a map that shows local rivers, branches of rivers, parks, and lakes, with cities and towns clearly marked. Ahead of time, look for an area that shows smaller streams feeding into larger streams. Some maps show shopping malls or details that would have large parking areas.

Homework Options

Reflection

- **Science Content:** Summarize the ideas you have about how water and land might interact in your community. *(Answers will vary with students' experiences but may include that water can change the land and that land can change the water by adding materials to it.)*

- **Science Process:** From your experience of working in groups, how do you think scientists make some decisions? *(Answers will vary but might include that scientists might share many ideas with other scientists before making a decision and that scientists don't always agree with each other's conclusions.)*

Preparation for 1.2

- **Science Process:** How would you model the flow of rainwater on a mountainside? *(Accept all reasonable hypotheses.)*

- **Science Content:** Think about a mountain and a valley below it. What do you predict will happen to some of rain that falls on the mountain during a thunderstorm? *(Students should be able to predict that some of the rain that falls on the mountain will run down into the valley.)*

SECTION 1.1 IMPLEMENTATION

◀ *1 class period* *

1.1 Understand the Question

Think about the Questions

The questions for this *Learning Set* are *How does water affect the land as it moves through the community?* and *How does land use affect water as it moves through a community?* It is a good idea to think about what you already know about how moving water affects the land and how the land affects the water. It is also important to think about what you are unsure about and what you would like to investigate.

Get Started

Think about these questions, and share ideas with your class about the source of your water and the lands it flows through. Listen carefully to all the ideas presented. You may want to write down some of the ideas you hear.

During the discussion with your classmates, you may have discovered that there are a few things you already know. You probably also discovered that there are many things you don't know yet. These are things you need to know to answer the questions. You are going to think of several questions that might help you to answer this *Learning Set's* questions and add them to the *Project Board*.

With your group, you are going to develop two questions that might help you understand how water changes as it moves through the land around communities. When you write your questions, keep in mind that your questions should

- be interesting to you,
- require several resources to answer,
- relate to the *Big Question* and the river ecology, and
- require collecting and using data.

Also, make sure you avoid yes/no questions and questions with one-sentence answers.

LT 15

LIVING TOGETHER

1.1 Understand the Question

Think about the Questions

20 min.

Students work in groups, like scientists, to think of questions that will help answer the Big Question *of the* Unit *about how water and land affect each other. They test their questions against a set of criteria and agree on two testable questions to include in the* Project Board.

○ Engage

Lead students to think more in depth about the *Big Question* by asking them what they already know about what happens when water moves over land. Ask them to give examples from their experiences.

TEACHER TALK

❝What experiences have you had where you have seen water and land interact with each other? Have you ever watched a heavy rainstorm?❞

*A class period is considered to be one 40 to 50 minute class.

"What happens when lots of water rushes along a gutter? Have you ever observed a layer of dusty rain spots on a window right after a rainstorm? How does the dust get onto the glass?"

Get Started

20 min.

Students think about what they should investigate to answer the question about how water flow affects and is affected by land.

unsure ab____ ___ you would like to inve___

Get Started

Think about these questions, and share ideas with your class about the source of your water and the lands it flows through. Listen carefully to all the ideas presented. You may want to write down some of the ideas you hear.

During the discussion with your classmates, you may have discovered that there are a few things you already know. You probably also discovered that there are many things you don't know yet. These are things you need to know to answer the questions. You are going to think of several questions that might help you to answer this *Learning Set's* questions and add them to the *Project Board*.

With your group, you are going to develop two questions that might help you understand how water changes as it moves through the land around communities. When you write your questions, keep in mind that your questions should

- be interesting to you,
- require several resources to answer,
- relate to the *Big Question* and the river ecology, and
- require collecting and using data.

Also, make sure you avoid yes/no questions and questions with one-sentence answers.

LT 15

LIVING TOGETHER

○ Engage

To get students going, give them an overview of what they will be doing. Inform students that they are going to develop two questions for a class project about how water and land affect each other.

"Writing questions does take some thought. First, by yourselves, use the four criteria on page 15 to write two questions about topics you would like to investigate about how water and land interact. Start by thinking about something that is of interest to you. Then, you will talk about these two questions in a small group. Don't worry, everyone in the small group will have written two questions, too. Then, as a group, pick two questions to talk about in a class discussion. The questions that the class decides on will be recorded on the *Project Board*."

△ Guide and Assess

If individual students have difficulty formulating questions, draw their attention to the list of criteria on page 15 in the student edition. Ask them to think about something about water or land in their community using these criteria.

It might help some students if you rephrase each criterion, such as:

- What is there about water and land in my city that I think is interesting?

- What resources will I need? That means, where will I find information that I need? Will I have to go to the library or on the Internet? Do I need special equipment?

- What does my question have to do with the *Big Question*? (Remind students of the *Big Question: How does water quality affect the ecology of a community?*)

- What kind of data will I need to collect to answer my question? Will I draw a picture? Will I take pictures with a camera? Will I have to make a table or draw a graph?

◇ Evaluate

As a class, have each group decide on two questions to add to present in a class discussion. Listen for how each group interacts. If a group has trouble agreeing on just two questions, tell them to select two and write a list of the others for possible later use.

Ask each group to present their questions.

△ Guide Presentations and Discussions

Ask each group to present their two submissions. Ask why the group selected these questions. Then, open the discussion to the class. You might begin by asking if other groups have the same questions or topics.

META NOTES

When a class discussion begins, students might have a hard time participating. If this is the case, model for students by taking one of the questions that has been submitted and begin to subject it to the four criteria on page 15 in the student text. This way, students will have a hint as to what to do and what words to use. They will feel that they can participate.

TEACHER TALK

"Now you have [eight to twelve] questions and you want to write just two on the *Project Board*. What would be a scientific way to decide on those two questions? What tool have you been using to write these questions? How can you use these criteria to narrow your choices? *(The criteria on page 15 is where you are trying to direct their attention.)*"

Update the Project Board

10 min.

Students work to fill in the second column of the Project Board, What do we need to investigate? *with two questions. Students also can add new information to the first column,* What do we think we know?

First, develop your own questions. When you have completed your two questions, take the questions back to your small group. Share all the questions with one another. Carefully consider each question and decide if it meets the criteria for a good question. With your group, refine the questions that do not meet the criteria. Choose the two most interesting questions to share now with the class. Give your teacher the rest of the questions so they might be used later.

How does water quality affect the ecology of a community? What advice should we give Wamego?				
What do we think we know?	What do we need to investigate?	What are we learning?	What is our evidence?	What does it mean for the challenge or question?

Update the *Project Board*

You will now share your group's two questions with your class. Be prepared to support your questions with the criteria you used. Your teacher will help you with the criteria if needed. Then your teacher will add your questions to the *Project Board.* Throughout this *Learning Set,* you will work to answer some of these questions.

Later in this *Learning Set,* you will conduct some investigations and use models to understand how water moves through the land, and how it affects the land it flows through. The investigations will require you to make careful observations and record all your results. The *Project Board* can help you to organize your work as you proceed.

LT 16

Project-Based Inquiry Science

△ Guide

Have each group present their two questions and give reasons or evidence that support their choice of questions. Questions that can be supported by evidence are questions that meet the criteria on page 15 in the student text.

Invite students to add new information that they have learned to the *What do we think we know?* column on the *Project Board.*

Assessment Options

Targeted Concepts, Skills, and Nature of Science	How do I know if students got it?
Water and land interact with each other.	**ASK:** What are three ways that land and water interact in your community? **LISTEN:** Students' answers will vary with their community and their experiences but some of the comments you might hear students relate would include that there are rivers or lakes near where they live. You might also hear that people visit these places, or that they have seen brown soil in a river after a storm, or that they have experienced water in a basement after a rainstorm.
Criteria and constraints are important in design.	**ASK:** What four criteria should you keep in mind when writing questions about a topic and designing an investigation? **LISTEN:** Students should be able to say that their questions should 1) be interesting, 2) need several resources, 3) relate to the *Big Question*, and 4) require collecting and using data.
Scientists often work together and then share their findings. Sharing findings makes new information available and helps scientists refine their ideas and build on others' ideas. When another person's or group's idea is used, credit needs to be given.	**ASK:** Why did you work together in a group before you wrote information on the class *Project Board*? **LISTEN:** Answers will vary but you may hear students describe that working together was part of the process for deciding what information was most important to put on the *Project Board*. **ASK:** In what ways do you act and think like a scientist when you work in a group? **LISTEN:** You should be able to hear students say that they act like scientists because they meet to consider a variety of questions and to decide on, or prioritize, which questions are the most important to investigate first.

LIVING TOGETHER

Teacher Reflection Questions

- Why might it be difficult for students to formulate questions for the *Project Board* that met the criteria listed on page 15 of the student text?

- What is one way to help members of groups to overcome possessiveness of their own ideas and come to a conclusion about the two questions to share with the class?

- How can students work more efficiently together in groups?

NOTES

SECTION 1.2 INTRODUCTION

1.2 Investigate

Model How Water Flows In and Around a River

Overview

Students review the water cycle and think about their own local water system. Then, students work in groups to build a model of how water moves across and through land. **NOTE:** *While students build a model watershed in this section, they will not be introduced to watershed as a vocabulary term until* Section 1.3. Students observe how water flows into and over their model and relate this to how water moves downward over landforms. Students will be able to predict that water flows from higher to lower elevations in a watershed.

◀ *1 class period*

*A class period is considered to be one 40 to 50 minute class.

Keep an eye out for slippery floors. Have plenty of paper towels on hand.

Targeted Concepts, Skills, and Nature of Science	Performance Expectations
Water in a watershed travels predictably, from higher to lower elevations.	Students should be able to predict that water moves from higher to lower elevations as long as natural conditions prevail.
Scientists use models to simulate processes that happen too fast, too slow, on a scale that cannot be observed directly (either too small or too large), or that are too dangerous.	Students should be able to build and use a model to demonstrate movement of water in a watershed. Students should move toward understanding that models or simulations represent natural phenomena, constructed to observe one feature or variable.
Predicting, observing, and explaining are important investigative skills.	Students should be able to demonstrate their investigative skills by observing the flow of water. They predict the flow of water from higher to lower elevations in all instances as long as normal physical phenomena persist.

LIVING TOGETHER

Materials	
1 per classroom	Roll of paper towels
1 per class	Class *Project Board*
1 per class	Large rectangular pan (jellyroll size)
1 per group	Large sheet of paper (butcher paper or other)
6-8 per group	Building blocks (different sizes; different heights)
1 per group	Spray bottle
1 per group	Pair of red and blue permanent markers or crayons
1 per classroom	Water (optional: colored for easier observation)
1 per group	Bits of clay or two-sided masking tape
(optional)	Large rectangular pan (jellyroll size)
1 per classroom (optional)	Projection of model from student edition
1 per class	Masking tape
1 per group	Overhead transparency

Activity Setup and Preparation

- Ahead of time, prepare a model to make sure the materials work to use as a demonstration as necessary.

- If the blocks or structures used to represent mountains or buildings are unsteady in the pan, put a small piece of clay or two-sided masking tape under each one ahead of time to hold each one in place.

- Test different kinds of paper for covering the blocks to see which type will work best for the students. The water should flow down and around the different objects under the paper. It should pool in some areas, run in others.

- Test how many squirts from the spray bottle will be needed to get water flowing.

- Make sure the water bottles produce a fine spray and not a jet of water. Test ALL spray bottles that will be used by students.

- Note any issues in the model that may be challenging to students, such as crumpling the paper and getting it to drape over the blocks.

- Prepare a materials station. Each group should have one pan, two different markers, six to eight different-sized blocks, and paper for draping.

- Make tape and clay available for the class. Caution students that if they need to use the tape or clay, that they use it before they begin to use the spray bottles.

- Do not include spray bottles or paper towels at the materials station.

NOTES

....

....

....

....

....

....

....

....

....

....

....

Homework Options

Reflection

- **Science Process:** Study the photograph on page 17. Use it to design a model on paper. Predict what will happen to water when it rains in this area. Draw arrows to show the direction water flows. Write a caption that includes your prediction. *(Students' drawings should include high and low places. Some might notice that the most obvious water in the photograph is at the lowest elevation. Captions should indicate that water will flow from high places down to the lake.)*

- **Science Process:** A scientist wants to study how water moves through a section of his/her state. What might she/he do to learn where water flows right now? *(Answers will vary as scientists use many kinds of models to predict outcomes. The scientist might make a physical model or a computer simulation showing the elevations as they are now.)*

Preparation for 1.3

- **Science Content:** Research the difference between runoff and groundwater. Look back at your model and state where water might have moved as runoff and where it might have soaked into the ground. *(Runoff is water from rain or melted snow that moves over the surface of the land. Groundwater is water that is below the surface of the ground. Runoff might have run over rocks or roads. Groundwater would have soaked into the ground that is covered with plants.)*

- **Science Content:** How does the model you made compare with the definition of a watershed on page 21? *(Students should begin to recognize that the model they built of blocks of various elevations and some pooling of water is a model of a watershed.)*

SECTION 1.2 IMPLEMENTATION

◀ *1 class period* *

1.2 Investigate

Model How Water Flows In and Around a River

Building **models** is one way scientists are able to re-create the real world in a lab. Scientists try to represent a phenomenon they are investigating.

model: a way of representing something in the world to learn more about.

simulate: to imitate how something happens in the real world by acting it out using a model.

elevation: the height of a geographical location above a reference point, usually sea level.

Scientists use models to simulate processes they can not closely examine in the real world.

They try to make the model as accurate as possible. In this investigation, you are going to build a model to **simulate** how a river flows. You will use your model to investigate how changes in the areas near the river might alter how the water moves. When you build the model, think about how it is similar to or different from a real river.

Design and Build Your Model

Look at the diagram on the next page. Your group will build a model that looks similar to what you see in the diagram. It doesn't have to be exactly what is pictured here. However, you need to have some high spots—areas of high **elevation**. You also need some low spots—areas of low elevation. Discuss with your group where these spots should be.

LT 17

LIVING TOGETHER

1.2 Investigate

Model How Water Flows In and Around a River

5 min.

Students will build a model of a watershed, although they will not be introduced to the term until Section 1.3. *Asking students to come up with their own ideas of models will prepare them to recognize some of the similarities and differences between the model they build and what actually happens in real life.*

○ **Engage**

Have students review the *Project Board* they created to remind them of the questions they decided to focus on. Ask students to describe what they have seen happen when rainwater falls and list their observations. Ask students how they might model what happens when rain falls in and around land. Create a list of their ideas.

*A class period is considered to be one 40 to 50 minute class.

TEACHER TALK

"Study the photograph on page 17. How would you describe the land? Is it flat or is it hilly? How is the photograph of land with hills and mountains different from land that is flat? *(Land with hills and mountains have high and low areas. Explain that these differences are differences in elevation.)*"

△ Guide

Show students the model that you prepared. Let them look at it from the side so that they can get an idea of differences in elevation. Let them describe differences in the height of various blocks you might have used. Explain that the tops of tall blocks are at a higher elevation than the tops of shorter blocks or the bottom of the pan when looked at from the side.

△ Guide and Assess

Check for students' understanding of the idea of elevation. If students have trouble with this concept, have them practice describing the elevation of structures in photographs on pages 14 and 17 in their text or even structures that they might be able to see from the classroom window. Have them point out areas that are at low elevations (*the river on page 14 and the lake on page 17*) and areas that have a higher elevation (*the tall building or hills*). Have them use the term *higher elevation* when pointing out the tops of taller buildings and the term *lower elevation* when pointing out the river and the lake.

META NOTES

Students may have a difficult time with the idea of elevation. Using a variety of examples in the classroom and even photographs can eventually help them develop a meaning for the term.

NOTES

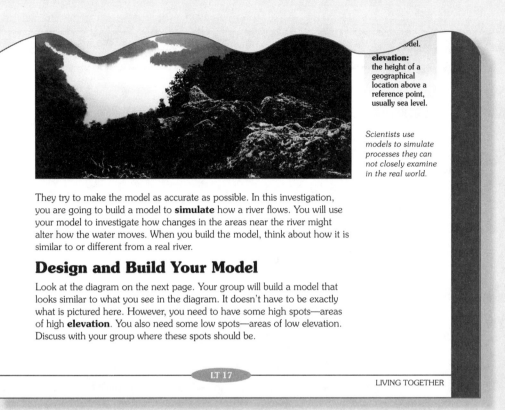

elevation:
the height of a geographical location above a reference point, usually sea level.

Scientists use models to simulate processes they can not closely examine in the real world.

They try to make the model as accurate as possible. In this investigation, you are going to build a model to **simulate** how a river flows. You will use your model to investigate how changes in the areas near the river might alter how the water moves. When you build the model, think about how it is similar to or different from a real river.

Design and Build Your Model

Look at the diagram on the next page. Your group will build a model that looks similar to what you see in the diagram. It doesn't have to be exactly what is pictured here. However, you need to have some high spots—areas of high **elevation**. You also need some low spots—areas of low elevation. Discuss with your group where these spots should be.

LT 17

LIVING TOGETHER

Design and Build Your Model

30 min.

Students design, gather materials, and build their own model. They consider what parts of the model represent high elevation areas, such as mountains and low elevations where a river might flow.

△ Guide

As students design and construct their models, watch for problems in their models similar to any you experienced while building your own model. Offer suggestions based on how you solved the same or similar problem.

Focus students' attention on what their goals are (*to construct a model using objects of different heights that will result in a flow of water*) by having them answer the two questions on page 18.

1. Student answers will vary but should include: Tall and short blocks, which change the elevation of the land, represent hills or mountains. Water in the spray bottle represents rainfall. The crumpled paper cover represents land. Rivers, ponds, and lakes will be represented by water as it runs or collects on the paper.

2. Student answers will vary.

Predict

5 min.

Students predict the outcome of their experiment based on the structure of their model. By making predictions, students formulate ideas about how water will flow.

META NOTES

Prediction in scientific research is important. In real-world situations, scientists make models and predict what they think will happen during an investigation. If a prediction is supported again and again by investigation, the scientific community can accept the idea.

If a prediction is disproved or not supported by investigation, then the prediction needs to be altered or changed in some way and tested again.

Materials
- large sheet of butcher paper
- building blocks
- spray bottle
- water
- large pan
- blue and red permanent markers or crayons
- overhead transparency sheet

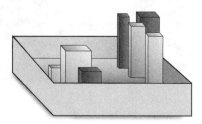

Use the building blocks to create areas of higher and lower elevation. Arrange the objects in your tray. Make sure they are spread around and that there is one big object at one end of the tray. Make sure you have a low area running through the middle of your model. This will represent the river.

Crumple a large piece of butcher paper. Be careful not to rip any holes in the paper. Uncrumple the paper.

Carefully cover the objects with the butcher paper by pressing the paper down around them. Use tape to secure the paper to the base of the pan.

Answer the following questions:

1. Which features of a river and the surrounding area do the objects in your model represent?

2. Why did your group decide to arrange the objects the way you did?

Predict

claim: a statement about what a trend means

Use a marker. On the butcher paper in your model, write an H on the high parts, and an L on the low parts. Look at your model carefully and observe the way the paper folds and dips around the objects. Think about what would happen if you sprayed water on the paper, as if it were raining. Imagine how the water would move along the paper. Think about where the water would flow. Where would the water pool or puddle? Where would the water drip into the pan?

On a piece of paper, draw a box to represent the pan used in the model. Write H and L to represent the areas you marked on the butcher paper. Make your sketch as accurate as possible. Use a blue marker or crayon to draw in the paths you think water would take if it rained on your model. Use arrows to show the direction the water will flow.

Draw in the places where the water will pool. Under your picture, write a few words describing what you think will happen when it rains on your model. You can use arrows to connect your words to the different parts of your drawing.

△ Guide and Assess

Determine if students have marked their models properly with H's and L's for high and low portions of the model.

Check to see that each student has drawn a paper plan that matches their model, including where the high (H) and low (L) elevations on their model are located. Urge students to mark arrows on their paper plan that represent their prediction of the paths water will take as it moves through the model and where water will pool or run.

To save time, ask students for their predictions as you check their models. The prediction should describe the path they think water will take. If they have not made a prediction, suggest two choices for them to pick from such as "Do you think that the water will flow up the taller structures on the model?" or "Have you ever watched water on a window during a rainstorm?" "Where does that water move? What direction do you think the water will move on your model?"

To begin, ask group members to describe what they saw to the rest of the group. The group will discuss reasons for any differences in their observations and any difficulties they had with their observation procedures. Then groups will categorize their observations. First, they should copy each observation onto a sticky note and group these sticky notes into categories by behavior. Then, they should label the categories using index cards. After categorizing and labeling their observations, groups will identify trends by writing descriptions of the behaviors in each category.

NOTES

....

....

....

....

....

....

....

....

....

....

....

Run Your Model

10 min.

When students run their experiments, they test the predictions they made earlier. Simulations can help students to visualize how land affects the flow of water.

META NOTES

Students should see alternative ways of categorizing the data and analyze the effectiveness of each method.

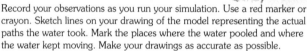

1.2 Investigate

Run Your Model

Now you will simulate how water moves on land when it rains. Use the spray bottle to lightly spray water on your model. The rainfall is supposed to be a light shower. Spray the water carefully so that all parts of your model get wet like they might in a real rainstorm. Everything should get wet equally.

Observe

Observe the way the water flows. When you spray at the top of the highest structure in your model, observe how and where the water flows. Look at where the water stops.

Record your observations as you run your simulation. Use a red marker or crayon. Sketch lines on your drawing of the model representing the actual paths the water took. Mark the places where the water pooled and where the water kept moving. Make your drawings as accurate as possible.

Make sure that when you spray water on your model, the water goes into the pan, *not* on the floor. Wipe up any spills immediately.

Explain

Consider the blue lines in your drawing that show what you predicted would happen when you added the water to your model. Compare the blue lines to the red lines that you drew to show how the water actually ran. Think about how accurate your predictions were. Did the water flow the way your predicted it would? Describe your results. Use a sentence like this: "I thought the water would flow ….. and when I sprayed water on my model I noticed that the water flowed ….. I think the water flowed the way it did because…" Use a *Create Your Explanation* page. Remember that a good explanation connects your **claim** to your evidence and science knowledge in a logical way.

Communicate Your Results

Investigation Expo

Scientists always share their understandings with one another. Presenting their results to others is one of the most important things that scientists do. You will share what you have found in an *Investigation Expo*. To prepare for this, you will use an overhead transparency.

LT 19

LIVING TOGETHER

⚠ Guide

Before passing out the spray bottles, have each group decide who will spray the water. Check to make sure the spray tops are on tightly. Demonstrate how to use the spray bottle to get a fine rainfall. Caution students that the water should not come out as a torrent. Advise them to spray above the top of the highest structure. They should continue to spray until the water flows down the model and collects in pools or streams.

"Think carefully about what you are going to do. Remember that several things are going to happen all at once. One of you will spray water onto your model. ALL of you will have to watch how the water moves and where it goes once it builds up on the model. At some point, there will be enough water so that it flows. Watch carefully to see where it flows first and in what direction it flows.

Remember that you have marked your paper plan with a prediction of where water would flow. After water flows over the model, you will mark your paper plan with the actual pathway the water took."

Observe

Observe the way the water flows. When you spray at the top of the highest structure in your model, observe how and where the water flows. Look at where the water stops.

Record your observations as you run your simulation. Use a red marker or crayon. Sketch lines on your drawing of the model representing the actual paths the water took. Mark the places where the water pooled and where the water kept moving. Make your drawings as accurate as possible.

Observe

5 min.

⬡ Get Going

Pass out the spray bottles. One student should spray. Everyone in the group should watch closely to track the path of the water and describe whether and where it forms rivers or pools.

☐ Assess

Watch as students spray water on their models to make sure they are creating a fine rainfall. Listen for voiced descriptions of where the water is moving. Collect spray bottles as students complete the simulation.

Caution students to wipe up spills on tables or floors immediately.

META NOTES

Because all members of the group are responsible for watching what happens to the water as it is sprayed onto the model, there may be a variety of views of what actually happens. Careful observation increases the accuracy of the observation.

NOTES

....

....

....

Explain

5 min.

Students start to articulate ideas based on their observations. They do this first independently and then in small groups.

Explain

Consider the blue lines in your drawing that show what you predicted would happen when you added the water to your model. Compare the blue lines to the red lines that you drew to show how the water actually ran. Think about how accurate your predictions were. Did the water flow the way your predicted it would? Describe your results. Use a sentence like this: "I thought the water would flow ..... and when I sprayed water on my model I noticed that the water flowed ..... I think the water flowed the way it did because…" Use a *Create Your Explanation* page. Remember that a good explanation connects your **claim** to your evidence and science knowledge in a logical way.

Make sure that when you spray water on your model, the water goes into the pan, *not* on the floor. Wipe up any spills immediately.

△ Guide

Draw students' attention to the questions (*Did the water flow the way you predicted it would flow? and so on*). Have students use the water flow diagrams they drew for the investigation to answer the questions individually.

Explain to students that they will present their results as a scientific explanation. They will compose a statement that connects a claim to evidence and science knowledge that supports the claim in a logical way. Provide a *Create Your Explanation* page for students to use to organize evidence, in support of their claim about how water moved in the model. Review that a *claim* is a statement or conclusion reached from one or more investigations. *Evidence* is the data or observations made during an investigation. *Science knowledge* is information about how things work as described by experts during previous investigations.

Then walk through points on page 19 that students should think about as they make their explanation. For instance:

TEACHER TALK

❝When you begin to think about how to explain what happened during your investigation to the class, start by making a list of your claim, your evidence, and the science knowledge that backs your claim. Talk with each other to make sure you have a clear statement for each one of these items. Collect the model and the drawing of the model that you made and marked. When you give your explanation, you can use the wording on page 19, such as *I thought the water would flow… I think the water flowed the way it did because…*❞

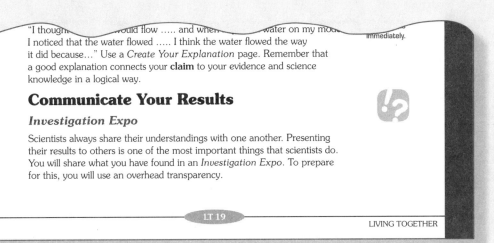

"I thought ......would flow ..... and when ......water on my mod...... immediately.
I noticed that the water flowed ..... I think the water flowed the way
it did because..." Use a *Create Your Explanation* page. Remember that
a good explanation connects your **claim** to your evidence and science
knowledge in a logical way.

Communicate Your Results

Investigation Expo

Scientists always share their understandings with one another. Presenting
their results to others is one of the most important things that scientists do.
You will share what you have found in an *Investigation Expo*. To prepare
for this, you will use an overhead transparency.

LT 19

LIVING TOGETHER

Communicate Your Results
Investigation Expo
10 min.

*During the class
discussion, students
begin to find patterns
in the observations
and results presented.
They will share their
observations, as would
scientists in the real
world.*

○ Engage

Explain that students will be presenting as a group. Provide a blank
overhead transparency and marking pen for each group to trace their
model's results. Give each group a few minutes to draw their prediction
and actual results on the transparency.

△ Guide Presentations and Discussions

Inform students that as each group presents its results, members of the
audience can politely ask questions about the model, the results, and the
group's conclusions. Begin a discussion after each group presents. You may
need to model asking questions. Then ask a student to ask a question. Make
sure the presenters respond to the student who asks the question and not to
you. During the discussion, encourage students to use respectful language
such as:

> **TEACHER TALK**
>
> **"**I agree with you because...
>
> I disagree because...
>
> Could you clarify...?
>
> How does that statement support your claim that...?
>
> What did you observe that makes you think...?
>
> Based on your observations, would you predict that...?**"**

△ Guide

Remind students that a claim is a statement about a trend or a pattern that
they see forming. They may not recognize it as a pattern until everyone in
the class discusses their results during the *Investigation Expo*.

Trace the diagram that you drew of your model onto an overhead transparency. Be ready to describe your investigation and clearly detail all your results. The answers to the following questions will be very helpful in preparing your presentation.

- Describe your predictions. (What patterns did you think you would see?)
- Describe why you think your prediction was accurate or inaccurate.
- Describe how the water moved in the model. (What patterns did you see?)
- How did the outcome compare to your prediction? Where did the water flow more quickly? How was the flow different from what you predicted? Did the water pool where you thought it would? How was it different?
- Explain why the water flowed and pooled where it did.
- Describe what you learned about water flow.

As you look at the overheads presented by other students, make sure you can answer these questions. Ask questions that you need answered to understand the results and the explanations others have made.

What's the Point?

In this section, you built a model to simulate how water flows across a landscape when it rains. Placing different-sized objects under the paper created the higher and lower elevations. You also drew a sketch of the model, and you predicted how water would run over the paper. When you ran the simulation you probably noticed that the water always moved from areas of high elevation to areas of lower elevation. Water cannot move uphill. You also noticed that water flowed and created puddles in several places as it flowed. These puddles represent lakes or ponds in the real world.

Water on land works that way too. If you watch where rain falls in one rainstorm, you will be able to predict the path water will take in the next rainstorm. This will be the case as long as the land stays the same from the first rainstorm to the next. But if the land changes, the water will flow and pool differently. You will need to consider how new construction in Wamego might change the land and affect how water flows.

Have students think about their predictions and compare them to their observations of how water flowed. (*Most predictions probably stated that water would flow downward.*)

To tie together the class discussion, ask students to look at the overhead transparency from all groups. Ask students what similarities they see between their group's model and the results of other groups.

◇ Evaluate

Conclude the class discussion by eliciting from the group that there are a number of patterns. Listen for students to articulate that although the models were different, there are overall patterns among them. The flow of water was, in each model, from areas of higher elevation to areas of lower elevation. A list of possible patterns may contain the following:

- Water flows from high to low elevations.
- Depressions create opportunities for water to pool into "lakes."
- Water flows in predictable patterns as long as the surface is not changed.

TEACHER TALK

"Now that you have had the opportunity to look at each group's model, what can you conclude about how water behaved in each of the models? Is there a pattern among them? How did the water flow in each model? What did you see that you might be able to use to sum up how water generally behaves?"

Encourage students to develop at least two statements that reflect the goals of the section, such as: 1. Water flows from higher to lower elevations. 2. One can predict that unless some condition changes, water will always flow from higher to lower elevations and not in the opposite direction. 3. A model can help us understand physical phenomena.

To close the discussion, ask students to comment on how working with a model helped them to understand how scientists develop new ideas and solve some problems.

NOTES

....

....

....

....

....

Assessment Options

Targeted Concepts, Skills, and Nature of Science	How do I know if students got it?
Water in a watershed travels predictably, from higher to lower elevations.	**ASK:** Describe the patterns you observed in your model. **LISTEN:** Students should say that water travels from higher elevations to lower elevations. **ASK:** What happens when water moves downward? **LISTEN:** Answers may vary but you should be able to hear students say that water flows downward and may form rivers. They might also say that some water collected at the lowest levels in lakes or ponds.
Scientists use models to simulate processes that happen too fast, too slow, on a scale that cannot be observed directly (either too small or too large), or that are too dangerous.	**ASK:** What is a model? **LISTEN:** Students should say that a *model* is a representation of something. A model helps you gain an understanding of how something works. A model makes it easier to think about the parts of something that is too large to study all at once. **ASK:** What did you model in this lesson? **LISTEN:** Students may say that they modeled how water moves in and around land. Remember that they have not learned the term *watershed* yet.
Predicting, observing, and explaining are important investigative skills.	**ASK:** What does it mean to be predictable? **LISTEN:** Students might say that being *predictable* means that under the same conditions, a person is able to say that something will act the same every time. **ASK:** What was predictable about water in this investigation? **LISTEN:** Students should be able to say that water always moves downward from higher to lower elevations.

Project-Based Inquiry Science

Teacher Reflection Questions

- What issues did students have in trying to make their watershed models different from the one depicted in the text or the one built by you?

- What problems did students encounter in making drawings of how the water in their model behaved? How can these problems be overcome?

- How can students be helped in situations where they have to make choices as a group, such as where they had to choose two questions to investigate?

NOTES

....

....

....

....

....

....

....

....

....

....

....

....

1.3 Read

What is a Watershed?

1 class period* ▶

*A class period is considered to be one 40 to 50 minute class.

Overview

In *Section 1.2*, students built a model that gave them evidence that water moves from higher to lower elevations. This was a basic setup for a watershed, although students were not given that term. In this section, students are introduced to the term *watershed*. They are introduced to the parts of a watershed and learn that slope affects how water flows through a watershed. Students learn that some water moves over land as runoff and some soaks into the soil to become groundwater.

Targeted Concepts, Skills, and Nature of Science	Performance Expectations
A *watershed* is the land area from which water drains into a particular stream, river, or lake.	Students should be able to identify the parts of a watershed and explain that as water falls on an area, it is absorbed in that area and/or drains into a river system. Everything on the land that drains into a river is in the watershed.
Water in a watershed travels predictably, from higher to lower elevations.	Students should be able to predict that water moves from higher to lower elevations in a watershed so long as natural conditions prevail.
Water movement shapes the land, carving out rivers and streams.	Students should be able to infer that the shape of the land is, in part, the result of the movement of water.

Materials	
1 per classroom	Projection of watershed
1 per group (optional)	Kitchen sponge
1 per group (optional)	Metal or plastic pan

Materials	
1 per group (optional)	Spray bottle with water
1 per classroom (optional)	Source of water
1 per group (optional)	State or local map, an atlas, and/or topographic map

Activity Setup and Preparation

Optional: To illustrate a basic concept of groundwater, set up a demonstration with an ordinary rectangular kitchen sponge, a pan, and water in a spray bottle.

Place a dry sponge in a dry pan. At one end of the pan, add a small amount of water and tilt the pan slightly so that the water flows into the sponge. Continue to add small amounts of water very slowly, letting the sponge absorb the water. At some point, the sponge will become saturated. Excess water will begin to flow out of the sponge. Have students use the sponge as an analogy for the movement of water through the pores in rock. Water moves slowly through the pores of the sponge as groundwater moves slowly through pores in rocks and soil.

This setup can also demonstrate the effects of slope. If tilted, the water will move through the pores of the sponge more quickly. In nature, gravity acts on water in groundwater rocks. As slope increases, gravity increases as well, causing water to move through groundwater at a faster rate than it would if the terrain were flat.

Homework Options

Reflection

- **Science Content:** Look at the drawing of the watershed on page 21. Make a copy of the drawing and mark the high (H) and low (L) elevations. Draw arrows to show the direction in which water would move in this watershed. *(Students' drawings should include high and low places. Arrows should indicate that water is flowing down the slopes.)*

- **Science Content:** Three small streams feed into a large river. How do you know that the three streams are part of the same watershed as the large river? *(The three streams are feeding into the large river.)*

Preparation for 1.4

- **Science Process:** Scientists called geologists use many kinds of maps in their research. Find out the differences between road maps, relief maps, and topographic maps. Write a paragraph that explains the important features of each kind of map. (*Road maps are two-dimensional. They show roads, town and cities, and give some elevations; topographic (topo) maps are two-dimensional but show much more detail in terms of structures on land. Topo maps also show every change in elevation by a series of lines. Relief maps are three-dimensional and actually have raised areas indicating differences in elevations that can be felt.*)

NOTES

....

....

....

....

....

....

....

....

....

....

....

....

....

Project-Based Inquiry Science

SECTION 1.3 IMPLEMENTATION

1.3 Read

What is a Watershed?

You just created a model of a river system. You included the river and all the land that drained rainwater into the river. All the water in your model came from rain on the land around the river. The water flowed into the river system. The river and all the land that drains water into the river is called a **watershed**. Everything that sits on the land that drains into a river is *in* the watershed.

For example, imagine a house sitting on top of a hill in the area around a river. That house would be in the watershed. The rain that falls around and on the house, including the driveway and garden, would drain into the river. Water might have to travel very far, but eventually all water ends up in a river or lake.

The diagram shows a watershed. Watersheds come in all shapes and sizes. They cross county, state, and national boundaries. No matter where you are, you are in a watershed!

In a watershed, water flows from higher to lower elevations. The dotted lines in the illustration mark the boundaries of the watershed. The boundaries are in areas of higher elevation. Land on one side of the boundary is in one watershed. Land on the other side of the boundary is in another watershed.

How Does Water Move in a Watershed?

How the water moves in a watershed depends on the shape of the ground. If there are dips in the ground, then water might pool there. When the elevation is high (like the parts marked H on your model), water will run off and move toward areas of the ground that are lower in elevation. The shape of the land determines how fast the water flows. When the **slope** of the land is steep, as shown in the picture at the top of the next page, water runs off very quickly. If the slope is gentler, the water will run off less quickly.

When it rains, water lands on the ground. Then it moves across the ground. The water will continue to run off the ground to collect in lower areas.

watershed: the land area from which water drains into a particular stream, river, or lake.

slope: a measure of steepness. It is the ratio of the change in elevation to the change in horizontal distance.

LT 21

LIVING TOGETHER

1.3 Read

What is a Watershed?
10 min.

Students define watershed *based on their own experiences building a watershed model.*

◯ **Engage**

Engage students by showing a projection of the watershed on page 21 and having them compare it to the sketches they made for the investigation in *Section 1.2*. Then have students read the opening paragraphs for the section, which formally introduces the term *watershed*. In this way, students realize that they already have some experience with the concept of a watershed and can now identify it with a technical term.

*A class period is considered to be one 40 to 50 minute class.

How Does Water Move in a Watershed?

20 min.

Ideas about how and where falling water moves once it hits the ground are extended.

...in the ... boundaries of ... boundar... are in areas of higher elevation. Land on one side of the boundary is in one watershed. Land on the other side of the boundary is in another watershed.

How Does Water Move in a Watershed?

How the water moves in a watershed depends on the shape of the ground. If there are dips in the ground, then water might pool there. When the elevation is high (like the parts marked H on your model), water will run off and move toward areas of the ground that are lower in elevation. The shape of the land determines how fast the water flows. When the **slope** of the land is steep, as shown in the picture at the top of the next page, water runs off very quickly. If the slope is gentler, the water will run off less quickly.

When it rains, water lands on the ground. Then it moves across the ground. The water will continue to run off the ground to collect in lower areas.

...ter drains into a particular stream, river, or lake.

slope: a measure of steepness. It is the ratio of the change in elevation to the change in horizontal distance.

LT 21

LIVING TOGETHER

○ Engage

Ask students how fast they think water flows from one place to another in a watershed. If they hesitate to answer, ask a student how fast water moved in their model watersheds *(sometimes slow; sometimes fast)*.

TEACHER TALK

"What is the reason for the different rates of flow? What do you think made water move faster in some places and not so fast in others? *(Water flowed faster in places where it was moving from very high or tall elevations to lower elevations.)*

How fast did it move when there wasn't much difference between the elevations?" *(Students should say that when there wasn't much difference in elevation, water moved more slowly.)*

Have students look at their models (*or the one that you made*) in *Section 1.2* from the side. Formally introduce the term *slope* and relate it to the steepness of the land. Explain that the degree of steepness is the slope of the land. To underscore the concept, have students describe a slope that is very steep and one with a lesser slope.

TEACHER TALK

"What gives your model its shape?" *(The different elevations. Some differences in elevations are steeper than others.)*

☐ Assess

Have students look at their models again. Listen as students use the term slope in describing parts of their models. Urge them to use other terms in their descriptions such as *high elevation* and *low elevation* as well.

⚠ Guide

Formally introduce the concepts of runoff and groundwater by reading the first three paragraphs on page 22 of the student text.

Ask students to think about what happens when rainwater falls on the ground.

TEACHER TALK

"What do you think happens to all the water that falls from clouds? What kinds of surfaces does it fall on?" *(Students should say that some of the rain falls on soil and plants and some falls on hard surfaces such as roads and sidewalks.)*

TEACHER TALK

"What are some of the hard surfaces that you see every day?" *(Students should name examples like roads, sidewalks, and parking lots as familiar hard surfaces.)*

Call on students' experience with seeing a summer rainstorm to foster an understanding of the term *runoff*. Explain that runoff happens on the top of the ground. Ask students what other hard surfaces would produce runoff. Guide students to understand that buildings are made of materials with hard surfaces that water rushes over. If students aren't forthcoming with additional suggestions, ask them what materials buildings are made of. *(Hard materials such as brick, granite, concrete, and steel.)*

TEACHER TALK

"When large amounts of water fall on a hard surface during a heavy rainstorm, what happens to the water?" *(Answers should include that it runs in sheets across these surfaces, sometimes with small waves, and flows into the street, often swirling in the gutter before flowing down the drain; eventually runoff flows into a stream.)*

Then focus students' attention on what runoff looks like as it runs across a hard surface.

TEACHER TALK

"How is rain that falls on a park or a farm different from runoff?" *(Answers should include that it soaks into the ground.)*

Refocus on the other possibility of water soaking into the ground. Explain that water that soaks into the ground becomes part of the groundwater. Explain that groundwater moves slowly under the surface, through the soil and eventually moves to a river or stream.

□ **Assess**

TEACHER TALK

"Where did you see runoff and groundwater in the model you created in *Section 1.2?***"** *(Answers should include that runoff formed as rainwater hit hard surfaces such as buildings (the blocks) and other hard areas. It had no place to go except along the surface. Rain that soaked into the paper in some places represented groundwater.)*

NOTES

runoff: water from rain or melted snow that moves over the surface of the land.

groundwater: water that is located below the surface of the ground.

Scientists call the rainwater that hits the ground and that moves on the ground **runoff**. In your model, you saw a lot of runoff because your paper did not let the water go through.

But not all the water that falls on the ground will run off. Some will run into the ground. The soil can absorb (soak in) some water. This is one place that your model was not a very good model of the river area.

The process of absorption is very important. Water that is absorbed into the soil is used by plants and animals that live underground. Once the water is absorbed by the ground, it moves through the ground, always downhill, toward the river. The water that is absorbed by the land and moves under it is called **groundwater**.

Stop and Think

1. In your own words, describe a watershed.

2. In which direction does water flow in a watershed?

3. What is the difference between surface runoff and groundwater?

4. Describe two ways that your model was

   a) the same as a real watershed.

   b) different from a real watershed.

What's the Point?

Watersheds include all the land that surrounds a river. The water in a watershed falls in the watershed and flows to the river. The water can run off the land and flow into the river. This water is called runoff. Sometimes the water is absorbed by the land and flows under and through the soil. This water is called groundwater. Although you cannot see the groundwater, it moves to the river just like the runoff. Plants use the water that is absorbed by the soil.

LT 22

Project-Based Inquiry Science

Stop and Think
10 min.

Help students use this review to make a firm connection between what occurred in their model and what happens in a real watershed.

META NOTES

During a class discussion, students' concepts may be modified and become more concrete.

Ask students to answer questions 1 through 4 on page 22 in preparation for a class discussion. Students' answers to question 4 may vary but will form a pool of possible answers for use during the class discussion.

1. A watershed is an area of land that drains into a stream, river, or lake.

2. Water flows from high elevations to low elevations.

3. Surface runoff is water that flows on the surface of the land. Groundwater is water that seeps into the ground but eventually moves into a river, stream, or lake.

4. a) Similarities to a real watershed: The paper on the model is like land; it flows from higher to lower elevations; absorption by the paper in the model is like ground absorbing water; water flowing over the paper is like runoff; the spray of water from a water bottle represents rainfall; differences in elevation in the model represent the various changing slope of land.

b) Differences from a real watershed: Paper does not absorb water the way soil absorbs it; land materials in a real watershed are made up of rocks, soil, cement, and grass, whereas paper represented the land in the model watershed.

Assessment Options

Targeted Concepts, Skills, and Nature of Science	How do I know if students got it?
A *watershed* is the land area from which water drains into a particular stream, river, or lake.	**ASK:** A harmless yellow dye was put in a small mountain stream. Several days later, the dye appeared further down the mountain in a river 10 miles away. What is the relationship between the mountain stream and the river?
	LISTEN: Students' responses should include that the stream and the river are part of the same watershed and that water containing the dye flowed down in elevation into the river, which is at a lower elevation.
Water in a watershed travels predictably, from higher to lower elevations.	**ASK:** During a heavy rainstorm, water rushed down the streets in a town and quickly raised the level of water in the river water. Would you describe the slope of the land the town is built on as steep or level with the river? Was the river fed by runoff after the storm or by groundwater? Explain your choices.
	LISTEN: Students should respond that the slope of the land is steep because the water moved quickly down the streets to the river. The river is fed by runoff because the level of the river rose quickly. It would have risen more slowly if fed by groundwater because the water would have moved more slowly through the ground.

Targeted Concepts, Skills, and Nature of Science	How do I know if students got it?
Water movement shapes the land, carving out rivers and streams.	**ASK:** How does water movement shape the land? **LISTEN:** Students should be able to explain how groundwater, surface runoff and watersheds create rivers, lakes, and streams.

Teacher Reflection Questions

- In this section, ideas that students began to form about watersheds were affirmed and extended by their experiences with models. What difficulties did students have with these ideas? What concepts should be revisited to further develop students' understanding?

- The goal of the *Project Board* is to support students' learning through public presentation and consideration. By sharing their ideas with others, students make their own ideas more concrete and are open to the thinking of others. How did the *Project Board* assist students in formalizing the targeted concepts?

- The content reading in this section may be difficult for middle school students. Some students may have difficulty assessing the high level of science content. What can you do to further support students in assessing the reading?

1.4 Explore

A Case Study: Watersheds in Michigan

2 class periods *▶*

A class period is considered to be one 40 to 50 minute class.

Overview

Students have learned about watersheds in *Sections 1.2* and *1.3* but they have not had to apply it to a new context. In this section, students learn that water flows from areas of high elevation to areas of lower elevation in a watershed in Michigan. Students use raised relief maps of Michigan and trace a two-dimensional map of movement of water within a small watershed of their own choosing, then through major watersheds in Michigan to the Atlantic Ocean.

Targeted Concepts, Skills, and Nature of Science	Performance Expectations
Water in a watershed travels predictably, from higher to lower elevations.	Students should be able to predict that water moves from higher to lower elevations in a watershed as long as natural conditions prevail.
Watersheds define the flow of water from an area of land into a river system and the flow of river systems into lakes and oceans. Nested watersheds are smaller watersheds that are a part of larger watersheds.	Students should be able to understand that a large watershed is made up of and fed by many smaller nested watersheds, also called sub-watersheds.
Students read raised relief maps and know the difference between three-dimensional and two-dimensional maps.	Students should be able to identify high and low areas on a raised relief map by feeling the differences in the elevations. Students should be able to tell that using a relief map is more useful than a two-dimensional photograph of the same area.
A *watershed* is an area of land on which water falls. Some of the water is absorbed and some is runoff as it drains across the land's surface and into a river system.	Students should be able to correctly define terms such as *runoff*, *watershed*, and *groundwater*.

Materials	
1 set per classroom	Laminated photographs of a variety of watersheds
1 per group	Washable transparency markers
2 per class	Raised relief map of the state of Michigan
1 per group	Conventional paper map of Michigan
1 per class	Transparency of two-dimensional relief map of Michigan
2 per group	Overhead transparency

Activity Setup and Preparation

Students will compare a geographic location on a paper map with the same point on a raised relief map to observe differences between two-dimensional and three-dimensional maps. Demonstrate how to locate a high point on the raised relief map.

Locate six or eight high elevations and trace watersheds for these points before students begin the activity so that you will understand any difficulties students may encounter in tracing watersheds.

Students should be able to see elevation markings on a paper road map such as:

- Grand Rapids, "El. 657 ft.,"
- Mount Pleasant, "El. 770 ft.,"
- Detroit, "El. 600 ft.," and
- Mount Arvon, "El. 1979 ft." (*Mount Arvon is the highest elevation in Michigan. It is located just east of L'Anse, MI, in the Upper Peninsula.*)

When it comes to group use of the raised relief map, make certain that each student in each group traces a different path to a river.

Homework Options

Reflection

- **Science Content:** Find out about the watersheds in your area. Draw a map of the boundaries of these watersheds. Mark any lakes and the rivers into which the watersheds flow. (*Answers will vary.*)

- **Science Content:** How does the information gathered from the map study help to answer the *Big Question: How does water quality affect the ecology of a community? (Student answers will vary. However, their answers should begin to reflect that they are developing an understanding that as water moves from one place to another, it might carry materials from one area to another and therefore, affect the land into which materials are being carried.)*

Preparation for 1.5

- **Science Content:** Use the raised relief map to focus on the Rouge River watershed. Trace the path that water from this watershed takes to the Detroit River. Think about how water from the Rouge River watershed might affect the Detroit River. *(Answers will vary but students may say that materials put into the river up on one of the branches flow into the Detroit River.)*

NOTES

....

....

....

....

....

....

....

....

....

....

....

SECTION 1.4 IMPLEMENTATION

1.4 Explore

A Case Study: Watersheds in Michigan

You have been looking at how water flows in a model of a watershed. You discovered that water flows from higher to lower elevations.

In this section, you will begin to study a set of real watersheds. The watersheds you will study are in Michigan. By exploring these watersheds, you will see how watersheds connect with and interact with each other. Understanding connections between watersheds will help you give good advice to the Wamego town council. Your teacher is going to show you a type of map of the state of Michigan that shows elevation. It is called a **raised relief map**. It will show you areas of Michigan that are higher and lower in elevation. You will be able to touch the map to feel the different heights. You might think the land in Michigan is flat, but it actually has a variety of elevations. There aren't any large mountains, but there are plenty of large hills. These areas are at a higher elevation than the areas around them.

raised relief map: a three-dimensional map that shows elevations.

Materials
- relief map of Michigan
- topographic map of Michigan
- map of USA
- washable transparency markers

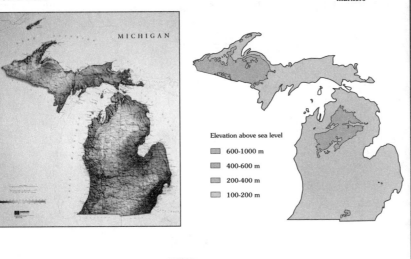

Elevation above sea level

- 600-1000 m
- 400-600 m
- 200-400 m
- 100-200 m

LT 23

LIVING TOGETHER

1.4 Explore

A Case Study: Watersheds in Michigan

10 min.

Students apply what they have learned about watersheds to a new context.

◯ Engage

Have students think back to the model in *Section 1.2* that made them first consider how water flows over and around land. Then, draw their attention again to the diagram of the watershed on page 21 in the student text.

*A class period is considered to be one 40 to 50 minute class.

"Let's look at the diagram of the watershed on page 21. Think about how water moved in your model. What was the most consistent thing you observed in the model-watershed activity? *(Water flows from higher to lower elevations.)* Did you see any models where water flowed differently?

Now think about a real-life situation. If a city is high on a hill and there is a river below it, in what direction do you predict groundwater around and under the city will flow? *(Downward, away from the city to the river.)*"

⃝ Engage

Project a transparency of a conventional two-dimensional road map of the Great Lakes area without using that name. Ask students to point out the lakes that surround Michigan. Some students may know these as the Great Lakes.

△ Guide and Assess

Ascertain if students understand that the Great Lakes are parts of watersheds by asking where they think the water in the Great Lakes comes from. Have students look at the map on page 25 in the student text.

"Let's connect what you saw in your model to this real-world situation. Where do you think the water in the Great Lakes comes from? *(Students should be able to relate the water in the Great Lakes to rainfall or precipitation and some might be able to give the correct answer—watersheds.)*

If the water in the lakes comes from watersheds, what can you conclude about the elevations of the Great Lakes and the elevation of the land around them?" *(The land around the Great Lakes is at a higher elevation than the lakes themselves.)*

△ Guide

Transition students toward the relief-map activity by asking them to describe different kinds of maps they might have seen. Then ask one member from each group to pick up a conventional map and a relief map of Michigan from the supply station.

To draw students' attention to the characteristics of a relief map, suggest that they run their hands over the relief map and the conventional map. Ask students to quickly tell one difference between the relief map and the conventional map. Students should be able to say that the relief map has raised areas and the conventional map does not have raised areas. Relate the raised areas on the relief map to elevations.

TEACHER TALK

"This may be the first time you have ever looked at a relief map. How is it different from a road map? What do you think the raised areas on the relief map signify? How is it like the model that you built?**"** *(Relief maps have raised areas that indicate elevation; conventional road maps mark some places with elevations in feet above sea level, but they are not raised.)*

META NOTES

Because this is a new situation, it may take students a few minutes to transfer the concept they learned from their model to a real-life situation and risk answering questions.

NOTES

....
....
....
....
....
....
....
....
....
....
....
....
....

Procedure

30 min.

Students use a relief map to track water flow in a real watershed.

META NOTES

On an overhead, show students how to follow and mark a possible river path to one of the lakes. It may be one that you have worked out previously.

Procedure

1. Compare the raised relief map with a paper map of the same area. The relief map is useful because you can touch it and feel the high and low spots on it. The relief map is a three-dimensional picture of the state of Michigan. The paper map represents the same area shown in the raised relief map. However, the paper map has only two dimensions.

2. Choose one point on your paper map. Compare it to the same spot on the raised relief map. Now look at a high elevation point of the plastic relief map and find the same spot on your paper map. How can you tell on the paper map that this point is an area of high elevation?

3. Use the raised relief map and work in small groups to find one area of Michigan that has a large hill. Starting at the top of the large hill you chose, have one member of your group use a transparency marker to draw on the relief map the direction that water will follow as it runs down the hill. Remember what happened in the watershed model you built earlier. Water moves downhill.

4. Continue to trace the water path you started to the nearest Great Lake and remember that water cannot run uphill. The members of your group can help the recorder identify the path. If the path seems to be going uphill, you need to find a new path. If you follow a path that is incorrect, wipe off the marker and return to the previous segment of the path. Your task is to find a path that does not go uphill and ends up in one of the Great Lakes.

   You have just traced one path that water might take. This might be the path of a river. All the land that drains water into this path is called a watershed.

5. Repeat these steps with each member of your group. Choose a different hill as your starting point each time. Each member of your group should have a chance to draw a water path.

Stop and Think

Answer the following questions. Be prepared to discuss your answers with your group and with the class.

1. How difficult was it to trace a path of water that does not go uphill?

2. Look at the lines you drew to mark the path of water from the top of a hill to one of the Great Lakes. What do these lines tell you about how the elevation of the land in Michigan compares with that of the Great Lakes?

LT 24

◯ Get Going

Make sure that students understand the activity procedures. Have students summarize how they will study the relief map. Their summary might be as follows:

1. Compare the relief map and the conventional map.

2. Choose a location on the conventional map. Find the same location on the relief map.

3. As a group, select a large hill on the relief map. With a marker, trace the direction water will follow down from that hill.

4. Track the water down through its watershed to the nearest Great Lake. Do not start back up any hills.

5. Each group member chooses a different hill and traces his or her own path to one of the Great Lakes.

Monitor the groups' progress by asking if some students have difficulty finding a clear river path downward. Ask them if they are sure their pathways always travel downhill. You might supply selective hints from the paths you previously tracked yourself. If too many students in a group are trying to find pathways that lead into Lake Michigan, you might ask if anyone tracked a watershed that leads into Lake Huron or Lake Erie. You can also ask if anyone found a watershed that seems to empty into more than one lake. If it looks like the majority of the class needs more time, give the groups extra time.

◇ **Evaluate**

Encourage students to relate their findings on the relief map activity in terms of *watersheds* and *elevations*. Listen for students to describe their findings in these terms. Listen for them to describe water moving from higher to lower elevations.

△ **Guide**

When each group appears finished, tell students to prepare to answer the *Stop and Think* questions at the bottom of page 24.

Have a class discussion on the questions. During the discussion, students might relate the relief map to their models from *Section 1.2*.

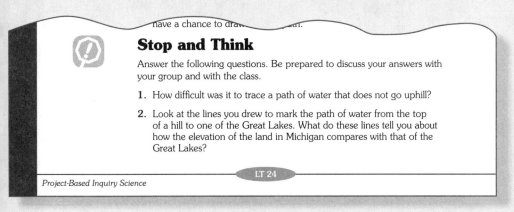

have a chance to dra_____

Stop and Think

Answer the following questions. Be prepared to discuss your answers with your group and with the class.

1. How difficult was it to trace a path of water that does not go uphill?

2. Look at the lines you drew to mark the path of water from the top of a hill to one of the Great Lakes. What do these lines tell you about how the elevation of the land in Michigan compares with that of the Great Lakes?

LT 24

Project-Based Inquiry Science

Stop and Think

10 min.

1) Some students may say that it was very difficult to find a path that did not go uphill at some point. This might be a result of the size of the relief map. Encourage students who may have been able to work around difficulties in using the raised relief map to share how they used the map.

2) The Great Lakes are at a lower elevation than that of the land in Michigan. Ask students to consider how this is different from or similar to their models.

TEACHER TALK

"Think back to your model watershed. Were there any places where water flowed uphill? *(There should not have been.)*

Where was there the most trouble with finding a clear pathway for water to flow? *(Probably toward the highest areas.)*

Did you find any places where water might move in opposite directions? *(When water moves in separate directions from about the same area of land, this area is called a divide.)*

Based on your model and the elevations you found on the relief map, why does water flow into the Great Lakes from land in Michigan? *(The land in Michigan is elevated above the lake and water flows from higher to lower elevations.)*"

NOTES

....

....

....

....

....

....

....

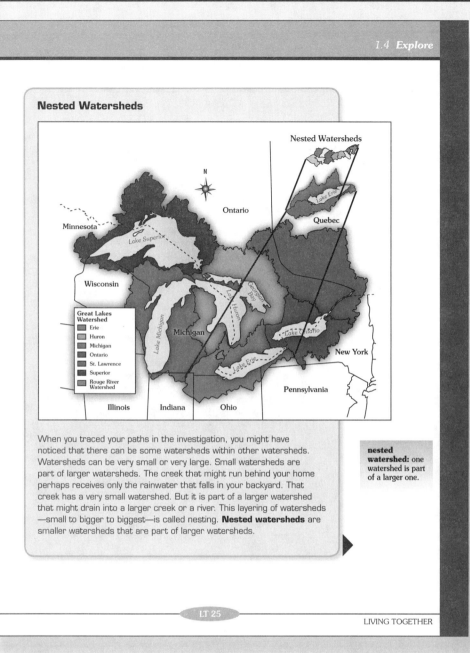

Nested Watersheds

When you traced your paths in the investigation, you might have noticed that there can be some watersheds within other watersheds. Watersheds can be very small or very large. Small watersheds are part of larger watersheds. The creek that might run behind your home perhaps receives only the rainwater that falls in your backyard. That creek has a very small watershed. But it is part of a larger watershed that might drain into a larger creek or a river. This layering of watersheds —small to bigger to biggest—is called nesting. **Nested watersheds** are smaller watersheds that are part of larger watersheds.

nested watershed: one watershed is part of a larger one.

Nested Watersheds
10 min.

Students are introduced to nested watersheds and learn that small watersheds flow into larger ones.

△ **Guide**

Build on students' experiences with the watersheds on the relief map by having them look at the map on page 25. Be sure that they observe and voice that the Great Lakes fill with water from all the surrounding areas, including Canada.

TEACHER TALK

❝In the last activity, each student traced a water path on the relief map. Each path represented a separate watershed. Were some of these watersheds right up against each other? What does that tell you about where watersheds are found? *(Every part of land on Earth is part of a watershed.)* Small watersheds feed into larger watersheds. These small watersheds are called *nested watersheds*.❞

The raised relief map of Michigan you used earlier shows you areas of higher and lower elevation. The map also shows you a string of lakes that surround Michigan. This string of lakes touches Michigan on three sides. These lakes are called the Great Lakes. They are the ending point for all the water that runs off the land in Michigan.

All the water on the land in Michigan moves to one of the Great Lakes. The Great Lakes are at a lower elevation than all the land in Michigan. The chart below shows the elevation of each of the Great Lakes. The elevation is measured at the lowest part of the lake. The numbers indicate how many meters above sea level the lakes are.

If you look at the numbers, you will notice that both Lake Michigan and Lake Huron are at a lower elevation than Lake Superior. So, because water runs downhill, water from Lake Superior enters either Lake Michigan or Lake Huron. Water that moves to Lake Huron continues through rivers and canals to the lakes it is connected to that are lower in elevation—Lake Erie and on to Lake Ontario. The water flows through rivers from Lake Ontario on to the Atlantic Ocean. Water that moves to Lake Michigan flows down the Mississippi and eventually ends up in the Gulf of Mexico.

Elevations of the Great Lakes

Great Lake	Elevation above sea level
Lake Superior	183 m
Lake Michigan	177 m
Lake Huron	177 m
Lake Erie	174 m
Lake Ontario	75 m

△ Guide

Help students analyze their findings with a series of questions:

TEACHER TALK

❝According to the map, where do watersheds in this part of Canada first empty into? *(Into the Great Lakes.)*

Think about water moving into the Great Lakes. For this to happen, how would you describe the elevation of all the land in watersheds around the Great Lakes, both in the United States and Canada? *(The land around the Great Lakes must be higher than the Great Lakes themselves.)*❞

What's the Point?

From your work with your watershed model, you discovered that water always flows from higher to lower elevations. You used this knowledge to trace the path of water from higher elevations to lower ones on a map of a real watershed. You also discovered that there are often many smaller watersheds nested in a larger watershed.

The watershed you explored was in the state of Michigan. You are using a map of Michigan because you are going to study a river in Michigan as a model of how the water quality in a river affects the ecology of the river and its watershed. This will help you understand how the water quality in Crystal River could affect the ecology in its watershed.

Water always flows from higher to lower elevations.

△ Guide

Draw students attention to differences in the elevations of the Great Lakes by having them turn to the chart on page 26.

META NOTES

As water travels from Lake Erie to Lake Ontario, its elevation drops about 99 m by the time it reaches Niagara Falls.

TEACHER TALK

"Look at the elevations of the lakes themselves in the chart on page 26. Lake Erie is near Lake Ontario. In what direction does the water flow— from Erie to Ontario or from Ontario to Erie? *(From Erie to Ontario.)* Think about what you have learned about the effect elevations have on how water flows. Why do you think it flows that way? *(Because Lake Erie is at a higher elevation than Lake Ontario.)*

Think about the overall direction that water from the Great Lakes is moving. As water moves through the path from Lake Huron to Erie to Ontario, it is moving in an easterly direction and eventually reaches the Atlantic Ocean. Water moving from Lake Superior moves into Lake Michigan and eventually down the Mississippi to the Gulf of Mexico."

Assessment Options

Targeted Concepts, Skills, and Nature of Science	How do I know if students got it?
A *watershed* is an area of land on which water falls. Some of the water is absorbed and some is runoff as it drains across the land's surface and into a river system.	**ASK:** During a heavy rainfall, think about why a major city would probably have more runoff or absorbed water **LISTEN:** Students should be able to say that a large city would have more runoff than absorbed water because of all the solid surfaces, such as streets, highways and parking lots. **ASK:** What is the relationship between absorbed water and groundwater in a watershed? **LISTEN:** Absorbed water soaks into the ground and travels through a watershed as groundwater.

Targeted Concepts, Skills, and Nature of Science	How do I know if students got it?
Water in a watershed travels predictably, from higher to lower elevations.	**ASK:** Lake A is 52 feet above sea level. Lake B is 111 feet above sea level. They are connected by a river. Why won't Lake A flow into Lake B? **LISTEN:** Students should be able to tell that Lake A is at a lower elevation than Lake B and therefore will not flow into Lake B, but Lake B will flow into the connecting river and then into Lake A.
Watersheds define the flow of water from an area of land into a river system and the flow of river systems into lakes and oceans. Nested watersheds are smaller watersheds that are a part of larger watersheds.	**ASK:** What does it mean that watersheds are nested? **LISTEN:** Students should voice the fact that all land is part of a watershed and small watersheds each feed into a larger watershed. **ASK:** What probably supplies most rivers, one large watershed or many nested watersheds? **LISTEN:** Most rivers are probably supplied by more than one watershed.
Students read raised relief maps and know the difference between three-dimensional and two-dimensional maps.	**ASK:** What does a raised relief map show that a conventional road map cannot show? **LISTEN:** Students should report that a raised relief map depicts higher elevations by actually molding parts of the map into raised shapes representing the higher elevations. Conventional road maps merely print the elevation of certain points across a state. **ASK:** How would a conventional road map indicate the elevation of Mount Arvon, which is 1979 ft. above sea level and the highest point in Michigan? **LISTEN:** Students should indicate that the high point would be written on a road map as: Mount Arvon El. 1979 ft.

Teacher Reflection Questions

- Did students have a difficult time applying the idea of water flowing from a higher to a lower elevation in a new situation? Would students benefit from an elevation drawing or a topographic map of the state of Michigan for determining the pathway of a water path during the activity? Maybe half the groups should have a topographic map and half should have the relief map.

- Are some students confused by the fact that the St. Lawrence River travels northeast on its way to the Atlantic but does not travel up in elevation? Do they confuse going north with ascending in elevation?

- Were there time management issues during the relief map activity? How could these be improved?

NOTES

1.5 Read

Introducing the Rouge River Watershed

1 class period ▶

A class period is considered to be one 40 to 50 minute class.

Overview

Students have learned about watersheds and studied specific watersheds in Michigan. In this section, students read about the Rouge River watershed. The Rouge River watershed is used as an example of one watershed that has changed over time due to the way people have used the river and the adjoining land. In this section, students will see the Rouge River as a part of a bigger system and will read about some of the history of the watershed. They also learn that watersheds throughout Michigan eventually drain into one of the Great Lakes. Students begin to understand that changes in the quality of the river water brought about by human activity affect people living in the watershed. Students begin to apply this information to the possible changes in the town of Wamego and the Crystal River there. They will use this information in subsequent lessons when they discuss changes in water quality brought about by four different land uses.

Targeted Concepts, Skills, and Nature of Science	Performance Expectations
Human activity can affect the ecology of a community. Humans use rivers for residential, commercial, industrial, and agricultural purposes. These activities affect water quality along a river.	Students should be able to recognize how land along a river is used and that these uses can have multiple effects on the quality of a watershed.
Watersheds define the flow of water from an area into a river system and the flow of river systems into lakes and oceans. *Nested watersheds* are smaller watersheds that are a part of larger watersheds.	Students should be able to say that a large watershed is made up of and fed by many smaller watersheds or sub-watersheds. Water from watersheds flows into larger bodies of water such as rivers and lakes, which eventually flow to the ocean.

Materials	
1 per class	Overhead projection of Michigan watersheds
1 per student	Laminated maps of Rouge River area
1 per group	Blue and red washable markers or crayons

Activity Setup and Preparation

Make an overhead projection of the map of Michigan watersheds shown in the student text.

Make sure you have sufficient maps of the Rouge River area that students can mark as they answer Question 2 under *Stop and Think* section.

Homework Options

Reflection

- **Science Content:** Research and describe ways in which people have impacted a watershed where you live. (*Answers will vary but should include the following common uses: residential, recreational, commercial, industrial, and agricultural. Some activities may have benefited a watershed, such as cleaning up one that has been damaged or neglected.*)

- **Science Process:** Make a flowchart that demonstrates the main parts of the watershed in which you live or a process, such as getting ready for school. (*Answers will vary. Flowcharts should include arrows, indicating that this is an ongoing process and indicating the direction in which the process or sequence is moving.*)

- **Science Process:** Identify local organizations involved with water-quality improvement where you live. Briefly describe the kinds of work they do that benefits the watershed. (*Many watersheds have groups who monitor water quality and periodically meet to collect trash that has accumulated within the streams of the watershed.*)

Preparation for 1.6

- **Science Process:** How can a model help you understand something that is very complex? *(In Section 1.6, students will view a teacher-run model of a stream table to prepare them to build their own.)*

- **Nature of Science:** How does collaborating on a design and building a model with others reflect the way scientists work on a problem? *(Answers will vary but should include that working together to design and run a model reflects the way scientists work because scientists often have to work with others who bring different ideas to help solve a problem. The final outcome will be the result of many different people's thoughts.)*

NOTES

....

....

....

....

....

....

....

....

....

....

....

....

LIVING TOGETHER

NOTES

Project-Based Inquiry Science

SECTION 1.5 IMPLEMENTATION

◀ *1 class period* *

1.5 Read

Introducing the Rouge River Watershed

The Rouge River running through a city.

You used a relief map of the state of Michigan to see how different watersheds nested within one another. You applied what you read about how water flows from higher to lower elevations to identify the different watersheds in the area. Throughout the rest of this Unit, you will be looking at one specific watershed in Michigan. You will examine the Rouge River watershed. (Rouge is the French word for red.) You can apply what you discover about the Rouge River to investigate other watersheds, including your own.

Where is the Rouge River

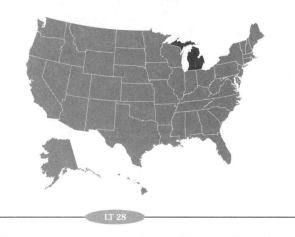

Project-Based Inquiry Science

LT 28

1.5 Read

Introducing the Rouge River Watershed

Students are introduced to the Rouge River and its watershed in Michigan, which will be the focus of the remainder of this Unit.

⃝ Engage

Photographs can be useful tools for making observations and drawing tentative conclusions about changes in the Rouge River watershed. Ask students to use the photograph to describe how they think some of the land is being used in this watershed. List students' ideas.

*A class period is considered to be one 40 to 50 minute class.

Where is the Rouge River Watershed Located?

10 min.

Based upon the photographs, hold a class discussion about what students perceive about how the Rouge River watershed is being used.

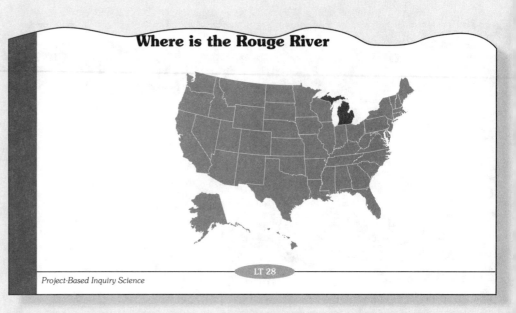

Where is the Rouge River

Project-Based Inquiry Science

LT 28

○ Engage

Show a projection of the map on page 29 to illustrate the watersheds throughout Michigan. Ask students to identify the markings on the map. *(They are nested watersheds and watershed boundaries. Certain cities are identified: Detroit, Lansing, and so on.)* Ask students to review what they already know about watersheds and, looking at the diagram, determine what Michigan's watersheds flow into. *(All Michigan watersheds flow into one of Lakes Superior, Michigan, Huron, or Erie. Some of this water eventually flows into Lake Ontario, but not directly from Michigan watersheds.)*

NOTES

....

....

....

....

....

....

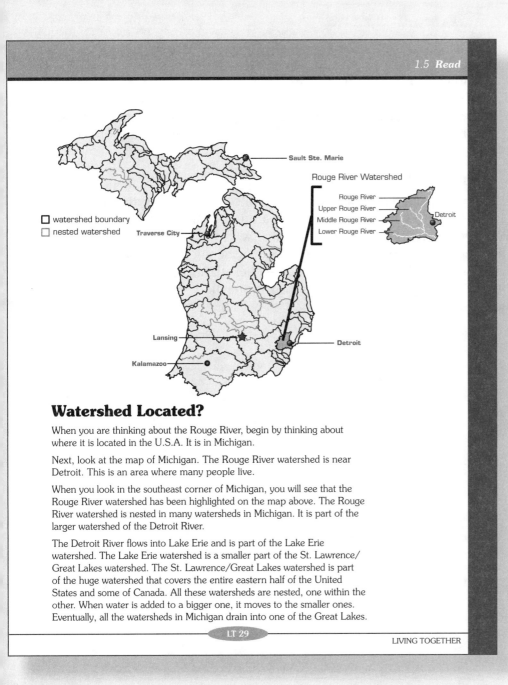

Watershed Located?

When you are thinking about the Rouge River, begin by thinking about where it is located in the U.S.A. It is in Michigan.

Next, look at the map of Michigan. The Rouge River watershed is near Detroit. This is an area where many people live.

When you look in the southeast corner of Michigan, you will see that the Rouge River watershed has been highlighted on the map above. The Rouge River watershed is nested in many watersheds in Michigan. It is part of the larger watershed of the Detroit River.

The Detroit River flows into Lake Erie and is part of the Lake Erie watershed. The Lake Erie watershed is a smaller part of the St. Lawrence/ Great Lakes watershed. The St. Lawrence/Great Lakes watershed is part of the huge watershed that covers the entire eastern half of the United States and some of Canada. All these watersheds are nested, one within the other. When water is added to a bigger one, it moves to the smaller ones. Eventually, all the watersheds in Michigan drain into one of the Great Lakes.

LT 29

LIVING TOGETHER

△ Guide

Ask students to summarize the main idea in the last paragraph on page 29 to reiterate that watersheds flow into larger bodies of water such as the Great Lakes. Then, explain that a tool called a flowchart could be used to graphically depict that concept. Explain that a *flowchart* is a tool that can help a person to remember a sequence.

META NOTES

A sequence or flowchart can help students keep track of the order in which events or a process occurs. A flowchart can be used to describe any watershed simply with each step leading to a larger body of water.

TEACHER TALK

"A *flowchart* is a simple way to connect things—to show the order in which a number of things, processes, or events occur. Let's make a flowchart using the last paragraph on page 29 starting with Rouge River watershed to show how watersheds feed into larger bodies of water."

△ Guide

With the help of a map showing the United States and Canada, have students identify major parts of the path that water travels from the Rouge River watershed to the Atlantic Ocean. With their input, create a flow chart that depicts this sequence (Rouge River watershed→The Detroit River→ Lake Erie→the Saint Lawrence River→the Atlantic Ocean). Students should be able to say that water moves from watersheds into progressively larger bodies of water.

NOTES

How Have People Used the Rouge River?

About 1.3 million people live, work, and play in the Rouge River watershed. Although the river with all its branches is 203 km (126 miles) long, that is still a lot of people. Throughout time, people have used the river for many different purposes. It has supplied drinking water and fish. It has been used for generating electricity. And it has been used as a dumping ground for waste.

Over 150 years ago, there were very few people living in the Rouge River watershed. Most of them lived in Detroit. The Rouge River was clean and had lots of fish in it, even though waste was dumped into it.

Then, about 100 years ago, the population in the watershed started to grow. Many factories for building cars, trucks, and airplanes moved into the area. With the factories came the people who worked in them.

Early industry along the Rouge River in Detroit introduced pollution that severely affected the quality of water.

With the increase in population, the demand to use the river water also increased. More cities and towns began to dump sewage into the river. More river water was needed to generate electricity. Some of the industrial waste was dumped into the river or buried in nearby areas.

Over time, the amount of sewage in the river made recreational activities impossible. Starting around 1950, people began to recognize that the quality of the river was very poor, and something had to be done. In the following years, sewage treatment plants were built. Cities were told they could no longer dump their waste into the river. Areas close to the river were set aside and protected as public parks. Although these measures helped to improve the river environment, illegal dumping of garbage and

LT 30

Project-Based Inquiry Science

How Have People Used the Rouge River?

10 min.

Students learn that humans use the Rouge River watershed for a variety of purposes and that these uses impact the watershed in a variety of ways.

◯ Engage

Help students build a picture of the history the Rouge River watershed and as it is today by having them read page 30. First, have students gather in their groups or teams. While students are gathering into their groups, write a list of categories on the board for students to use in a discussion after they read information about the watershed:

- Changes in population: (*increased over 150 years from few to more than 1 million people*)

- Effects of increased population: (*more people usually means more waste, more industry, greater demand for water*)

- Uses of the river over time: (*drinking water, factories for cars, trucks, airplanes; waste dumping*)

- Ways to help a watershed: (*sewage-treatment plants instead of dumping; cleanup by citizen/environmental groups; laws to prevent illegal dumping*)

△ Guide

Have one member of each group read aloud a portion of page 30 to emphasize how the Rouge River watershed has been used and has changed over time. At the end of the reading, ask volunteers to describe one thing they heard in the reading about how the Rouge River watershed has changed. Record their suggestions under one of the categories you have listed.

NOTES

_____

_____

_____

_____

_____

_____

_____

_____

_____

waste continued. In the 1980s, groups of citizen volunteers decided to remove all the garbage still found along the river banks. These groups have been working on this cleanup ever since. Every year they meet for at least one day to clean up the areas along the river.

Reflect

Look back at the pictures you saw at the beginning of this Unit. These pictures showed many different ways people use a river. All these pictures were taken along the Rouge River, in Michigan. Look at each picture and answer the following questions. Be prepared to share your answers with your group and the class.

1. Does the river flow through a city or town, or through a farm? Can you tell by looking at the pictures?

2. Where do you think the different pictures you saw were taken? Where do you think the industrial parts of the river might be? On a map of the watershed, color that area blue. Where do you think the river is used for recreation? Color that area red.

3. Compare your answers to the answers of other members of your group. Did you all pick the same areas? What else do you need to know to decide if the areas of the map are correctly identified?

4. How do you think the changes in the quality of the river water affect the lives of people living in the watershed?

5. List three examples of human activities that affect the quality of a river. Use what you learned about the Rouge River to explain what might happen to the river, and how the people in the watershed are affected by the changes in water quality.

6. Do you know of groups that are involved in cleaning up the river nearest your home? Do a little research to see if there is a group near you.

Human activities do not always damage the water quality of a river.

LT 31

LIVING TOGETHER

Reflect
15 to 30 min.

Students will think about the types of places the Rouge River flows through, how people have used the river, and discuss whether the quality of the river reflects the way the river has been used.

○ Engage

To get students thinking about the effect of the uses of the river on the quality of the watershed overall, have them turn back to pages 6 and 7 to determine these possible uses from the photographs on those pages. List each photograph with a brief description (*boating, storm drain, farming, houses, and so on*).

△ Guide

Have students focus on what is going on in each picture and volunteer a use or description. Encourage students to envision themselves in the scene and describe how the river is being used at that point (*recreation, industry, agricultural, and commercial*). Record students' descriptions next to the appropriate word in the list. It is all right if students suggest interpretations different from those of other students.

△ Guide and Assess

Prepare students to answer the six *Stop and Think* questions individually. Caution them that they will next discuss the answers in their groups, then as a class.

Lay out laminated maps and markers for each group to use in completing *Stop and Think* Questions 2 and 3.

TEACHER TALK

"Now that you have listed some of the uses and changes that have happened along the Rouge River, let's think a little more about what this means for the watershed. At first, you will need to turn back to pages 6 and 7 to answer the first two *Stop and Think* questions. Look at each photograph again. Think about what each scene might be telling you about how a watershed is used every day. Then ask yourself two questions: *What in the photograph tells me how the river is being used?* and *How does this use affect the quality of the river?* Then, answer the remaining *Stop and Think* questions.

You will then discuss your observations within your group. Finally, be prepared to discuss your answers with the class. Be ready to provide evidence (*from the photographs*) to support your description of how the river has been used and how these uses have affected the river."

Provide students with about five minutes to answer questions individually, then, ask them to gather into their groups. Finally, have students discuss their answers to each question as a class. Listen for the following responses or guide students toward these responses.

1. Most students will say, that based on the variety of pictures, the river flows through many different areas—used for industrial, recreational, residential, agricultural, and other purposes.

2. Answers will vary. Many students may say that industrial areas are at the end of the watershed where it flows into a river. Many may suggest that recreational areas are further back up in the watershed where there are fewer people.

3. Answers may vary. Students may say that they need to know more information about the area before they can decide where certain uses would be likely to occur. Ask where they might perform research to obtain this information.

4. Answers will vary about how changes in the quality of the water affects people's lives. Many will say that industrial areas make the quality less than what it should be.

5. Choices of human activities that affect watershed quality will vary. Some will say water consumption by local human population, dumping wastes, industrial uses, and agriculture.
   NOTE: This question is important as it sets students up for *Section 1.6*.

6. Answers will vary. Most states have websites concerning their watersheds. Information is provided about who to contact for each watershed and what volunteers are doing or can do to clean up a watershed.

NOTES

....

....

....

....

....

....

....

....

....

Apply
10 min.

The Rouge River is used throughout the Unit as an example of issues about water quality and its effects on humans and other animals living near and in water.

META NOTES

Making connections to or being familiar with a local watershed can reinforce the content of this Unit. Throughout this Unit, focus students' attention on making connections between the watershed they are studying (Rouge River) and the status of a watershed in their local community.

Apply

Think about the land around Crystal River. How will it be different with buildings on it? When people build buildings, they cover the land and change the types of land cover that originally existed there. Buildings, houses, and roads all cover the ground. Trees, grass, and soil are moved and may or may not cover the ground the same way they did before. Water can soak into soil, but it flows as runoff when it lands on buildings and roads.

Think back to when you have been outside when it was raining. What happened to the rain when it hit the ground? How did the type of land cover affect what happened to the rain? What kinds of land cover are there in your community?

Think about the watershed you built. You decided that as long as you did not change the structures in the watershed, the water would continue to run off in the same pattern. The direction of the water flow and the places where the water pooled would not change as long as you did not change the objects or the paper. If you had used soil instead of paper, you would have noticed that some water soaked into the ground.

Now you are going to consider the differences among four different land uses. During your class discussion, use a table like the one below to organize what you are learning about each land use.

Residential	Commercial	Industrial	Agricultural

What's the Point?

In the United States, many rivers have been used by people for dumping waste, draining storm water, recreation, and industry. The Rouge River is one example of such a river. There are many other rivers across the United States that have been used in the same way. When people use a natural resource like a river, they modify the environment around it. Human activities in any part of the watershed impact the quality of the river because all the water in the watershed sooner or later ends up in a river. Sometimes water travels many miles, over and under the land, to get to the river. It can carry with it many different materials that can make the river unsafe or unclean.

LT 32

○ Engage

Draw students' attention to the *Big Question* across the top of the student page. *(How Do Flowing Water and Land Interact in a Community?)* Then, ask them to think again about the challenges faced by the Crystal River, which runs past the little town of Wamego.

△ Guide

Ask students to describe the land where they live and any signs of how it might be changing. If they have trouble getting started, you might ask some guiding questions. "Are there individual houses with yards or are there large apartment buildings? Is the land flat or hilly? Are there factories that depend on having a large source of water? Are there any trees or woods nearby? Are there groups of stores or tall office buildings? Are there shopping malls surrounded by large parking lots? Are there farms with large fields?"

Have students begin to think about how these things might be affecting the watershed in which they live.

Ask students to describe evidence that land where they live has been changed. *(The land was scraped flat for a new mall. An empty field was laid out with street, houses were built and trees were planted.)* If evidence isn't forthcoming, ask questions to elicit an example of something students would recognize. You might ask if anyone has noticed the building of a warehouse or a new development of houses and roads where woods existed a year before. Explain that each of these is an example of evidence that land is changing.

> ### TEACHER TALK
>
> **❝**Picture a densely populated city. Now think about what it was like before any buildings were built on it. Why do people build cities? How is it different from suburbs? How is it different from farmland? What kinds of activities take place in each of these areas that would affect a watershed? Make sure you support your ideas.**❞**

△ Guide and Discuss

Stimulate a class discussion with questions and thoughts about why land changes. Using what they have learned about different types of land uses, have students make suggestions about filling in the table on page 32 with this information. Inform students that as they think of more things pertaining to each kind of land use, they will be able to add to the table.

> ### TEACHER TALK
>
> **❝**In this section, you saw that the watersheds in Michigan directly or indirectly affect all of the Great Lakes. You began to think about how different kinds of activities can impact the quality of water in a watershed and eventually impact the quality of a river and the places into which that river flows. In the next section, you will investigate these situations in more detail by modeling how different kinds of land uses affect water quality.**❞**

> ### META NOTES
>
> During discussions, listen for students to give evidence or reasons for why land in an area has changed.

> ### META NOTES
>
> Students often confuse evidence with opinion when supporting their ideas. *Evidence* is based on observations of facts. *Opinions* are often personal interpretations of situations. Keep students focused on clearly observable facts.

> ### META NOTES
>
> Continue to draw students' attention back to the *Big Question* throughout the *Learning Set* as they accumulate information.

Assessment Options

Targeted Concepts, Skills, and Nature of Science	How do I know if students got it?
Human activity can affect the ecology of a community. Humans use rivers for residential, commercial, industrial, and agricultural purposes. These activities affect water quality along a river.	**ASK:** How do activities along the Rouge River affect the quality of the river water? **LISTEN:** Students should be able to iterate that each activity adds something to the water as it moves through the watershed. **ASK:** Why might an increase in population in a watershed decrease the quality of the available water? **LISTEN:** Students should be able to say that as the population in the watershed increases, water quality decreases because more water is used and more wastes of all kinds are produced.
Watersheds define the flow of water from an area into a river system and the flow of river systems into lakes and oceans. *Nested watersheds* are smaller watersheds that are a part of larger watersheds.	**ASK:** Why is water in the Great Lakes affected by activities in watersheds throughout Michigan? **LISTEN:** Students should recognize that watersheds throughout Michigan affect water in the Atlantic Ocean because water from some of these watersheds flows to the Atlantic Ocean.

Teacher Reflection Questions

- Do students understand that water running through a watershed can never truly be cleaned up completely once it is contaminated with wastes from human activities such as industry and recreation?

- Can students ascertain from photographs how humans impact river water quality?

- Did students understand the process of making a flowchart to represent the sequence in which water flows from some Michigan watersheds to the Atlantic Ocean?

1.6 Investigate

How Does the River Affect the Land and How Does Land Use Affect the River?

◀ *4 class periods* *

*A class period is considered to be one 40 to 50 minute class.

Overview

Groups are assigned a specific type of land use to model—residential, commercial, industrial, or agricultural. Groups plan their model on paper and then construct the model using the supplied materials. Students predict how the water will affect the assigned land use and how the land will affect water that flows through the modeled land. In an *Investigation Expo*, students communicate their results and discuss how water quality is affected by how people use land. Students also discuss how the land might be changed by how water moves sediments through the watershed.

Targeted Concepts, Skills, and Nature of Science	Performance Expectations
Human activity can affect the ecology of a community. Humans use rivers for residential, commercial, industrial, and agricultural purposes. These activities affect water quality along a river.	Students recognize how land along a river is used and that these uses can have multiple effects on the quality of a watershed.
Water and land interact with each other.	Students conclude that water and land interact when they observe that water can change the shape of land and materials from land can change the quality of water in a river.
Water flow transports and redistributes materials in a stream.	Students experience that materials on land, such as sediments, are moved by water and deposited elsewhere, thereby changing the land.

Targeted Concepts, Skills, and Nature of Science	Performance Expectations
Scientists use models to simulate processes that happen too fast, too slow, on a scale that cannot be observed directly (either too small or too large), or that are too dangerous.	Students build and use a model that demonstrates that water changes land and that land changes water in a watershed. Students should move toward understanding that models or simulations represent natural phenomena that test one variable.
Predicting, observing, and explaining are important investigative skills.	Students should be able to demonstrate their investigative skills as they predict and then observe the way water moves through various land-use models.
Scientists must keep clear, accurate, and descriptive records of what they do so that they can share their work with others, consider what they did, why they did it, and what they want to do next.	Students should be able to keep descriptive and accurate records of how they designed and built their land-use models.

Materials

1 per classroom	Stream table pan with earth materials for demonstration**
2 per group	Spray bottles, one with clear water and one with water colored with a small amount of blue food coloring
1 per group	Popsicle stick or tongue depressor for creating stream path
1 per group	Stream-table pan with runoff weep-hole at end
1 per group	Earth materials** or soil already measured and bagged
1 per group	Containers to collect runoff from stream table weep holes
1 per group	Wood block or books to prop up the stream tables
1 per group	Colored marker set
1 per group	Presentation supplies
For each classroom	Source of water generous amounts of newspaper and/or paper towels Large plastic bags for disposing of soil and other earth materials Waste-water bucket Kitchen sieve to straw materials after use and reuse

Materials

**Provide various materials for students to use as they build different land-use models. These materials may be small rocks, sand, tall and short wood blocks for houses and large factory buildings, small flat rocks or pieces of flat plastic to act as hard surfaces such as parking lots or roads. Collect small plants for trees and pieces of sod for lawns.

Activity Setup and Preparation

Demonstration

- Set up two model stream tables according to the diagram below. These will enable students to view possible different effects of water on the land and land on the water that eventually moves out of the pan. Bank the soil material firmly to one end of the stream table, but not the full length of the stream table. Before you demonstrate the tables for the class, test the most effective way to apply water to get the desired result. Keep track of how much water needs to be applied to get a result. Test both the curved stream and the straight stream in the flat position and in the tilted position.

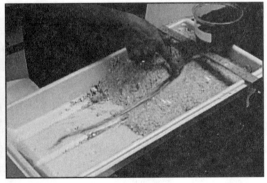

- During your practice runs, test the effect of the shape of the stream bed (*curved or straight*) by using your finger to "draw" a river in the soil. Then spray the stream table with water until water collects in the stream bed and begins to flow. Make a note of how much water was needed to get the desired effect. (*As an alternate approach, you might let water slowly drip from a reservoir at the higher end of the stream table until a river forms and flows.*)

- Before students build their models, determine approximately how much soil students will need to perform their investigations. Measure this out into gallon resealable bags ahead of time, each bag to contain enough for one stream table.

Student Models

- Students will build their stream-table models based on your demonstration models. However, they will add other materials to create a certain type of land use, the effects of which they have been assigned to evaluate (*residential, commercial, industrial, and agricultural*). Provide them with materials that they can use to make these models, but also allow them to use other materials after checking with you.

- Students who have been assigned an industrial site to model, where there may be toxic runoff from manufacturing processes, might use a second color in the spray bottle. If they use yellow as a second color, then it may turn green when it meets with the blue water from the river, giving them another way to identify changes that land can make to water.

Homework Options

Reflection

- **Science Content:** Summarize your observations of how the flow of water in your stream table affects the surrounding land and how the land affects the flow of water in your stream table. What implications does this have for a real river like the Rouge River? *(Answers will vary with the kinds of materials used to depict the land use. Agricultural land will probably be composed mostly of sod. Most of the water sprayed on it will become groundwater. Any models with hard surfaces will have more runoff.)*

- **Science Process:** What was one variable that was tested during the demonstration? *(One variable was the effect of the shape of a stream on the movement of materials. Another possible variable was the effect of gravity on the movement of materials when the pans were tilted at an angle.)*

Preparation for 1.7

- **Science Content:** Are the effects of residential and commercial land uses different from the effects produced by industrial and agricultural land uses? If so, how are they different? *(The purpose of these questions is to get students thinking about how different uses may have different specific effects on water in a watershed.)*

- **Science Process:** How do the materials in runoff from a residential area differ from runoff from agricultural land? *(Students should be able to analyze the differences in runoff from the two types of land use.)*

SECTION 1.6 IMPLEMENTATION

1.6 Investigate

How Does the River Affect the Land and How Does Land Use Affect the River?

You have been reading about watersheds and how land use can affect the quality of water in a watershed. Now you will look at how water can affect the land that it moves across. Then you will consider the many ways that people use the land in a watershed. You will build your own models of four of these uses: residential, commercial, industrial, and agricultural. Each of these uses changes the land in a different way.

First, you will look at a demonstration of a model. Your teacher has built a river model for you to observe. In the model, your teacher is going to use some soil and water to demonstrate how a river might change the land around it. The model will be run several different ways. Each different model will demonstrate a different way for the water to flow down the river.

Demonstration

A stream table can be used to simulate processes occurring in an active river or stream. Since it is not always possible to study these processes in a real river, scientists use stream tables in the laboratory to do that. They build a model of the river or stream and the land around it. Then they move water through it to simulate water processes. This helps them understand the causes and effects of water processes.

A simple stream table consists of a large pan covered with Earth's material, such as sand, rocks, or grass. A line is drawn in the sand to represent the river. Water drips into the sand from a container at one end. The stream table may be slanted to help the water flow downhill. At the other end of the pan is a drain hole, connected to a bucket, to collect the water.

Scientist use stream tables to simulate processes in an active river or stream.

LT 33

1.6 Investigate

How Does the River Affect the Land and How Does Land Use Affect the River?

Students observe a demonstration and prepare for building a model stream using a stream table.

---

*A class period is considered to be one 40 to 50 minute class.

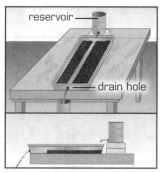 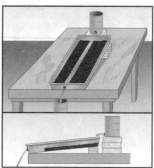

reservoir

drain hole

flat   slanted

○ Engage

Begin by asking students to review what they know about how water behaves in a watershed. Record students' ideas. To get students thinking about how water and land can affect one another, ask students to describe what they might have seen on news programs about floods.

Then, introduce the demonstrations you are about to do. Indicate that you are going to use a scientific tool called a *stream table* to demonstrate how water affects land as it moves through parts of a watershed. Then, tell students that after the demonstrations, they will work in their groups to build similar models of their own.

Identify the parts of the stream table as you set it up so that students will know how to set up their stream tables.

Demonstration

20 min.

Students are introduced to erosion *and* deposition. *Students also are introduced to* stream tables, *which they will use to model how land use affects these processes.*

△ Guide

Before you begin the demonstrations, sketch the four demonstration setups so all students can see. Identify each part with a label. Draw four different ways the models will be used. Label whether the river will be *straight* or *curved*. Label whether the model will be *flat* or *elevated*.

Have students prepare diagrams of their own and inform them that they will take notes during the demonstration. Reiterate the importance of keeping notes during any experiment. Introduce the *Demonstration Notes* page. Emphasize the importance of recording observations in science. Let students know that they will be using their notes to share observations in their groups.

TEACHER TALK

"In science, it is important to keep notes about every part of a project. Scientists take notes on what they read, predictions they make about problems they are working on, ways to do experiments, and results they obtain during experiments. If you record your ideas in the same way, you can compare what you observed during your experiment with what you predicted might happen. To help you remember and think about your ideas, you will record your predictions and observations on a *Demonstration Notes* page. Then, you will use your notes to compare your outcome with those of other groups.**"**

Stream Table 1: The Flat Straight Stream

Set up a flat stream table and let it remain flat. Make a straight river through the soil.

Ask students to write, on their *Demonstration Notes* paper, their predictions of what will happen to the land when water flows down the river. At this point, let water flow down the river.

Ask students to record their observations. Help students to describe how the water is moving. In addition, ask if they notice whether the water is picking up any solid materials from the banks of the river. Listen as students describe their observations. Students should be describing how the water moves through and over the land. If they describe it as not moving or moving very slowly, ask for a reason. Their answer could be that it is because the table is flat.

During the demonstrations, identify the places where soil is picked up *(erosion)* and where soil is dropped *(deposition)* by the stream. However, hold off in using those particular terms.

META NOTES

In this section, students will observe four different stream tables. They will observe the changes in Earth materials as water flows. They will be observing erosion and deposition.

TEACHER TALK

"One thing that you noticed in the first stream table is that the water moves very slowly. The slow-moving water did not move much earth material. What can you do with the model to make it represent the real world better and help us to see what happens when water and land interact?**"**

△ **Guide**

Stream Table 2: The Elevated Straight Stream

Raise one end of the stream table to allow the water to move more quickly.

Ask students to record their prediction on their *Demonstration Notes* page of what will happen to the water and land in the stream table now that it is elevated.

Using the reservoir or bottle of water, pour water down the straight river. Students should record their observations of the water as it now moves down the river. They should describe what is happening to the land as the water moves in the river.

Stream Table 3: The Curved River Pattern

Ask students to predict what will happen if the river were slightly curved. Make a large *C*-shaped curve in the river. The curve has to be large enough to change the pattern of erosion and deposition but not so big that the water cannot move down the river.

Ask students to predict how they think the *C*-shaped river will change how the water and land interact. Use the reservoir or spray bottle to create the C-shaped river. Use the same water delivery system as in the other parts of the demonstration.

Discuss with students how the *C*-shaped river differs from the straight river at the same elevation. Encourage students to identify the places on the river where land is moved (*erosion*) and where the land is then deposited (*deposition*).

Stream Table 4: The Very Curvy River Pattern

Ask students to suggest other river systems that might demonstrate the interaction between water and land. Students should suggest that it would be interesting to observe a more curvy river. Ask students to predict how a more curvy river will change the way the river and land interact. Encourage them to look for trends between the straight river and the *C*-shaped river when they make their prediction. Repeat earlier procedures in setting up and testing the very curvy river.

Ask how the very curvy river differs from the straight river and the *C*-shaped river in how it erodes and deposits land.

1.6

Different models of stream tables.

Natural processes, such as flooding, can change the structure of land and water in a watershed.

Your teacher will set up the stream table in four different ways, as shown in the diagrams.

Sketch the different models. As you watch the water flow through the model, pay very close attention to the way the land on both sides of the river changes. Pay attention to

- how the soil moves,
- where along the bank the soil moves, and
- where the soil ends up.

Make notes about what you observe for each of these situations. You might want to mark your sketches based on what you observed.

Stop and Think

Look at your sketches and the notes you took about the river models you observed. What did you notice about how the soil was moved by the river? Answer these questions. Be prepared to discuss your answers with your group and the class.

1. When the river was straight and the pan was level, how did the soil move along the river?

2. When your teacher made the pan more slanted by lifting the water end of the pan, how did the water move compared to the level pan? How did that change affect the soil that the river moved?

3. Your teacher also made rivers that were more curved. How did that change the way the soil moved along the river?

Stop and Think

15 min.

△ Guide

Ask students to use evidence from their observations to describe how soil moved in a river, depending on the shape of the river.

Then have students work in their groups to develop one answer for each *Stop and Think* question.

Call the class together and have one group answer question 1. Then ask other groups whether their conclusions were the same. Ask them to suggest reasons for different conclusions. Answers may vary but they should all contain the following information.

1. The soil didn't really move along the flat, straight river.

Erosion and Deposition

5 min.

Students are introduced to the terms, erosion and deposition.

META NOTES

In *Project Based Inquiry Science*, students learn concepts through an iterative process. In this section, students have observed the processes of erosion and deposition. They have also observed how water flows differently in various types of "rivers." They now have experienced the processes and can attach concept words to the processes. During the demonstration or in the *Stop and Think* discussion, it is appropriate to define deposition and erosion using examples from the demonstrations. Students learn the concepts more deeply when terms are introduced after students have had experiences related to the concepts.

Erosion and Deposition

Before you start building stream models, you need to know something about how water moves in a watershed and the effect that has on the stream or river. During a heavy rainfall, water in a stream flows fast and will pick up a lot of dirt. As water flows against the bottom and sides of the river channel, it removes more dirt, sand, soil, and debris. Scientists call the "removal of dirt" **erosion**. When water slows down, the collected dirt in the river drops out of the water. Scientists call the debris and soil "dropping out" of water **deposition**.

erosion: a process in which Earth's materials are loosened and removed.

deposition: the setting down of Earth's materials in another area.

land use: how people use Earth's surface.

Design and Build Your Model

When people settle in areas near a river, they change the land around it. Trees, grass, and soil are removed to make room for buildings, houses, and roads. Other areas are used for farming or to grow crops. The way people use Earth's surface is called **land use**. Your class will model the effects of changes in land use on river processes.

Your teacher will assign your group one of the four types of land use to model: residential, commercial, industrial, and agricultural. Read the section describing the land-use model assigned to your group. Use the pictures included in each section to help you imagine the land use that you will model.

LT 35

LIVING TOGETHER

2. Water flowed when the pan was elevated at one end and it flowed faster. The steeper the slope, the faster the water moved. Faster water moved soil in the river bed, then dropped the soil at the bottom of the pan when the water stopped moving.

3. The curves caused the water to pick up soil in different places and drop it elsewhere. Water picks up soil on an outside curve (*water is moving faster*) and drops it off on the inside of a curve (*water is moving more slowly on inside curves*).

△ Guide

Now that students have experienced something concerning how water can move soil in a river system, you can introduce the terms *erosion* and *deposition*.

1.6

erosion—the process where soil or loose materials are picked up by water (or wind) and carried elsewhere.

deposition—the process whereby soil or other particulate materials are dropped out of water or wind in another area.

Ask students if they have ever seen a flood on a news program. Have students volunteer to describe the color of the floodwater. Ask students to describe where all the material in the floodwater comes from and where it goes using the terms erosion and deposition.

○ Engage

Define the term *land use*. Then ask students, "How had the land around

Design and Build Your Model

When people settle in areas near a river, they change the land around it. Trees, grass, and soil are removed to make room for buildings, houses, and roads. Other areas are used for farming or to grow crops. The way people use Earth's surface is called **land use**. Your class will model the effects of changes in land use on river processes.

Your teacher will assign your group one of the four types of land use to model: residential, commercial, industrial, and agricultural. Read the section describing the land-use model assigned to your group. Use the pictures included in each section to help you imagine the land use that you will model.

LT 35

LIVING TOGETHER

you—where you live or go to school- changed? What are some of the different ways land is used where you live?" Have students be very specific in their answers *(a new highway down the street, a new shopping center where a local farm used to be, a new housing development, a parking lot where a house used to be)* or use a local map to show where changes were taking place.

△ Guide

Explain that each group will be assigned a specific land use to model. They will use a stream table and design a land model to show the effects of that land use on a watershed. They will test their model and then share their data with other groups to compare how each kind of land use affects water in a watershed. The results of the teacher demonstration done earlier can serve as a control for the students' tests.

If necessary, review the parts of a stream table with students. Review how to use the stream table.

Create eight student groups and assign each a land use to model. Explain that there will be two groups modeling each land use. Two groups will test how residential land use affects water in a watershed. Two groups each will test how industrial land use, how commercial land use, and how agricultural land use affect water in watersheds.

META NOTES

It is important to have at least two examples of each type of land use. Students will then be able to compare the results for their land use to another group. If necessary, groups should be made up of no more than three students so that there are at least eight groups in the class for discussions or investigations.

Design and Build Your Model

40 min.

Models are used to explore many different phenomena in science. Being able to analyze a situation and design a model of it is an important skill. These models will help students understand the effects of different land uses and land cover on water quality.

You will use a stream table to model effects on the watershed. Begin by building a model as close as possible to one of the models your teacher built. All the groups should begin with a similar model.

After you complete your investigation, you will share the data your group collected with other groups. In this way, you can compare the effects each land use produces in the river. It is important to be careful when building your model and collecting data so you can compare your results with those of other groups later.

When you build your model in the stream table and watch how water flows through it, pay attention to the areas of erosion and deposition you can see. On a diagram of the stream table, draw and label areas of erosion and deposition. If you are not sure where you are seeing this happen, ask your teacher to help.

Model Residential Use

Suppose a developer decides to build houses next to the river. The land she builds on is covered with grass. The builder builds five houses, equally spaced from each other. Each house includes a two-car garage. Sidewalks connect one house to the other. The builder removes all the soil and grass that was originally on the site. She plans to add a large grass lawn to each house once they are built.

LT 36

○ Get Going

To help students begin their design process, you might remind students of the importance of criteria and constraints when solving a challenge. With the class, identify the criteria and constraints of this investigation. If students are not able to identify the criteria and constraints, remind them that criteria are goals to be satisfied to accomplish the investigation. *Constraints* are factors that limit how you can solve a problem. The *criteria (criterion)* for this problem is to find out how an assigned land use affects a watershed by modeling it. The materials they will use (*stream table and water*) are the constraints of the problem.

Model Commercial Use

Some land along the river has been designated as a space for building a small mall. The mall will include a movie rental store, a grocery store, a coffee shop, a plant nursery, and a day care. The mall needs 100 parking spaces and will take up the area of about ten houses.

Model Industrial Use

Many industrial factories are built along rivers. The water from the river is often used in the factories for manufacturing and then returned to the river. In your model, you are going to build a paper mill. Large trees are brought to your factory to be ground up and made into the white paper you write on. The trees are very big. You need to build a road wide enough for trucks to get to the factory in your model.

The factory building will take the area of about ten houses. The pictures shown may help you to imagine how an industrial land use might look.

Have students discuss variables while designing their models. Ask, "What will be the independent variable and what will be the dependent variable for your land use investigation?" Tell students to begin by asking themselves, "What will make my group's model different from other models?" *(Your assigned land use.)* This should help you determine what your independent variable is. Ask students, "Will your investigation have a control?" *(The teacher's demonstration can serve as the control.)* "What parts of your investigation will stay the same as everyone's else's?"

As students identify criteria and constraints, record them on the board so that students can refer to them as they design their models and procedures.

△ Guide and Assess

On the board, list suggested steps to be taken before a model is run. Write this out ahead of time to save time. Suggest to students that the order in which an experimental procedure is done can give them a path to follow, although steps often get changed.

1. Design an experimental model on paper.

2. Have the design checked or modify it and have the revised design rechecked.

3. Write out the procedure steps to test the model.

4. Collect materials for the model.

5. Build the model according to the approved design.

6. Have the assembled model checked by your teacher.

7. Answer the *Predict* questions on page 39 in the student text before running the model. (*Teachers should encourage students to use the words* erosion *and* deposition *where needed and should check their understanding of the words in context.*)

8. Run the test on the model.

9. Record observations to be used as evidence in discussions later.

Model Agricultural Use

You might not be familiar with how land looks when it is used for farming or raising animals. You can use the pictures to help you imagine what this type of land looks like. Build your stream table to look similar to a farm. On your farm, there are a few buildings and a house. The crops have just been planted for the coming year. The model you will build should have a lot of exposed soil. There is a river running through the area. There is also a lake nearby. Families living near the farm use the lake for swimming.

Procedure: Build Your Model

Make a list of the ways the land in your model will be used. Be sure to include all the details from the description in your list. Each member of your group should sketch how the land is going to look after all buildings and features of your land use are built. Remember to include all the details from the list you created.

When all the members of your group have completed their drawings, compare them to one another and allow each group member time to describe their plans. Listen carefully for details you may not have included in your drawing.

Now you are ready to build your model. Your teacher will provide you with materials to use. Use your group's drawings to build a model representing the land use you have been assigned.

LT 38

Project-Based Inquiry Science

Procedure: Build Your Model

20 min.

Have students discuss their plans and designs for their models. Then, direct them to start implementing these ideas.

Inform students that while they are designing their model, they should keep in mind how their land use affects erosion and deposition in the watershed. Monitor students as they design their models. Inform students that they should check their designs with you before they assemble materials and build their models.

Check to see that each group builds its model according to their design. If they have to change the design somewhat, and this is entirely possible, remind them to record these changes because it might affect their observations.

Predict

10 min.

Predicting requires students to think about what might happen. Later, they compare their predictions with their observations and think about what they mean.

Predict

Before you try your model, predict how water will flow in your land-use model when you spray water on the different types of land cover. Make a prediction about how water will move on each of the land covers you built in your model. Even if you are not sure, try to guess based on what you already know about that land cover.

As you make your predictions, think about the following questions:

- What will happen to the water that hits the grassy areas? What will happen to the water that hits the sandy areas?
- What makes the water move differently through the different land covers in your model?
- How will the amount of runoff vary among the different land covers?
- How will the time the water takes to get through the different land covers and to the river vary?
- What will the water look like when it enters the river compared to when it fell as rain?
- How clean will the water be that drains through the different land covers?

Procedure: Run Your Model

Spray water on your stream table. Observe where the water flows as it moves through the model.

Recording Your Observations

As you observe how the water flows, record your observations on a data chart. You might also want to record notes on a sketch of your model. Be as specific as possible. Things you might want to check could be: how much of the water you spray ends up in the river; how fast the water flows over the areas where the land cover has changed; does the water form puddles anywhere? Add your own ideas for other interesting patterns you observe.

Analyze Your Results

1. What makes the water move differently through the different land covers in your model?

2. Is the time the water takes to get to the river different in your model compared to the models your teacher demonstrated?

3. How does the amount of runoff vary among the different land covers?

LT 39

LIVING TOGETHER

Call students' attention to the *Predict* questions on page 39 in the student text before they run their model.

Predict questions need to be answered before students run their model. Students might prepare for answering the *Predict* questions as a homework assignment. Then, as you are approving design plans, ask students the

Procedure: Run Your Model

Spray water on your stream table. Observe where the water flows as it moves through the model.

Predict questions. Once students have made their predictions, let them spray water on their model. Let them observe where the water flows, where runoff occurs, and how long it takes the water to move through the land cover, and what the water looks like as it enters the river, after it has been sprayed over the land cover.

TEACHER TALK

"Please remember that only one person should spray water on the model, but everyone in the group needs to be watching what happens to the water. Keep track of where the water goes. Remember, water moves in a certain way depending on the land use you modeled. Models that represent a city with buildings, roads, and parking lots will have more runoff and less groundwater seepage into a river than would agricultural land use.

Your observations are your evidence. Everyone should record their own observations so they can be shared with each other later. Discuss your model while you are making your observations.**"**

Recording Your Observations

As you observe how the water flows, record your observations on a data chart. You might also want to record notes on a sketch of your model. Be as specific as possible. Things you might want to check could be: how much of the water you spray ends up in the river; how fast the water flows over the areas where the land cover has changed; does the water form puddles anywhere? Add your own ideas for other interesting patterns you observe.

◯ Engage

Students might record data more quickly and accurately, if, ahead of time, they have a checklist or table of items to mark off or if they have a drawing of their model to mark as the experiment proceeds. On a drawing, students might mark "P" where water pools or collects, "R" where runoff pours into a river, or "G" for where they see water soaking into the ground.

△ Guide

Remind students to record observations, such as:

- How much of the water that is sprayed on the model winds up in the river?

- How fast does water flow over the land area model?

Procedure: Run Your Model

10 min.

Students begin to implement their planning and predicting by running their models.

Record Your Observations

10 min.

Students use data charts to record and organize their observations.

- Where is the water pooling and where is the water being absorbed by Earth?

- Where are you seeing the processes of *erosion* and *deposition*? How are the examples you are seeing the same as and different from the models in the teacher demonstration?

Reiterate that the observations students make during the experiment provide them with the evidence they need to explain what happens when water and land interact, so they will need to watch carefully and quietly.

☐ Assess

Monitor students as the experiment runs to see if everyone is able to record data. Ask questions about what you observe and listen for answers that reflect what you see taking place. If a student's observation differs from what you observe, ask the student to explain because he or she may have observed something different.

Another place where students may have difficulty is in following the procedures that they originally outlined. Get them back on track by asking what procedure the group had decided to follow and whether they are in fact, using it. However, explain that during an experiment, it is not uncommon for a scientist to see a better way to do something. Explain that if the group agrees, they can change a procedure, and the revised procedure needs to be recorded by all members of the group.

There should not be a need to refill the spray bottle, and so once each group is finished spraying their model, the bottles can be collected or returned to the appropriate materials place. As each group finishes, students can begin to discuss their data within the group.

Some students may finish the activity quickly. Inform students that if they finish their experiment earlier than other groups that they should move on quietly to the *Analyze Your Results* questions. Ask them to think ahead to prepare to discuss their results with the other group in the class that is experimenting with the same land use.

Analyze Your Results

10 min.

Answering these analysis questions helps students prepare for the Investigation Expo *during which they will present their models and communicate their results.*

Analyze Your Results

1. What makes the water move differently through the different land covers in your model?

2. Is the time the water takes to get to the river different in your model compared to the models your teacher demonstrated?

3. How does the amount of runoff vary among the different land covers?

LT 39

LIVING TOGETHER

Ask students to answer the analysis questions by themselves, and then discuss these answers within their group, because individual members of a single group may have observed something missed by other members of the group.

1. Water moves differently over hard surfaces, such as pavement, than it moves over areas where it is absorbed, such as grass.

2. Answers will vary with the land use and land cover tested.

3. Water runs off pavement or hard surfaces faster than it runs off grass because soil absorbs water and hard surfaces do not absorb water.

4. Runoff from hard surfaces into a river will appear cleaner than water that is absorbed or runs into the ground. Water that runs over bare soil will appear cloudy or brown with soil that has been eroded.

5. Answers will vary with the model that has been constructed. Models that represent a city with buildings, roads, and parking lots will have more runoff and less groundwater seepage into a river.

> **META NOTES**
>
> A group discussion of experiment outcomes and conclusions acts as a way to engage students for the next segment of this section.

NOTES

....

....

....

....

....

....

....

....

....

....

Communicate Your Results: Investigation Expo

10 min.

Students create presentations to show the results and observations of their land-use model investigations.

Explain

5 min.

Students share their observations and results.

4. How clean is the water that drains through the different land covers?

5. Is the groundwater that enters the model river the same across the length of your model or does it vary?

Communicate Your Results

Investigation Expo

Use the *Analyze Your Results* questions as a way to discuss the results of your investigation in your group.

For the *Investigation Expo*, create a poster with a diagram of your land-use model. Make your diagram as detailed as you possibly can. Include all your land covers as well as your results. Indicate on your diagram places of erosion and deposition, and places where there was a lot of runoff in your model.

During the *Investigation Expo*, you are going to describe to your class how your model worked. You need to include enough details in your presentation so that your classmates will understand how the land cover in your model changed how the water moved. Answer the following questions in your presentation:

- How did the water move in different parts of the stream table?
- How do you think the land cover you modeled might affect how the water is absorbed by the ground compared to vegetation (plant life) or bare soil?

As you listen to the presentations of the other groups, observe how water flows for each land use. Compare the places were erosion and deposition occur in the different models. Compare the amount of runoff produced by different land covers. How do the residential, commercial, industrial, and agricultural land use each affect the amount of runoff produced, compared to that of the bare soil in your teacher's model?

Explain

As the water flowed through the land, you probably noticed that it flowed differently when it was covered by pavement (small laminated pieces), houses, or vegetation. When people have not changed land, water flows in predictable ways. It may flow over the land and run off into a river. If there is more vegetation, the water may be absorbed by the land, be used by the plants, or run through the land as groundwater. No matter how it moves, the water always ends up in the river. When the land use is changed, the

Explain to students that they will present the results of their experiment in an *Investigation Expo*. Inform each group that they will make a poster that shows a diagram of their land-use model and the results of their experiment. The poster should include where erosion and deposition took place and where there was a lot of runoff. List the items that you expect to see in students' posters. Students can refer to this checklist as they develop their posters and presentations.

The following *Teacher Talk* will help prepare students for the *Investigation Expo*.

"To prepare for the *Investigation Expo*, your group will need to think about what your model was about and be able to tell what happened during the experiment. You will need to describe the results and the conclusions you made based on them.

To present your work, your group will draw a poster that shows your land-use model with all its parts labeled. Then, you will have to agree on a statement that tells everyone else in the class how your group's land use model affected the way water moved.

Practice your presentation. Decide who in your group will explain the model that was used, who will explain where erosion and deposition took place, who will point out where runoff took place or why it didn't take place, and who will read what your group concluded about the way your land use affects water in a watershed.

Finally, during your presentation, answer the two bulleted questions on page 40 in your book."

△ Guide

Give students time to work in groups on their posters and prepare their presentations.

Assist students by walking through the parts that they need to construct an explanation of their experiment and its outcome. List the three parts (*claim*, *evidence*, and *science knowledge*).

Using a *Create Your Explanation* page, have students work in their groups to construct their explanations for the land-use investigation based on evidence collected during the investigation.

"From earlier investigations in the *Launcher Unit*, you may recall that an *explanation* begins with a claim and it is backed by evidence and science knowledge. A *claim* is a statement or conclusion reached through observation and/or investigation. *Evidence* is the data collected during an experiment. *Science knowledge* is knowledge based on what experts have previously investigated. In your group, think about what your claim is as it relates to the land use you experimented with. Then, summarize the evidence that you observed in your investigation. Finally, think about what you learned from observing the original stream table demonstrations. This represents your science knowledge. Once you have all three of these parts, you can assemble an explanation for use in your presentation."

META NOTES

Students will already have some idea of what an explanation is. In science and *PBIS*, an explanation has specific components: a claim, evidence, and science knowledge.

META NOTES

Recall that two different groups investigated each land use. They therefore, have the same experiment question. Have the two groups present one after the other so that you and the class can compare the experiment plans and results.

patterns of water movement are also changed. More water may run off. The land may absorb less water because it is covered. That could have consequences for plants because they need water at their roots. When water is not absorbed by the ground, it cannot reach the roots of the plants.

1. Think about the stream tables. What is one variable that might affect the amount of deposition in a river? Explain how that variable affects the amount of deposition in the river.

2. What is one variable that might affect the amount of erosion in a river? Explain how that variable affects the amount of erosion in the river.

Explain

Use *Create Your Explanation* pages to help you organize your claims and evidence as you develop your explanations.

Communicate

Share Your Explanation

Your class will meet to discuss each group's explanation. As a class, select or create a set of explanations that explains what affects the deposition in a river and another set that explains what affects the amount of erosion around the river.

What's the Point?

You have built and observed models of a stream with different land uses. As water flowed through your model, you observed how the water changed the shape of the land, and where erosion and deposition were occurring. The movement of soil is critical to how rivers work. The amount of eroded soil carried by the water and the places where it is deposited can greatly affect the balance of a river system.

In addition to changing the pattern of erosion and deposition, different land uses also change water quality. It is important to understand how human land use changes the river and the water so people can look for ways to minimize its impact.

When you changed the land cover in your model, the patterns of erosion and deposition changed. How people use the land around the river can change the water quality.

LT 41

LIVING TOGETHER

Explain

5 min.

Students share the claims and evidence of their investigation and construct explanations.

△ Guide and Assess

As groups construct their explanations, listen for logical connections. If you hear opinion statements being expressed, you might ask the group, *"What evidence did you observe that supports that statement?"* If they cannot demonstrate evidence from the investigation, explain that the statement is an opinion and different from evidence based on data.

If the majority of the class expresses difficulty in constructing an explanation, stop the class and hold a class discussion to clarify any problems.

Let students know that they will be presenting their explanations to the whole class.

Have students think about the variables in their experiments. Remind students to answer the two questions on page 41 as they prepare their presentations.

1. Deposition occurs when water slows down. When water slows down, it is no longer able to keep the particles of soil suspended. Gravity pulls the particles of soil downward and they drop to the bottom of the stream.

2. The speed of the water is the variable that affects erosion, but slope also affects the speed. If a river is straight, erosion will increase if the river is on a steep slope. Curved rivers undergo more erosion on the outside of the curve where the river moves faster. Soil gets deposited in the inner curves of a river because the water moves more slowly there. It cannot hold onto the particles of soil. Gravity pulls them down.

△ Guide

Explain to students that they also will construct a poster that represents their group's investigation of land use. The poster will be used during the groups' presentation.

Distribute materials for the poster and explain your expectations for the poster. The poster should include:

1. A title that includes the land use investigated

2. A drawing of the investigation setup with labels

3. Arrows that indicate the direction of water flow

4. Labels that tell where water pooled or formed groundwater or runoff

5. The prediction of how this land use affects water flow

6. A summary or concluding statement that explains the effect of that particular land use on water flow and uses the terms erosion and deposition

7. The poster is to be used during the group's presentation to the class.

> **META NOTES**
>
> A *poster* is a physical product that should reflect what a student has learned.
>
> Teachers should always provide students with their expectations for a product.

LIVING TOGETHER

Communicate Share Your Explanation

40 min.

In presenting models and test results, students act the way scientists do. This helps students to articulate their ideas and refine them when there is feedback from other students.

Communicate

Share Your Explanation

Your class will meet to discuss each group's explanation. As a class, select or create a set of explanations that explains what affects the deposition in a river and another set that explains what affects the amount of erosion around the river.

△ Guide Presentations and Discussions

Students come to closure, reviewing what they have learned about how water and land interact and how land use changes what is in the water. They make initial connections between water flow and the ecology of a watershed and revisit the Project Board.

Have each group present their posters and their conclusions about how land use affects water. If two groups experimented with the same type of land use, have these groups present back-to-back so that students might compare outcomes.

Encourage students to ask questions during each group's presentation, but to use language that is appropriate, such as:

"Why did you …"

"I don't agree because…"

"I don't understand how you reached that conclusion because your evidence showed…"

> **TEACHER TALK**
>
> **❝**When other groups show their posters, look for the same kind of information in their presentations. This will help you to compare your results with how water moved and where erosion and deposition took place in their land use models.**❞**

Have students add information about the effects of land use on water and any new questions they thought of to the *Project Board.*

NOTES

....

....

....

Assessment Options

Targeted Concepts, Skills, and Nature of Science	How do I know if students got it?
Human activity can affect the ecology of a community. Humans use rivers for residential, commercial, industrial, and agricultural purposes. These activities affect water quality along a river.	**ASK:** What are two kinds of materials that could change water quality near farmland? **LISTEN:** Students might describe loose soil that is eroded from farmland into rivers and fertilizers that are absorbed into groundwater. **ASK:** What is one reason why industrial and commercial uses of land might result in more runoff than land used for agriculture? **LISTEN:** Students should be able to explain that industrial and commercial uses usually mean buildings, streets, and parking lots, which are hard surfaces that water runs off of.
Water and land interact with each other.	**ASK:** Why does erosion change land? **LISTEN:** Students should be able to say that erosion changes the land because soil is moved from one place to another. **ASK:** Where does more erosion take place in a curving river? Explain. **LISTEN:** More erosion takes place on the outside of the curve because water is moving faster on that side of the river.

NOTES

....

....

....

Targeted Concepts, Skills, and Nature of Science	How do I know if students got it?
Water flow transports and redistributes materials in a stream.	**ASK:** During a flood or after a heavy rain, rivers rush along and often become rusty brown in color. Why? **LISTEN:** Students should be able to say that the color is due to a large amount of material being carried away or eroded by the moving water. **ASK:** Where does eroded material go when it is carried off in a river? **LISTEN:** Students should be able to relate that eroded materials carried by a rushing river are eventually deposited or drop out of the river further downstream when the river slows down.
Scientists use models to simulate processes that happen too fast, too slow, on a scale that cannot be observed directly (either too small or too large), or that are too dangerous.	**ASK:** How did the land use model help you understand that water can erode materials? **LISTEN:** Students should relate that the moving water in the model caused soil to erode or be moved from one place to another.
Predicting, observing, and explaining are important investigative skills.	**ASK:** Why was it important to have all members of each group watch the land-use model as it was tested? **LISTEN:** Students' answers will vary but may include that not every member of the group will see everything, but that when all members watch carefully, one member may see something that others have missed.

Targeted Concepts, Skills, and Nature of Science	How do I know if students got it?
Scientists must keep clear, accurate, and descriptive records of what they do so that they can share their work with others, consider what they did, why they did it, and what they want to do next.	**ASK:** How did your group act like scientists as you prepared and conducted your experiments and presentations about how land use affects water quality in a watershed? **LISTEN:** Answers will vary. Students should relate that within each group, all members observed the experiment as it took place and took notes. All members discussed how their particular land-use model affected the quality of the water sprayed on the model.

Teacher Reflection Questions

- How did the investigation help students understand the difference between erosion and deposition of earth materials?

- What difficulties did students have in relating their stream table models to real-life instances that they experience every day, such as city streets, shopping mall parking lots, and huge housing developments?

- How can the stream table setups within the classroom be less messy while still giving students a hands-on experience?

NOTES

1.7 Read

More about the Effects of Land Use on a River

1 class period ▶

*A class period is considered to be one 40 to 50 minute class.

Overview

Students built stream table models in the last section and observed how the flow of water through and over land was affected by the way the land was used. They learned that moving water can pick up soil (*erosion*) in one place and drop or deposit (*deposition*) it elsewhere as the water slows down. In this section, students learn that large and small activities that take place in residential, commercial, industrial, and agricultural land use can affect water quality. They learn that runoff can carry significant amounts of harmful substances into a river and that movement of soil through erosion and deposition can affect the quality of water in an area.

Targeted Concepts, Skills, and Nature of Science	Performance Expectations
Human activity can affect the ecology of a community. Humans use rivers for residential, commercial, industrial, and agricultural purposes. These activities affect water quality along a river.	Students should be able to provide examples of how land used by humans along a river can have an impact on the quality of water in a watershed.
Water and land interact with each other.	Students should be able to recognize a cause and effect relationship between moving water and land. They should be able to say that erosion by moving water changes land and that runoff and eroded soil change the quality of water moving into a river.
Water flow transports and redistributes materials in a stream.	Students should be able to say that when water moves materials from one place to another, it redistributes these materials. In doing so, sometimes harmful materials enter the watershed.

Targeted Concepts, Skills, and Nature of Science	Performance Expectations
Scientists use models to simulate processes that happen too fast, too slow, on a scale that cannot be observed directly (either too small or too large), or that are too dangerous.	An illustration is a kind of model. Students should be able to use an illustration to observe some physical phenomena and answer questions about what is depicted.

Materials	
Optional	Projections of illustrations from page 43 and 44 in the student text

Activity Setup and Preparation

Make separate overhead projections of the residential/commercial use illustration on page 43, the industrial land use illustration on page 43, and the agricultural land use illustration on page 44 from the student text.

Homework Options

Reflection

- **Science Process:** Select a land use that you did not model in *Section 1.6*. Pretend that you are a scientist studying possible changes in the watershed where this land use is located. Write a summary predicting how the land use might be affecting a river in the area. Use the terms erosion, deposition, and runoff in your summary. Provide two pieces of evidence for your claims. *(Answers will vary with the land use that is chosen but should include the three terms and a prediction based on evidence.)*

- **Science Content:** Find out what is classified in your state as Household Hazardous Waste (HHW). Write a definition for HHW. Using the term groundwater, explain why HHWs need to be disposed of legally and safely. *(All states have had to institute disposal options for HHW because many can leak into soil, contaminate groundwater, and create threats to human health.)*

Preparation for 1.8

- **Science Content:** In preparation for *Section 1.8*, think of two ways in which people cause or contribute to groundwater pollution. *(Answers will vary because there are many ways that people pollute groundwater intentionally and unintentionally. Any two of the following can be used to answer the question: Chemical spills from factories and automobiles, runoff containing leaked oil, lubricants and antifreeze from cars in parking areas, illegal dumping of toxic materials, fertilizers in gardens and on farm fields, untreated sewage, and salt and sand trucks during the winter. Accept any other reasonable material that may pollute groundwater.)*

- **Science Content:** Many people pollute their environment without thinking. What are two examples of non-point pollution that you have witnessed or participated in? *(Answers will vary but may include watching salt/sand combinations being distributed on roads during the winter; littering; plastic bags blowing around the street or caught in a tree; watching effluent pour from a factory into a stream or holding pond.)*

NOTES

SECTION 1.7 IMPLEMENTATION

◀ *1 class period* *

1.7 Read

More about the Effects of Land Use on a River

Erosion, deposition, and runoff happen naturally. Often, they do not cause problems in rivers. But the changes people make to land affect erosion, deposition, and runoff.

In class, you observed and experimented with a land-use model. The soil represented the land. The stream you made with your finger represented a river. As you sprayed water onto your model, you observed how the water flowed over and through the land. As the water moved, it changed the land.

As you built your model, you changed the way the surface looked. Houses and roads covered what was soil or grass before. You explored the effects of these changes on your model river. You looked at how the land use affected erosion and deposition as well as runoff.

In this section, you will read about ways in which humans have changed the land and examples of how that can affect the nearby rivers. This will help you understand the ways that land use changes the water that ends up in the river. As you read these descriptions, pay particular attention to the land uses you did not model. You will be able to discuss each of the land uses with your class.

Effects of Residential and Commercial Land Use

Both residential and commercial land use affect the river in similar ways. When you built your model for residential or commercial land use, you might have used plastic sheets to represent paved surfaces or roads. You also used plastic blocks to represent buildings. When you ran the model, you probably noticed that water was running off the plastic surfaces. The runoff eventually ended up in the river.

What you saw in your model is very similar to what happens in the real world. Rain runs off the surfaces of buildings, rooftops, and roads. The water falling on paved roads and concrete cannot penetrate the ground. Instead, it flows fast towards the river, increasing erosion. Because the water cannot penetrate into the ground, it is not available as groundwater for plants and trees that need it. Plants and trees may suffer from lack of

LT 42

Project-Based Inquiry Science

1.7 Read

More about the Effects of Land Use on a River

Students are informally introduced to the idea that the everyday activities of humans can affect water quality in their local watershed.

○ Engage

Write the terms *erosion*, *deposition*, *runoff*, and *groundwater*, on the board. Use a class discussion to help students review what they have learned and investigated thus far about how erosion, deposition, and runoff can affect water in a river.

Urge students to use the four terms to describe how water can change from its source (*such as rainfall*) to when it arrives in a river.

*A class period is considered to be one 40 to 50 minute class.

LIVING TOGETHER

"In the earlier section, you learned by modeling how different land use affects which materials and how much material reaches a river in a watershed. Let's reinforce what you saw in your land use models by studying illustrations of these different land uses.**"**

Effects of Residential and Commercial Land Use

10 min.

Students extend their ideas about erosion, deposition, runoff, and groundwater in this section.

Effects of Residential and Commercial Land Use

Both residential and commercial land use affect the river in similar ways. When you built your model for residential or commercial land use, you might have used plastic sheets to represent paved surfaces or roads. You also used plastic blocks to represent buildings. When you ran the model, you probably noticed that water was running off the plastic surfaces. The runoff eventually ended up in the river.

What you saw in your model is very similar to what happens in the real world. Rain runs off the surfaces of buildings, rooftops, and roads. The water falling on paved roads and concrete cannot penetrate the ground. Instead, it flows fast towards the river, increasing erosion. Because the water cannot penetrate into the ground, it is not available as groundwater for plants and trees that need it. Plants and trees may suffer from lack of

LT 42

Project-Based Inquiry Science

META NOTES

Research has shown that students have extensive misconceptions about scale where groundwater storage is concerned. Instead of thinking about groundwater moving through pores in soil, many think that water is most commonly stored underground in large facilities bordering on the size of a house or a skyscraper. *(Journal of Geoscience Education* 9/05, pp. 374–380.)

◯ Engage

Project an image of the residential and commercial land-use illustration from page 43 in the student text to use as the basis of a class discussion. Explain that a drawing is also a kind of model.

△ Guide

To make clear that students understand the terms being used, ask them to explain what the term *residential* means and what the term *commercial* means. Write the two terms side-by-side on the board and ask for examples of each from the local community.

Focus the discussion by asking students why these two are alike and why they are different in the illustration. Write their suggestions on the board. *(Alike: buildings, streets. Different: commercial has large hard surfaces (pavement) due to parking lots and wider streets; houses have lawns).*

Direct students' attention to the arrows in the illustration. Ask what the arrows on the diagram mean. *(Accept all reasonable responses. Arrows mean the direction in which water is moving. Arrows across green areas mean that the water is more than likely being absorbed by vegetation and ground cover and will probably reach the river through groundwater. Arrows from the parking lot at the mall mean runoff, with water moving more quickly to the river.)*

Ask students to compare the effects of the groundwater and the runoff on the river. *(Groundwater will have little effect on the river and may be clean. Runoff rushes to the river and can stir up sediment in the river, making the river muddy. It might also carry oil from the parking lot and contaminate the river water. The harder the surface, the greater the runoff.)*

Another way to ask how groundwater and runoff compare in their effects is to ask how vegetation and pavement affect erosion and deposition. *(The roots of vegetation absorb water and increases groundwater, but pavement increases runoff.)*

Have students discuss how their stream table models compare with what they see in the illustration.

NOTES

....

....

....

....

....

....

....

....

....

....

....

Effects of Industrial Land Use

10 min.

Students discuss and learn about the effects of Industrial land use and apply understanding to observations of local examples.

water. The intense runoff muddies the water in the river because fast-flowing waters can pick up and carry a lot of dirt along the way.

When you created grassy areas in your model, you also noticed that the water did not run off these surfaces. The water was absorbed by the grass and went into the ground. In a real watershed, the vegetation covering the ground acts like a sponge. It absorbs the rain and snow falling to the ground and releases the water over time. In this way, vegetation prevents erosion. Because the water it absorbs is released slowly, the dirt in the water is deposited along the way before the water reaches the river. The water entering the river is therefore much cleaner.

Effects of Industrial Land Use

Large industrial factories are often built along rivers. This is because many factories need large quantities of water during **manufacturing** processes. Rivers can provide the water they need. The river can also be used to transport goods and materials in and out of the factory by boat.

> **manufacturing:** the making or producing of anything.

When modeling industrial land use, you used plastic sheets to represent roads and plastic blocks to represent buildings. The effects that these surfaces have on erosion, deposition, and runoff are very similar to what you observed for residential and commercial land uses.

But industrial uses are different in several ways from other land uses. Factories often draw water from the river to run their processes. This water is then returned to the river at the end of the manufacturing cycle. During the process, water can pick up substances or change in other ways. Which changes occur

LT 43

LIVING TOGETHER

○ Engage

Project an image of the industrial land-use illustration from page 43 in the student text to use as the basis of a class discussion. To get students to relate to the idea of industrial land use, ask volunteers to name some local industries and list them on the board. Examples might be something such as: Acme Athletic Shoes, the Midtown Power Company, and the Big Box Hardware Company, or the Extra Crispy Potato Chip factory.

During the demonstrations, identify the places where soil is picked up (*erosion*) and where soil is dropped (*deposition*) by the stream. However, hold off in using those particular terms.

△ Guide

Prepare students for a class discussion by having them study the industrial-use illustration. Then, have each student write a question they would ask the class about groundwater or runoff activities in this picture. You may need to model a few questions. Have volunteers ask their questions of the class, not of you.

△ Guide and Assess

Make sure that students' questions are on the topic and that answers focus on the questions asked. Facilitate by modeling questions and refocusing answers as needed.

◇ Evaluate

Evaluate students' understanding of the differences in effects of two land uses by asking how the effects of the factories are different from the effects of houses in the residential illustration? (*Answers will vary. The factories seem to be putting out materials into the air, into the groundwater, and directly into the river.*)

NOTES

Effects of Agricultural Land Use

10 min.

Students are exposed to the idea that land use affects the water that living things depend upon.

---

depend on the type of factory and how the water is used. The water exiting a factory can be very different from the water entering one. When this water is discharged back into the river, the changes that occur can be harmful to aquatic organisms or wildlife in the area.

Effects of Agricultural Land Use

Today, many large areas of land are still being cleared to plant fruits and vegetables. As trees and grasses are removed to make room for farms and houses, the soil is exposed. Without the roots of the plants to anchor it, these soils can be eroded away easily. The eroded soil particles carried by the water and discharged into the river affect the water's quality.

The increased erosion of soils used to grow crops does not happen only once, when the land is cleared. When crops are harvested, the soils become exposed again until the next crop grows. More erosion occurs every time the soil becomes exposed. Over time, this increased erosion can change dramatically how a river system works. In some places, the river might widen. In other places, the depositions might change the river's shape or depth.

Soil and dirt from farmlands are not the only things that get washed into the river from agricultural use. Substances added to the soil to help the crops grow and keep insects away are also carried by the water into the river. Some substances are carried in runoff. Others are absorbed into the soil and carried underground in groundwater.

Reflect

Write a description of what happens to the water when it rains in your neighborhood. In your writing you should answer the following questions:
- Where does the rainwater go once it hits the ground?
- Where do you see that the rain is absorbed?
- Where does runoff occur?
- Where do you think most of the water ends up?

LT 44

Project-Based Inquiry Science

◯ Engage

Project the image of agricultural land use from page 44 in the student text. Begin by asking volunteers to identify the parts of the illustration. Students should report that they see buildings (*house, barn, silo*), a hill (*which means an elevation difference*), crops, and a river with fish in it.

◯ Get Going

Get students thinking about everything that is going on in the agricultural land-use illustration, by inquiring how students would enhance the drawing that would expand their understanding of how river water can become changed near a farm. Listen for answers that reflect the use of water for animals and crops.

Have individual students read the paragraphs to focus on the differences between one land use and another. Inform students that when they have looked at each type of land use, they will be better equipped to answer the *Reflect* questions.

- Paragraph 1: When ground cover is removed or scraped away, soil erodes and is carried to a river where it changes the quality of the river.

- Paragraph 2: Erosion increases as land is cleared and again when crops are harvested. The river changes as more eroded soil is washed in.

- Paragraph 3: Other substances from farming move into groundwater and into the river.

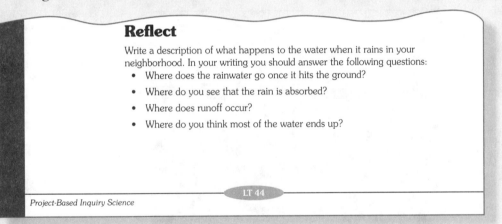

Reflect

Write a description of what happens to the water when it rains in your neighborhood. In your writing you should answer the following questions:
- Where does the rainwater go once it hits the ground?
- Where do you see that the rain is absorbed?
- Where does runoff occur?
- Where do you think most of the water ends up?

LT 44

Project-Based Inquiry Science

Reflect

10 min.

Students reflect on what they have read and how it applies to their neighborhoods.

○ Get Going

After students have answered the *Reflect* questions by themselves, ask for volunteers to answer each question. You might transition them to a class discussion by asking questions such as: "What evidence is there that vegetation was removed?" or "How much runoff do you think there is versus groundwater absorption?"

△ Guide and Assess

During the class discussion, answers should contain the following information in some form. By this time, students should be comfortable using the terms *runoff, groundwater, vegetation, erosion* and *deposition*. Listen for the use of these terms in their answers. If you do not hear them, you might ask students to restate an answer by using these terms.

What's the Point?

You read about the specific problems different land uses can create. Sometimes the uses create similar problems. The residential and commercial land uses are similar. However, industrial uses are very different from residential uses. When considering the differences between land uses, it is important to pay close attention to how the water is being used and how the land is being changed.

Runoff can carry different substances into water. Parking lots and streets contribute oils, grease, and dirt to runoff. These substances eventually make it into the river.

Erosion and deposition are always issues in land use. The movement of soil is critical to how rivers work. Water can carry soil and deposit it in other places. When soil enters the river, it can affect the quality of the water.

Groundwater is also critical because the water can carry different types of materials through the ground and into the river. Groundwater is especially important in agricultural land uses. Water moves through the ground and carries substances from the farm into the river.

Many water bodies play an important role in the towns and cities they run through. This river in Holland is used for both recreation and industrial transportation.

LT 45

LIVING TOGETHER

- If rain falls on pavement, most of it ends up pooling or running into a river. If rain falls on vegetation or soil, much of it is absorbed into groundwater; some will eventually end up in a river.

- Rain is absorbed where there is vegetation (*grass, trees, and exposed soil*).

- Runoff occurs when rain hits hard surfaces or pavement.

- If it hits hard surfaces, it could become runoff; if it hits vegetation, it is absorbed into groundwater.

△ Guide

Ask students to summarize what they learned in this section. Listen for students to say that different kinds of land use creates different kinds of problems for water as it falls on land. Students should be able to make a statement that relates land use and changes that occur in rivers. They should say that materials that erode from land are carried by water and often move in groundwater, eventually emptying into a river. They should also be able to say that materials such as oil and other fluids used in automobiles are picked up in runoff and carried to a river, thereby contaminating the river water. Industrial sites also produce materials that are carried either in runoff or via groundwater to streams and can be harmful.

Assessment Options

Targeted Concepts, Skills, and Nature of Science	How do I know if students got it?
Human activity can affect the ecology of a community. Humans use rivers for residential, commercial, industrial, and agricultural purposes. These activities affect water quality along a river.	**ASK:** How do you think the many parking lots and extensive road systems in and around New York City affect the quality of water in the Hudson and the East Rivers, which flow around the city? **LISTEN:** Students' answers should suggest that runoff from the streets and parking areas, carrying oil and other substances from cars, would carry into the Hudson and East Rivers and affect the quality of the water there.
Water and land interact with each other.	**ASK:** How might water and land interact in an industrial area to change the quality of water? **LISTEN:** Students should be able to say that water that reaches a river from an industrial area probably carries with it materials that can harm water quality. Water from an industrial area that is absorbed into the ground also harms water quality because it changes groundwater.

Targeted Concepts, Skills, and Nature of Science	How do I know if students got it?
Water flow transports and redistributes materials in a stream.	**ASK:** How might fast-moving runoff from a large parking lot overlooking a river redistribute material in the river? **LISTEN:** Students' answers might include that fast-moving runoff stirs up materials already in the river and moves them around.
Scientists use models to simulate processes that happen too fast, too slow, on a scale that cannot be observed directly (either too small or too large), or that are too dangerous.	**ASK:** How can an illustration of land use be used to explain what happens to a nearby river? **LISTEN:** Answers will vary. Students' responses might include that an illustration can show the basic parts of the land use. Arrows can be drawn to show the direction of the flow of water toward the river. Some might say that a caption or labels can expand understanding. **ASK:** How does a model like the stream table land-use model differ from an illustration as a model? **LISTEN:** The stream table model was a model that could be tested. Data or evidence was derived from the test. An illustration does not provide data on which to base a claim.

Teacher Reflection Questions

- In the *Reflect* segment, students were to apply the concepts on this section to their own community. Which concepts did students have trouble applying? Which concepts were easier for them to apply?

- In *What's the Point?*, students were asked to make connections to *the driving question*. How did you guide them in making connections?

- How did you help students access content in the text?

SECTION 1.8 INTRODUCTION

1.8 Read

What are Some Sources of Pollution in a River?

◀ *1 class period* *
*A class period is
considered to be one
40 to 50 minute class.

Overview

Students have learned that various land uses can change the water that
is part of their watershed. They have learned that runoff from residential,
commercial, agricultural, and industrial land use can carry different
substances into water that eventually winds up in river water. This runoff
can erode soil, which is carried to rivers. *Deposition* is the dropping of
soil from the river once it slows down. Depending on the size of the soil
particles and how swiftly the river water is flowing, eroded soil is deposited
at various places. Groundwater forms when water is able to seep into soil.
It is especially important for agricultural areas. Groundwater, too, eventually
reaches a river. Both runoff and groundwater can be contaminated by
substances or pollutants as they move over and through the ground.

In *Section 1.8,* students will read about point-source and non-point sources
of pollution. *Point-source pollution* occurs when pollutants are put directly
into bodies of water. These can be easily identified. Examples of point-
source pollution are untreated sewage, chemicals that leak from barrels into
groundwater and then into rivers, and chemicals from industrial smokestacks
that are washed by rainfall into water supplies. *Non-point sources* of pollution
are not so easily identified. They might be oil in runoff from cars and trucks on
city streets, excess fertilizer from lawns and farms, salt and sand used by cities
to treat icy roads, and trash that has been dumped by the side of the road.

Targeted Concepts, Skills, and Nature of Science	Performance Expectations
Human activity can affect the ecology of a community. Humans use rivers for residential, commercial, industrial, and agricultural purposes. These activities affect water quality along a river.	Students should be able to provide examples of how land used by humans along a river can have an impact on the quality of water in a watershed.

Targeted Concepts, Skills, and Nature of Science	Performance Expectations
Water flow transports and redistributes materials in a stream.	Students should be able to say that when water moves materials from one place to another, it redistributes these materials. In doing so, sometimes harmful or excess materials enter a watershed and cause changes there.
Pollutants to a watershed may occur from point source or non-point sources of human activities.	Students should be able to distinguish between point-source pollutants and non-point source pollutants and identify examples of each.

Materials	
1 per classroom (optional)	Projection images from students' text
1 per class	Class *Project Board*
1 per group	Set of images of various land uses along the Rouge River

Activity Setup and Preparation

There is no investigation for this section, but equipped with new information about point and non-point sources of pollution, students can study the illustrations of common settings on pages 46 and 47 in the student text. Students can also re-study photographs of activities along the Rouge River and add to their ability to observe about how pollutants enter a watershed and what effects they can bring about.

With expanded information derived from observation and investigations in previous sections, students can add evidence to the *Project Board*.

Homework Options

Reflection

- **Science Content:** Describe two non-point sources of pollution that you might see in your neighborhood or on your way to school. (*Answers will vary but may include: car exhaust, a car dripping oil or gasoline, a truck spreading salt on an icy street, or someone spreading fertilizer on a lawn.*)

- **Nature of Science and Science Process:** On the way to school one day, you see a tanker truck pulled up by a small stream. The tanker is unloading a liquid into the stream. What might you do? *(Answers will vary but there are laws against this activity. Suggestions may include making a note of any name on the truck or its license plate, calling local law enforcement agents, or calling the local environmental protection agency. Even if the truck is gone when law enforcement agents arrive, scientists can test the water to find out what was poured into the stream and, if the dumping company is known, they can be fined.)*

Preparation for Back to the Big Question

- **Science Content:** Write a rough draft of a report to the Wamego town council telling them, based on evidence you have learned thus far, some of the possible point and non-point types of pollution they will have to watch for if FabCo and increased housing is built in town. *(Answers will vary but the draft may include that a new industry and increased housing may change Crystal River water because of increased runoff from the factory; parking areas and new streets will have to be built; there will be increased oil on the streets because there will be more cars; and more use of salt on roads, increased litter, and more fertilizer on lawns, which will affect groundwater.)*

- **Nature of Science:** Explain how opinion can be a problem when a scientist is interpreting evidence collected from a scientific investigation. *(Answers will vary but should include that evidence should be based on facts observed or measured during an experiment. A scientist who adds to or colors his or her facts changes the evidence and it can no longer be used to make recommendations.)*

1 class period* ▶

1.8 Read

What are Some Sources of Pollution in a River?

5 min.

Students are introduced to the term pollution *to describe substances that change the quality of water in a watershed.*

META NOTES

The *What's the Point?* text on page 50 of the student book identifies the big ideas in this section. This text may help you to focus your students' learning through class discussions. Visit *What's the Point?* on page 50 before any class discussions.

SECTION 1.8 IMPLEMENTATION

1.8 Read

What are Some Sources of Pollution in a River?

pollution: substances added to air, water, or soil that cause harm to the environment.

You built and ran models of land use using a stream table. You observed the effects of the different kinds of land use on erosion, deposition, and runoff in a watershed. During the classroom discussion, you might have discovered something more. Human activities also change the quality of the water. As the water flows through the watershed, it carries with it stuff it picks up along the way. This could be dirt and soil eroded from land. Runoff and groundwater can also pick up substances like chemicals, small particles, and pieces of trash that can affect the quality of the water. Scientists call these substances that end up in the river **pollution**. Pollution can cause harm to human health or the environment. Most of the time, these substances result from human activities. Normally, they are not found in natural environments. They can be very dangerous because living organisms may not be able to handle them.

Stop and Think

Look at the pictures below and on the next page. You might have seen scenes like these in your neighborhood, or town. People are walking down the street. People are washing their cars. Some are taking care of their lawns. Someone is pouring something down a sewer drain. Workers are fixing something under the street.

LT 46

Project-Based Inquiry Science

○ Engage

Divide the opening paragraph into three sections to give students a chance to focus on various aspects of changed water quality.

- Have a student volunteer to read the first section (ending with *"Human activities…"*) for the purpose of reviewing what students have done and to bring the students' attention to the human influence on water quality.

---

*A class period is considered to be one 40 to 50 minute class.

- Have a second student read to the bold term, **pollution**, to develop an idea of how water moving through a watershed picks up materials that can affect water quality.

- Have the third student read through to the end of the paragraph, which emphasizes the potential harm that can come from polluted water.

◇ Evaluate

Evaluate students' understanding of pollution as presented in this paragraph by asking questions such as "What causes pollution? What things are found in polluted groundwater?, How do you know if water is polluted? and How might changes in water quality affect the quality of your life?"

Stop and Think

Look at the pictures below and on the next page. You might have seen scenes like these in your neighborhood, or town. People are walking down the street. People are washing their cars. Some are taking care of their lawns. Someone is pouring something down a sewer drain. Workers are fixing something under the street.

LT 46

Project-Based Inquiry Science

Stop and Think
15 min.

Students analyze common human activities that result in changes to water quality. This prepares them for learning the differences between point and non-point sources of pollution in rivers.

TEACHER TALK

❝ Think about what you've learned about materials and water moving through watersheds as you look at the two illustrations on pages 46 and 47. Study the two pictures. Think about where you live. Different people may see different activities going on. First, answer the questions individually. Then, you will talk about them as a class. **❞**

Possible Answers:

1. The images represent residential and commercial land use.

2. Residential:

   a) Man dumping oil down a drain — the drain may lead directly to a river. This is illegal in most states.

   b) Car leaking oil into a gutter — the oil eventually will be washed into the drain and go to the river.

   c) Person washing a car — the soapy water will enter groundwater and some may flow down to the drain.

   d) Woman littering — the litter may be washed into the drain and move to the river.

   e) Watering the lawn and sidewalk — some of the water will enter the groundwater; water landing on the sidewalk will be runoff.

   f) Woman raking and bagging leaves — removing litter from lawn, although when leaves break down and their organic matter enters groundwater, they act as a natural fertilizer.

3. Commercial:

   a) Store owner pouring dirty water into drain — probably goes directly into the river.

   b) Car leaking oil onto street — the oil eventually will be washed into the drain and go to the river.

   c) Woman littering — the litter may be washed into the drain and move to the river.

   d) Hole in street to possibly repair a broken sewer line — sewage may leak directly into the river.

4. Both areas will have runoff because of paved areas; grassy areas will absorb water. The commercial area has no grassy area for groundwater to form.

5. In grassy areas, water may pick up chemicals (fertilizers) from gardening. In both, runoff carrying discarded oil and contaminated wastes and litter cause pollution in rivers.

△ **Guide**

Consider projecting an image of each illustration. To facilitate class discussion, begin by asking students to identify the pollution they see going on in each image and list these.

META NOTES

Anticipate that students may interpret the illustrations differently from what you predict or plan.

1.8

1.8 Read

Answer the following questions as you look at the pictures.

1. What kind of land use does each picture represent?

2. Identify all of the examples of human activities shown in each picture. Describe how each activity might cause pollution.

3. How might each land use affect the local runoff and groundwater?

4. Think about what might happen in each place when it rains. Describe what could be on the ground that might cause pollution in a river.

Sources of Pollution in Rivers

Once pollutants get into soil or water, they are carried by surface and groundwater to rivers. In this way, they are distributed over large areas, sometimes miles away from their source.

Depending on how the pollution enters a body of water, pollution sources are divided into two groups: **point-source pollution** and **non-point-source pollution**.

point-source pollution: pollution that originates from a single point or location.

non-point-source pollution: pollution that comes from many sources over a large area.

LT 47

LIVING TOGETHER

Sources of Pollution in Rivers
10 min.

Students learn about two categories of pollution based on where the pollution originates: point-source pollution and non-point source pollution.

○ Engage

Prepare students for learning about the current classifications of pollution —point and non-point sources but without initially using these terms. Instead, ask, "What are some sources of pollution in a river?" List their ideas in two columns *(point and non-point)* without telling them what the columns mean.

Then, have students read *Sources of Pollution in Rivers* on pages 47-49 in the student text. After students have read the terms and their descriptions, go back to the initial two lists you recorded and have students identify what class of pollution their ideas fall into *(point or non-point)*.

Point Sources of Pollution and Non-Point Sources of Pollution

10 min.

Students study and discuss sources of pollution and apply understanding to examples in their community..

Point Sources of Pollution

Point-source pollution comes from a specific point or location. From this location, the pollution is discharged directly into rivers, lakes, or oceans. Scientists can easily identify the source of this type of pollution. They analyze the water at different points in the river or lake. The closer to the point source they measure, the higher the amount of pollutant they find.

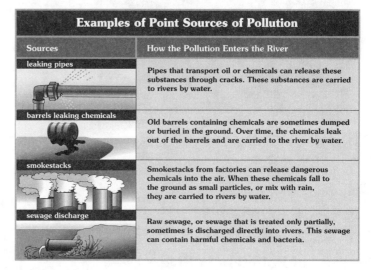

Examples of Point Sources of Pollution

Sources	How the Pollution Enters the River
leaking pipes	Pipes that transport oil or chemicals can release these substances through cracks. These substances are carried to rivers by water.
barrels leaking chemicals	Old barrels containing chemicals are sometimes dumped or buried in the ground. Over time, the chemicals leak out of the barrels and are carried to the river by water.
smokestacks	Smokestacks from factories can release dangerous chemicals into the air. When these chemicals fall to the ground as small particles, or mix with rain, they are carried to rivers by water.
sewage discharge	Raw sewage, or sewage that is treated only partially, sometimes is discharged directly into rivers. This sewage can contain harmful chemicals and bacteria.

Non-Point Sources of Pollution

Non-point-source pollution comes from many sources and locations. Scientists cannot easily identify all the sources of this pollution. For example, one non-point source of pollution is runoff containing fertilizer used on lawns or farmland. Because the runoff has material from so many different farms or lawns, it would be difficult to pinpoint the source of the fertilizer. Another non-point source of pollution is urban runoff from roads and parking lots. Water running off these surfaces can carry oil leaked from cars or salt used to melt ice to a river. This type of pollution is often carried to the river by runoff over large areas. Non-point sources of pollution are

○ Engage

Have half of the class look at the illustrations in the *Point Sources* chart and half of the class study the illustrations in the *Non-point Sources* chart. Then, have the class discuss and share reasons why the examples have been classified as they are. They might also want to go back and reclassify the two lists on the board or add to them.

△ Guide and Assess

To confirm that students understand the difference between the two categories of pollution, facilitate a class discussion by having them consider the following two scenarios. In scenario 1, some people live in cities on rivers, so it probably doesn't take long for a pollutant in some runoff to reach the river. In *Scenario 2,* many people live far from the nearest river. They live in a housing development where everyone has a beautiful green lawn because the lawns are fertilized and kept weed-free with pesticides. If a pollutant gets into groundwater several miles from the nearest stream, it may move through a large area of groundwater and take a very long time to reach the nearest small stream or river. Which class of pollution is represented in these scenarios? Are there point and non-point sources in both examples?

META NOTES

Many students may have difficulty imaging non-point source pollution. Because the "point" at which pollution takes place is not apparent and often the pollutant is not visible to the eye, students assume that there is nothing there. Unlike a smoke stack or leaking pipe, non-point source pollution doesn't provide clues as clearly as does point-source pollution. Throughout the remainder of the Unit, remind students that not all pollutants or polluters leave evidence that is easy to see or to trace.

NOTES

....

....

....

....

....

....

....

....

....

....

Reflect

10 min.

Students apply what they have just read to situations they have already looked at by revisiting the photographs of the Rouge River watershed in the Unit Introduction and in Section 1.6.

much more difficult to control. It is hard to determine who or what is responsible for this pollution. Non-point sources of pollution can originate from a very large land area such as an entire watershed.

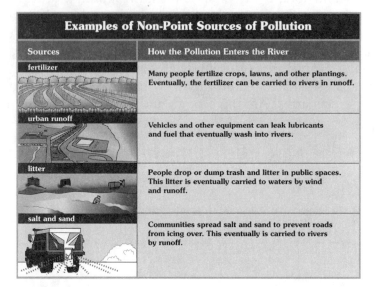

Examples of Non-Point Sources of Pollution	
Sources	**How the Pollution Enters the River**
fertilizer	Many people fertilize crops, lawns, and other plantings. Eventually, the fertilizer can be carried to rivers in runoff.
urban runoff	Vehicles and other equipment can leak lubricants and fuel that eventually wash into rivers.
litter	People drop or dump trash and litter in public spaces. This litter is eventually carried to waters by wind and runoff.
salt and sand	Communities spread salt and sand to prevent roads from icing over. This eventually is carried to rivers by runoff.

Reflect

Look back at the photographs you reviewed early in the Unit. The photographs show scenes of the land use you were assigned and that you modeled with your stream table. Discuss with your group the types of pollution that may result from your land use.

• Record all of the pollution sources your group identifies in the photos, including both point sources and non-point sources.

• Record any pollution sources you think might be there because of certain activities or events shown in the photos.

• For each pollution example you record, determine if it is a point source or non-point source of pollution. Make sure you write the reason why you think so.

LT 49

△ Guide

Before the class period, review the photographs from pages 6, 7, and pages 36 through 38. Select photos that will enable students to demonstrate their understanding of point vs. non-point source pollution sources.

Divide the class into their teams or groups. Assign each group one or more photographs that you have selected from earlier in the text. Ask groups to prepare a presentation that lists what they think are sources of pollution in the photograph, what they think is the reason for the pollution, and whether

Your teacher will lead a class discussion where each group shares their photos and their work. Listen carefully as other members of your group discuss their observations and conclusions. With your class, review and discuss the observations and conclusions drawn by other groups investigating a different land use. How are the pollution sources similar or different for each land use? Come up with a list of types of pollution you agree upon. List the types of pollution you found for each of the land uses your class has investigated.

Update the *Project Board*

The questions you focused on in this *Learning Set* were *How does water affect the land as it moves through the community?* and *How does land use affect water as it moves through a community?* Return to the *Project Board* to update any questions or ideas you have posted. You now have many items to post in the *What are we learning?* column. Be sure to cite (in the *What is our evidence?* column) the evidence you collected to support what you now know about pollution and land use. Discuss with your class what you recorded on the *Project Board* that can help you answer the two questions. You might make up and discuss several new ideas or understandings that should be recorded on the *Project Board*.

What's the Point?

There are many different ways that land use can add pollutants to a watershed. All of the different ways of polluting are grouped into two different types: point sources and non-point sources of pollution. Point sources of pollution, the kind that happen in many industrial areas, are very harmful. However, they are a lot easier to stop than non-point sources of pollution. Non-point sources of pollution can be more difficult to find. They do not come from a specific place. Agricultural areas and residential areas create a lot of non-point sources of pollution through fertilizers.

Project-Based Inquiry Science

Update the *Project Board*
10 min.

Students revisit the class Project Board *to add information that shows the evolution of their ideas as they have learned new information about water in watersheds.*

it is a point-source or non-point-source pollution. Students are also to give reasons (*evidence*) for their choice.

Have each group present their photo and conclusions. Encourage the rest of the class to communicate their ideas and ask questions of the presenters.

Inform the class to be polite during presentations. Tell them that they might not always agree with the presenters' conclusions. If they disagree with the presenters' conclusions, they should voice their differences politely, but be prepared to give reasons for their viewpoint.

△ **Guide**

Remind students that in column 3 of the *Project Board*, they are to answer the question *"What are we learning?"* In column 4, they are to provide evidence that supports what they say they have learned.

TEACHER TALK

❝You have learned a lot about point and non-point source pollutants. You will need this information as you move ahead through the Unit. Scientists keep track of everything they read, think about, and do in an experiment to answer their questions. The *Project Board* gives us a place to keep track of what you are learning. What you have learned in this *Learning Set* needs to be added to the *Project Board*. In addition, let's recall the *Big Question: How does water quality affect the ecology of a community?* How does what you are learning relate to this *Big Question?*❞

Ask students to state claims for column 3 and evidence supporting the claims for column 4. You might have to help them get started by writing the first claim statement. Remind students that they can have more than one claim. Examples of four possible claims and their evidence might include:

Claim 1: Human activities on land can affect the water quality of a river. If land is changed and roads and large areas are paved, water quality can be affected.

Evidence for Claim 1: Changing the form of land affects erosion and deposition of rivers. Our stream table model showed that for paved areas, erosion increased due to faster runoff.

Claim 2: Deposition of materials occurs when water movement slows down.

Evidence for Claim 2: In the stream table model, you observed that as water flowed faster, it carried more soil. Then, as it slowed down, it released the soil. This happens on curvy rivers where fast-moving water eroded the outside curves and then dropped soil on the inside curves where water slowed down. This was hard to see on the model, but you also read about this.

Claim 3: How land is used affects how water flows over or through land to a river.

Evidence for Claim 3: From the stream table model, you observed that when there is a lot of pavement, there is more runoff and it can move faster. You decided that this causes more erosion and carries more water to a river. Fast moving water can stir up soil in the river and erode soil there. In places where there is grass and trees (*vegetation*), water is absorbed into the ground. It becomes part of groundwater and moves more slowly into a river. Land without vegetation or pavement is bare soil. It can be eroded easily by fast-moving water and is moved to rivers.

Claim 4: Human activities can cause changes in water that result in pollution.

Evidence for Claim 4: In the reading, you learned how human activities such as careless dumping of materials, accidental spills, manufacturing, farming, and dumping trash can cause pollution in a watershed. Some of these substances go directly to a river and some of them move through groundwater over a long period of time.

Assessment Options

Targeted Concepts, Skills, and Nature of Science	How do I know if students got it?
Human activity can affect the ecology of a community. Humans use rivers for residential, commercial, industrial, and agricultural purposes. These activities affect water quality along a river.	**ASK:** What is one way that humans have affected the movement of water in a watershed? **LISTEN:** Students have many choices here. Listen for such human activities as pavement, and any of the four types of land use that students experimented with.
Water and land interact with each other.	**ASK:** How does pollution in a river near a factory or farm tell you that land and water interact? **LISTEN:** Answers will vary but students might say that pollutants found in a river or stream got there by way of groundwater or runoff from pavement. Those substances are not naturally part of a river.

Targeted Concepts, Skills, and Nature of Science	How do I know if students got it?
Water flow transports and redistributes materials in a stream.	**ASK:** How can runoff redistribute soil or sediment in a river? **LISTEN:** Students should be able to say that water in the form of runoff from pavement moves fast and can stir up and lift soil or sediment in a river, causing erosion. Later, as the water slows down, the sediment drops out or is deposited to the bottom of the river at another place along the river.

Teacher Reflection Questions

- What difficulties did students have in distinguishing point and non-point sources of pollution in the illustrations? Were they able to make connections between water flow in a watershed and the spread of pollution? What changes should be made to the lesson to help students find examples of point-source and non-point source pollutants?

- How did the *Project Board* assist students in making connections between concepts they have learned so far? In what ways is the *Project Board* valuable for recording information when students need to answer the *Big Question* and address the *Big Challenge?*

- How has students' conceptualization of a watershed changed since the beginning of this *Learning Set?*

Learning Set 1

Back to the Big Question and the Challenge

◀ *1 class period*

A class period is considered to be one 40 to 50 minute class.

Overview

Using information and evidence from their investigations on how water flows through a watershed and how specific types of land use can affect water resources, students can begin to discuss and agree on preliminary recommendations for the town council of Wamego. Students update their *Project Board* with their recommendations.

Targeted Concepts, Skills, and Nature of Science	Performance Expectations
Explanations are claims supported by evidence, accepted ideas, and facts.	Students should be able to begin to draft a recommendation based on their explanations containing claims and evidence from their investigations and science knowledge.
Scientists make claims (conclusions) based on evidence obtained (trends in data) from reliable investigations.	Students should have a clear understanding of the definitions of and relationship between *evidence* and *claims*.
Water in a watershed travels predictably, from higher to lower elevations.	Water flows from higher to lower elevations in watersheds.
Water flow transports and redistributes materials in a stream.	Students should understand that the water flowing from watersheds into a river or stream carries materials from the ground.

Materials	
1 per class	Class *Project Board*
1 per student	*Create Your Explanation* page

Homework Options

Reflection

- **Science Content and Process:** Write a rough draft to the Wamego town council explaining how one type of land use currently affects the Crystal River. *(Answers will vary but should reflect the characteristics of residential, commercial, industrial, or agricultural land use. Answers should include descriptions of amount of runoff or possible pollutants that might be produced as a result of that land use.)*

- **Nature of Science and Science Process:** Write a rough draft of a recommendation about where the town might place a new factory without harming the water used by the residents of Wamego. What might be some implications to other towns? *(Answers will vary but students should consult the map of the town on page 11 in the student text in preparing their draft.)*

NOTES

BACK TO THE BIG QUESTION/CHALLENGE IMPLEMENTATION

◀ *1 class period**

 Learning Set 1

Back to the Big Question and the Big Challenge

How Does Water Quality Affect the Ecology of a Community?

The questions you were investigating in this *Learning Set* were: *How does water affect the land as it moves through the community?* and *How does land use affect water as it moves through a community?* The answers to these questions might have seemed obvious at first. Now, though, you see how complicated the system of watersheds can be. You have seen how the quality of water depends on many factors. You have seen how watersheds are linked to each other.

Watersheds are an important part of understanding how water flows. Soil and chemicals might get into the runoff and be carried to the river. You have investigated how watersheds work. You also saw how runoff moves in a watershed. You noticed that runoff always moves from higher to lower elevation. Sometimes the water pools in low areas. Other times the water continues to move through the watershed to the river.

You investigated several different types of land use. You used a stream-table model to model the land cover that could be found with each land use. The stream tables helped you see erosion, deposition, and runoff for each land cover. You noticed that the water picked up some soil and carried it to the river.

In the final section, some new words described what you saw in your model. Your models provided the beginning of your understanding. Then you read about the different land uses. Some of the issues in the land uses were similar to each other. Other issues, like those in the industrial and the agricultural uses, were very different. In all the different land uses, runoff and erosion were very important processes. Also, each land use can introduce point-source and non-point-source pollution. You identified pollution sources for each land use. You are beginning to see the connection between land use and water quality. This connection will be very important as you consider what effect the changes in Wamego might have on Crystal River.

LT 51

Learning Set 1

Back to the Big Question and the Challenge

10 min.

Students begin to tie together what they have investigated about land use and water quality and find that the two influence each other.

○ Engage

First, review what students have accomplished in this *Learning Set*. They have learned what a watershed is and have investigated how the use of land affects the quality of water moving through the watershed.

△ Guide and Assess

On the board, list the terms *runoff, groundwater, erosion, deposition, point-source,* and *non-point source pollution.* Ask students to apply their

*A class period is considered to be one 40 to 50 minute class.

Explain and Recommend

10 min.

A recommendation is a claim in which a suggestion is made based on evidence and science knowledge.

Explain and Recommend

The *Big Question* for this Unit is *How does water quality affect the ecology of a community?* Recall that you will be answering smaller questions as you move through this Unit. Doing this will help you eventually answer the *Big Question* and address the *Big Challenge*. At this point, you are not ready to answer the *Big Question* completely and thoroughly and address the challenge. But you probably have some ideas of how watershed structure, erosion and runoff, land use, and pollution sources are a part of a good answer to the *Big Question* and the *Big Challenge*.

Discuss with your group your new knowledge and what you recorded on the *Project Board* that can help you answer the *Big Question* and address the *Big Challenge*. Identify answers to the *Big Question* and recommendations you might make to the Wamego town council. Using a *Create Your Explanation* page, state each of your answers and recommendations as a claim. For each claim, record the evidence you have and your science knowledge. Then write an explanation that connects the evidence and your science knowledge to your claim.

Communicate

Share Your Explanations and Recommendations

Share your claims, recommendations, and explanations with the class. If your classmates disagree with any of your claims or recommendations, discuss the evidence you have and the science knowledge that support them. Try to come to agreement. You can help one another revise your claims or recommendations to better match your evidence and science knowledge.

Update the *Project Board*

Add new claims and the evidence that goes with them to the *What are we learning?* and *What is our evidence?* columns. Add recommendations the class agrees on and the evidence that supports them to the *What does it mean for the challenge or question?* and *What is our evidence?* columns.

knowledge of how watersheds can change. Don't discourage the use of these listed terms, but listen for students to describe concepts and not merely give definitions.

○ Engage

Tell students that they are going to meet in their groups for ten minutes to utilize the evidence they have from *Learning Set 1* investigations and science knowledge. In their groups, they will begin forming an explanation that will be the basis of recommendations to the Wamego town council at the end of the Unit. Because they have just completed their first *Learning Set*, they do not have a complete picture on which to base their recommendations, but will be a start.

△ Guide

Remind students that a *recommendation* is a kind of claim that suggests what to do when a certain situation occurs. Record this general formula for students to refer to as they begin to plan their recommendations to the town of Wamego. Review each example so that students understand how to use the formula.

When a particular situation occurs, *do* or *try* the following or *expect* to see the following take place:

- *When* you are crossing the street, *[do]* look both ways to make sure the traffic has stopped.

- *When* the temperature is very warm and the dog is panting hard, *[try]* to remember to keep his water dish filled with cool water.

- *When* you see lightning in a summer sky, *[expect]* to see rain and hear thunder.

Students should also make use of a *Create Your Explanation* page to organize their thoughts. Because there are many things for students to coordinate as they begin pulling concepts together, assist by tying together what they need to do.

TEACHER TALK

"Your group investigated how water and specific land uses affect each other. The results of your investigation has given you evidence on which you can base a claim, such as, *As water rushed over bare agricultural land, the runoff eroded loose soil.* Use a *Create Your Explanation* page to construct an explanation. Your explanations should contain science knowledge.

Remember that your explanation needs to contain a claim supported by evidence and science knowledge in a logical way. Watch to make sure that there are no opinions in your explanation or claim."

○ Get Going

Tell students that they have ten minutes to construct their explanations in their groups and then each group will present their explanations and recommendations to the class.

Communicate

20 min.

Share Your Explanations and Recommendations

Communicate

Share Your Explanations and Recommendations

Share your claims, recommendations, and explanations with the class. If your classmates disagree with any of your claims or recommendations, discuss the evidence you have and the science knowledge that support them. Try to come to agreement. You can help one another revise your claims or recommendations to better match your evidence and science knowledge.

META NOTES

It might be easier for students to keep track of claims, evidence, explanations, and recommendations if they have a separate *Create Your Explanation* page for each investigation.

META NOTES

As students proceed through *Learning Sets 2* and *3*, they will have opportunities to revise their explanations and recommendations.

△ Guide

Encourage students to ask questions of the group presenting their explanations. Model the kinds of questions students might ask. This will help the presenters talk in terms of claims based on evidence and help listeners recognize how to use evidence and science knowledge effectively to come up with conclusions that the class can agree upon.

◇ Evaluate

Listen to verify that students present claims that are based on evidence and science knowledge and not on opinions. Listen for clearly stated recommendations that relate the evidence students present to the *Big Question* for the Unit: *How does water quality affect the ecology of a community?*

After the class has agreed on claims and recommendations that they can make at this point in the Unit, draw their attention to updating the *Project Board*. Ask students what claims they can add to the *What are we learning?* and *What is our evidence?* columns. Have them add evidence to the *What does this mean for the challenge?* column.

Update the Project Board

5 min.

Update the *Project Board*

Add new claims and the evidence that goes with them to the *What are we learning?* and *What is our evidence?* columns. Add recommendations the class agrees on and the evidence that supports them to the *What does it mean for the challenge or question?* and *What is our evidence?* columns.

Teacher Reflection Questions

- During this implementation, students were asked to agree on recommendations based on evidence from the investigations completed during this *Learning Set*. How difficult was it for students to agree as a class on these recommendations or did they agree too easily?

- Were students motivated by the investigations in this *Learning Set?* What can you do to increase this motivation the next time you teach this *Learning Set?*

- How can you increase student participation in discussions? What can you do to have all the students participate, not only the ones who consistently participate?

NOTES

NOTES

LEARNING SET 2 INTRODUCTION

Learning Set 2

How Do You Determine the Quality of Water in a Community?

◄ *8 class periods**

*A class period is considered to be one 40 to 50 minute class.

Students design investigations to test for the presence of six water-quality indicators. Claims and evidence are presented to the class. Conclusions are modified and refined through class discussion in preparation for drafting recommendations for the Wamego town council.

Overview

In *Learning Set 1*, students were introduced to the concept of watersheds. They learned that land use can affect the quality of water that drains into a river and ultimately, into oceans. In *Learning Set 2*, students begin a more concentrated study of the driving question *How do you determine the quality of water in a community?* when they learn about specific indicators that can be used to monitor water quality in a community's watershed. Students come to realize that these indicators reflect how land use can affect water quality. They are also introduced to how water quality can affect life forms in a body of water.

Students begin the *Learning Set* by discussing the problem of how to test water quality in a watershed. Initially, in *Section 2.1*, they believe that merely checking the clarity, smell, and condition of organisms tells them the quality of the water. In *Section 2.2*, they move in a more scientific direction when they design and present plans for investigating the effects of fertilizer on plant growth.

In *Section 2.3*, students' knowledge of what and how to test for water quality is expanded. They learn how to test water for pH. They consider whether organisms can continue to survive in a stream when the pH of the stream changes. In *Section 2.4*, students observe how temperature and turbulence affect the amount of dissolved oxygen in a body of water and how this, in turn, affects organisms living there. In *Section 2.5*, students read about the effects of changes in temperature, turbidity, and the presence of fecal coliform bacteria on water quality. By the end of the *Learning Set*, students have a better idea of how and why it is important to monitor water quality in a community and they can begin drafting recommendations to the town of Wamego.

> **LOOKING AHEAD**
>
> • In Section 2.2, students test the effects of fertilizer on duckweed growth. This experiment must be run over 5 to 10 days. In Section 2.3, the cabbage juice indicator must be prepared one to two days in advance, refrigerated until the day it is used, and brought to room temperature on the day it is used.

Targeted Concepts, Skills, and Nature of Science	Sections
Scientists often work together and then share their findings. Sharing findings makes new information available and helps scientists refine their ideas and build on others' ideas. When another person's or group's ideas are used, credit needs to be given.	2.1, 2.2
Scientists plan investigative questions and communicate ideas.	2.1, 2.2, 2.3, 2.4
Some common water-quality indicators that can be measured are pH, temperature, turbidity, levels of dissolved oxygen, nitrates, phosphates, and fecal coliform bacteria.	2.2, 2.3, 2.4, 2.5
Land use can affect water in a watershed.	2.2, 2.5
The use of fertilizers can affect water quality in a watershed.	2.2
In a fair test, one variable is manipulated and one variable is measured to see what happens as a result of changing the first variable. All other variables are kept constant.	2.2
pH is a measure of how acidic or basic a solution is and is one indicator of water quality.	2.3
Some aquatic organisms are sensitive to changes in pH.	2.3
Most aquatic organisms use dissolved oxygen for respiration.	2.4
Dissolved oxygen is an indicator of water quality.	2.4
The amount of dissolved oxygen in water increases as temperature decreases and as turbulence increases.	2.4
Water temperature is affected by temperature, turbidity, and fecal coliform bacteria.	2.4, 2.5
Thermal pollution, the result of human activity, causes water temperatures to increase, reduces water quality and can harm organisms in an aquatic ecosystem.	2.4, 2.5
Turbidity is a measure of how opaque water is and may increase due to disturbances in land structure or in the river bed as a result of human activity.	2.4, 2.5
Fecal coliform bacteria are found in the digestive tracts of animals.	2.5
Human activity can affect the ecology of a community. Humans use rivers for residential, commercial, industrial, and agricultural purposes. These activities affect water quality along a river.	2.1, 2.2, 2.3, 2.4, 2.5

Students' Initial Conceptions and Capabilities	• Most students will have an idea of what a scientific investigation is and how it proceeds, but they may not realize that investigations are usually carried out within a group and that results are shared among members of the scientific community. (Mead & Metraux, 1957.)
	• Students may not understand that scientists do investigations to test ideas, not to produce a desired result. (Carey, et al., 1989; Schauble et al., 1991; Soloman, 1992.)
	• Students may not understand the need to change only one variable (the independent variable) during an experiment. (Wollman, 1977a, 1977b; Wollman and Lawson, 1977.)
	• Students often do not grasp that the most important part of an experiment lies in analyzing results, not just collecting data. (Steven, 2003.)

Understanding for Teachers

Clean water resources are essential for life. It is a worldwide problem. Outside the developed countries of the world, between 2 and 3 billion people are without adequate or any sanitation. Without clean drinking water resources, illness and disease are common. It does not take long for these same problems to appear in developed areas either if the systems that protect drinking water supplies, such as sewage treatment, break down. Breakdown may come in the form of an actual break in sewer lines, which usually results in a call to boil water used for drinking until the problem has been solved. Breakdown may also come when someone intentionally or through ignorance, pours harmful or potentially harmful materials into a water resource.

No river or pond is completely free of pollutants. It is nearly impossible to keep pollutants out of air and water. The Environmental Protection Agency (EPA) monitors major watersheds throughout the country and makes this information available online. Individual states must set what are called *total maximum daily loads* for individual pollutants that provides information that shows maximum allowable levels of materials that may exist in specific bodies of water without causing harm.

Sediment, nitrates and phosphates, and harmful organisms (pathogens) from non-point sources are the greatest sources of problems in North America. Seventy-five percent of problems with water pollution come from erosion, materials washed out in rainfall, and runoff from farms and large livestock feeding areas called feed lots. In addition, according to the United States Department of Agriculture, runoff carries away nearly one quarter of all fertilizer applied to cropland.

By the end of *Learning Set 2*, students should understand that it is desirable for communities to monitor the quality of their water supplies and to take steps to correct problems as soon as they are noticed. There are several ways that communities monitor water quality beyond merely checking its appearance. The investigations and demonstrations in *Learning Set 2* enable students to expand their science knowledge about water quality. They also prepare students for *Learning Set 3* in which students will concentrate on the effects that water has on life forms that occupy ecological niches in watersheds.

Learning Set 2

How Do You Determine the Quality of Water in a Community?

8 class periods ▶

LEARNING SET 2 IMPLEMENTATION

Learning Set 2

How Do You Determine the Quality of Water in a Community?

You have investigated watersheds. You read how they connect water in small areas. You also saw how those smaller watersheds are nested in larger ones. You then reviewed some water samples from a watershed. There were differences among these samples. This indicates that the water is different at different locations in a watershed.

The water from this mountain reservoir makes its way to local residents' drinking glasses.

Even though the samples were different, the water in the watershed is all connected. Eventually, water from separate areas will mix together. It will all flow into one larger waterway. Thus, all of the samples you reviewed earlier might mix together farther downstream.

Suppose this mixture flowed through your community, and it was your only water source. Think about how you would then judge the quality of this water. Your class began to discuss how land use in the watershed could affect what ends up in the water. Also, what is in the water can affect the plants, wildlife, and people in the community. A community should want to know exactly what is in the water. In this *Learning Set,* you will work to answer the question *How do you determine the quality of water in a community?*

LT 53

LIVING TOGETHER

○ Engage

Have students recall the kinds of land uses they modeled in *Learning Set 1* or, if the models are still available, assemble them in a place in the classroom where students can view all of them together. Make sure students recall the terms *residential, commercial, industrial,* and *agricultural uses.*

Ask students to think about what it would be like if all of the land uses they were introduced to in *Learning Set 1* were brought together in one place. Record students' suggestions. Some students may quickly realize that many towns and cities are, in fact, made up of many, if not all, of the land uses.

*A class period is considered to be one 40 to 50 minute class.

2.0

❝Think about what might happen if all the water from each type of land use fed into your community's watershed. What might be in the water from homes, from factories, businesses streets, and farms? What might happen to the water in the watershed? Let's record some of the things you think would change in the watershed.**❞**

Introduce students to the driving question for this *Learning Set—How Do You Determine the Quality of Water in a Community?*

△ Guide

Have students read the introduction to *Learning Set 2* on page 53 in the Student Edition for the purpose of reviewing concepts learned in *Learning Set 1*. Tell students that recalling this information will give them a setting for studying about water quality.

Emphasize that by the end of *Learning Set 2*, they will be able to answer the driving question and they will be closer to answering the *Big Question* for the Unit: *How does water quality affect the ecology of a community?*

NOTES

....

....

....

....

....

....

....

....

....

SECTION 2.1 INTRODUCTION

2.1 Understand the Question
Water-Quality Indicators
Overview

1 class period ▶

A class period is considered to be one 40 to 50 minute class.

Students are introduced to the driving question of *Learning Set 2: How do you determine the quality of water in a community?* When challenged to determine what would happen if their watershed was their only source of water, students begin to think about the quality of that water. They consider ways to determine the quality of water. Students return to the *Project Board* to record what they think they know about water quality. They also record questions about water quality that they think they might want to investigate. This activity prepares them for activities they will do in subsequent sections where they are introduced to and investigate several measures of water quality.

Targeted Concepts, Skills, and Nature of Science	Performance Expectations
Some common water-quality indicators that can be measured are pH, temperature, turbidity, levels of dissolved oxygen, nitrates, phosphates, and fecal coliform bacteria.	Students should be able to indicate that pH, temperature, turbidity, levels of dissolved oxygen, nitrates, phosphates, and fecal coliform bacteria are indicators of water quality that can be measured.
How land is used can affect the water quality in a watershed.	Students should be able to relate land use to water quality in a watershed.
Scientists plan investigative questions and communicate ideas.	Scientists plan investigative questions and communicate ideas to learn more about water quality.

Materials	
1 per class	Class *Project Board*

Homework Options

Homework options provide students with some short, relevant work options. Each of these assignments encourages reflection on the current section or prepares students to make the connections between one section and the next. A variety of homework option tasks are included throughout the Unit.

Reflection

- **Science Content:** What can happen in a community where microbes are found in the water? How might these microbes affect the community? How are they spread? *(Students may connect their responses to information learned in the communicable disease Unit.)*

- **Science Content:** In *Learning Set 1,* you determined that there was a relationship between land use and the condition of a watershed moving through a watershed. Think about this relationship. Write two sentences describing how industrial land use and residential land use might affect water quality. Give a reason for each statement. *(Answers will vary. Students may say that industrial land use damages water quality because it often pollutes water with chemicals. Residential use also can affect water quality by careless dumping of materials like motor oil into drains.)*

Preparation for 2.2

- **Science Process:** On page 54 in the student text, four students discuss different ways to check the quality of a community's water supply. Each has a different opinion. How do you think their meeting is like one that scientists might hold when faced with a problem? *(Answers will vary but probably will include that four scientists may each have a different opinion on a problem. They listen to each other's thoughts and out of this may come new ideas or a way to investigate the problem.)*

1 class period* ▶

2.1 Understand the Question

Water-Quality Indicators

Students are given the opportunity to share their experiences. This motivates them for what they are about to learn.

2.1 Understand the Question

Water-Quality Indicators

microbe: an organism that can be seen only with a microscope.

You may have heard about unsafe water in a community. Sometimes a community may close beaches because it is not safe to swim in the water. The water may contain dangerous **microbes** (or microorganisms). Residents may be asked to boil water before drinking it. In other cases, people may be told not to fish in certain waters. Communities must decide when the water in their area is unsafe.

Get Started

Below are some students' ideas about how a community might decide if water is safe. Read and think about each student's ideas.

Rahim: "You can just look at water and know whether you should swim in it or drink it. If it is clear, or mostly clear, it's probably fine."

Sarah: "Water that is not high quality probably smells bad. I think scientists use smell to help determine if water is of high quality."

Lucia: "Scientists test water with chemicals, like chlorine, to see what is in the water and make it clean. I think I have heard of scientists testing for acid in water."

John: "I don't really know . . . but if you look at an area of a watershed and see a lot of dead plants and animals, and people are sick, the water is probably low-quality water."

Do you agree or disagree with each student? What might you say to each student? What might you investigate to determine if a student's idea is correct? Discuss your answers to these questions with your class.

Project-Based Inquiry Science

LT 54

○ **Engage**

Begin by asking students if they have ever heard a warning not to drink the water that comes into their homes without first boiling it for a few minutes. Make a list of students' responses about their experiences with drinking-water warnings. *(Some may relate alerts that they have heard on television news broadcasts, on the radio, or in newspapers. Some may have seen warning signs at a beach similar to the one on page 54 in the Student Edition about unsafe swimming conditions.)*

*A class period is considered to be one 40 to 50 minute class.

Get Started

Below are some students' ideas about how a community might decide if water is safe. Read and think about each student's ideas.

Rahim: "You can just look at water and know whether you should swim in it or drink it. If it is clear, or mostly clear, it's probably fine."

Sarah: "Water that is not high quality probably smells bad. I think scientists use smell to help determine if water is of high quality."

Lucia: "Scientists test water with chemicals, like chlorine, to see what is in the water and make it clean. I think I have heard of scientists testing for acid in water."

John: "I don't really know . . . but if you look at an area of a watershed and see a lot of dead plants and animals, and people are sick, the water is probably low-quality water."

Do you agree or disagree with each student? What might you say to each student? What might you investigate to determine if a student's idea is correct? Discuss your answers to these questions with your class.

LT 54

Project-Based Inquiry Science

Get Started
20 min.

Students want to know why they are doing things. The scenario *at the beginning of the Learning Set* provides *the student's with a reason to be concerned about community water quality.*

META NOTES

Initially, you may have to use wait-time for the class to begin asking questions of other students. In your mind, count to ten before resorting to calling on a student. Students eventually will become involved in the discussion.

△ Guide Discussions

Introduce the topic of testing water quality by planning a class discussion. Inform the class that they should read page 54 in the student text and think about the questions at the end of the page. Select four students at random to play the rolls of the four students shown on page 54. Have them sit across the front of the room, much like a panel. Have a fifth student introduce the purpose of the panel discussion To find out how a community might decide if its water supply is safe. Each of the four can read the script in the text, in turn, about how he or she thinks water quality can be tested. Then open the discussion to questions and opinions from the class. If the class is reticent about asking questions, be prepared to ask "Do you agree or disagree with the method (student name) has suggested?" to start the discussion.

- Remind students to give a reason for their opinion in a class discussion.

- As methods for detecting and testing water quality are suggested, make sure they are recorded for later use on the *Project Board*.

META NOTES

If, during a class discussion, students are beginning to ask too many yes/no questions or questions that utilize one word answers, begin to model asking questions that require more extended answers such as "How did...," "Why do you think that...?" "What did you mean when you said that...?"

Update the Project Board

10 min.

Students work to update the What do we need to investigate? *column of the* Project Board. *Students also can add new information to the first column,* What do we think we know?

Update the *Project Board*

Your class started a *Project Board* to help you keep track of your understanding and questions about water quality and ecology. At the end of *Learning Set 1*, you updated the *Project Board* with information about watersheds and water flowing through watersheds.

Your class has shared their ideas about what you need to do to decide if water is safe. You have also come up with ideas about investigations you could do. Now it is time to update the *Project Board* again. You may have new questions or new ideas about how to determine the quality of water. Record these ideas, predictions, and explanations in the *What do we think we know?* column. You may have some information about water-quality testing that you have read about before or someone has told you about. Suggest it for the *Project Board*. As you move along in your discussion, you may find that you have disagreements about some things you think you know about water-quality tests. Suggest things you need to learn more about. These questions could be turned into investigations you could perform to better understand water-quality testing. Record questions and ideas for investigation in the *What do we need to investigate?* column.

What's the Point?

Your class might have a lot of ideas and questions about how water quality is determined. Some of you might have some experience or knowledge in this area. Your initial conversations and ideas helped you to become aware of your understanding of the topic. You were able to share ideas and questions about water quality during the *Project Board* session. Once these different ideas are out in the open, your class can pursue investigations that focus on these ideas and questions.

LT 55

LIVING TOGETHER

TEACHER TALK

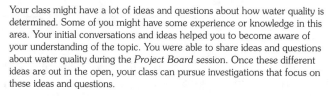

"The discussion seems to have raised a lot of new thoughts and questions. It's probably time to update the Project Board. Let's see where we left off. Where might we record what we've been saying about testing water quality? What makes water of good quality? What makes water harmful? What have you decided that you want to learn more about? Which of these questions might be turned into investigations."

◯ Get Going

Record students' ideas in the appropriate columns of the *Project Board*. There is no need to determine now if these ideas are right or wrong, doable or not doable. Write items that students disagree about, or can't seem to make a decision about, in the second column. Explain that scientists often have similar problems.

△ Guide

Write a date on all entries. This reminds students that the *Project Board* is a dynamic organizer and is used for a growing list of information and insights into a problem.

◇ Evaluate

Make sure that all students have an opportunity to voice an opinion about items entered on the *Project Board*.

Assessment Options

Targeted Concepts, Skills, and Nature of Science	How do I know if students got it?
Scientists often work together and then share their findings. Sharing findings makes new information available and helps scientists refine their ideas and build on others' ideas. When another person's or group's idea is used, credit needs to be given.	**ASK:** How did the students in the text reflect the way scientists sometimes work together? **LISTEN:** Answers will vary, but you should be able to hear students saying that each student scientist was allowed to express his or her ideas of how to test for water quality. Classroom discussion is somewhat like letting the whole scientific community think about the ideas. During the classroom discussion, other students might have added new ideas or information to the ideas expressed.

> **META NOTES**
>
> The *Project Board* can be revisited many times as ideas become modified. Trained scientists keep notebooks and other records of the progress of their thinking. Periodically they revisit these notes to see how their ideas are fitting together. This is like periodically checking the *Project Board* to see if it has information that can be used to answer the *Big Question*.

NOTES

....

....

....

Targeted Concepts, Skills, and Nature of Science	How do I know if students got it?
How land is used can affect the water quality in a watershed.	**ASK:** What did you learn about land use in *Learning Set 1* that you can use to find out how safe water might be in your community? **LISTEN:** Answers will vary but students may say that the way land is used in the community can contribute different materials added to water and therefore affect the quality of the water available in a community.
Scientists plan investigative questions and communicate ideas.	**ASK:** In what ways do you act and think like a scientist when you work in a group? **LISTEN:** You should hear students indicate that sharing ideas is one way that groups act and think like scientists.

Teacher Reflection Questions

- In this section, students determined that they will investigate water quality. What difficulties did students experience in thinking of ways to test for water quality?

- How can students be guided to make more of a connection between the information they put on the *Project Board* from *Learning Set 1* and the *Big Question?*

- The goal of this introductory activity is to activate students' initial ideas and prepare them for further investigation. Did they all feel free to suggest what they thought would work as a test for water quality? If not, how can you get more of them to express their opinions without feeling they will be judged?

NOTES

....

....

....

SECTION 2.2 INTRODUCTION

2.2 Investigate

Plant Growth as an Indicator of Water Quality

◀ *6 to 10 class periods* *

Overview

Students have learned the basic structure of a watershed. In *Section 1.8*, they started to think about point and non-point pollution sources that can affect water quality in a watershed. In this section, students learn that plant growth can be an indicator of the amount of certain commonly used substances (*fertilizers*) in water. They do so by studying how fertilizer affects the growth of a common aquatic plant, duckweed. Students plan and design an experiment in which duckweed is grown in various concentrations of fertilizer. Students observe that duckweed is more prolific when grown in water with higher concentrations of fertilizer. Later, students relate this information to what they learn about the effects of phosphates and nitrates (*ingredients of fertilizers*) and levels of dissolved oxygen on water quality.

*A class period is considered to be one 40 to 50 minute class.

Targeted Concepts, Skills, and Nature of Science	Performance Expectations
The use of fertilizers can affect the water quality in a watershed.	Students should be able to indicate that fertilizers can affect water quality in ways that can be measured.
How land is used can affect the water quality in a watershed.	Students should be able to relate land use to water quality in a watershed.
Scientists plan investigative questions and communicate ideas.	Students should be able to voice that scientists plan investigations and communicate ideas to one another.
In a fair test, one variable is manipulated and one variable is measured to see what happens as a result of changing the first variable. All other variables are kept constant.	Students should be able to distinguish the dependent and independent variables in their experimental plans.

Targeted Concepts, Skills, and Nature of Science	Performance Expectations
Scientists often work together and then share their findings. Sharing findings makes new information available and helps scientists refine their ideas and build on others' ideas. When another person's use to or group's idea is used, credit needs to be given.	Students should be able to relate the development of their individual experimental plans to a single class plan as being similar to the process scientists use to share the results of an experiment at a meeting or when they publish their results in a journal.

Materials

1 per classroom	Roll of masking tape
1 per group	Poster board
15 per class	Large, equal size-beakers or jars, labeled
1 per classroom	Wax marking pencil
60 per class	Duckweed plants
1 per group	Paper clips or tweezers
1 per classroom	Source of distilled or deionized water to fill jars
1 per classroom	Liquid plant fertilizer (8 oz)
1 per classroom	250-mL graduated cylinder
1 per classroom	Roll of clear plastic film
12 per class	Rubber band
1 per group	*Plant-Growth Data and Observations* page
1 per group	Grow lamp with light fixture setup for duckweed growth
1 per group	Colored markers set
1 per classroom	Graduated cylinder, 10-mL

Activity Setup and Preparation

Decide where you will want groups to display the posters showing proposed experimental designs so that the whole class will be able review each plan.

NOTE: It would be helpful for the teacher to try all of this experiment ahead of time, especially making up the solutions and checking to see how far away the light source should be for optimal outcome and to avoid scorching the plants.

- Timewise, this activity requires:
  - o one day to set up
  - o five to ten days of observations
  - o one final day to analyze collected data

- Student groups will plan and design experiments, then submit them to class discussion. The outcome is to be one experiment that the whole class agrees upon.

- Provide each group with a *Plant Growth Experiment Planning Page* to track the important parts of the experimental design process.

- Set aside room in a well-lit area of the classroom, but not an area of direct sunlight, as this can damage the plants. A grow light or a fluorescent fixture will improve results. Try the experiment ahead of time to determine how much student participation you can allow for setting up.

- Label three sets of four jars or beakers with a wax marking pencil at the 300-mL mark by pouring 300 mL distilled water into each jar and marking a line at the 300-mL meniscus. Pour out the water and let the jar or beaker dry.

- Label each jar with fertilizer concentrations: 0%, 0.5%, 1%, and 2% so that you have 3 jars marked 0%, three marked 0.5%, and so on.

- Use these directions to prepare the following concentrations of fertilizers: 0%, 0.5%, 1%, and 2%.

  a) 0% solution: Fill three jars or beakers, labeled 0%, with distilled or deionized water. No fertilizer is to be added to these three jars.

  b) 0.5% solution: Put 1.5 mL liquid plant fertilizer in a 10-mL graduated cylinder and add distilled or deionized water to the 300 mL mark. Do this for each of the three jars or beakers labeled 0.5%. Rinse the cylinder thoroughly, and dry.

c) 1% solution: Put 3 mL liquid plant fertilizer in a 10-mL graduate cylinder and add distilled or deionized water to the 300 mL mark. Do this for each of the three jars or beakers labeled 1%. Rinse the cylinder thoroughly, and dry.

d) 2% solution: Put 6 mL liquid plant fertilizer in a 10 mL graduate cylinder and add distilled or deionized water to the 300 mL mark. Do this for each of the three jars or beakers labeled 2%. Rinse the cylinder thoroughly and dry.

- When time permits, make a reserve stock of each of the same solutions to use during the experiment to replenish the solutions because they may evaporate.

- Using tweezers or an unfolded paperclip, gently transfer five duckweed plants into each of the 12 beakers. You should not use your fingers because they may add oils or other substances from your hands to the solutions. Choose green, healthy plants that have two fronds apiece and are approximately the same size.

- Cover the beakers with clear plastic film and secure with a rubber band.

- Place the beakers in indirect light. If possible, put the beakers under a 24-hour fluorescent or grow light, but not so close that the plants become overheated.

Homework Options

Reflection

- **Science Content:** What do you think the experiment that your class has designed is about? *(Students should be able to tell you that the experiment is a test that will show the effects of different amounts of fertilizers on plants that grow in streams and rivers.)*

- **Science Process:** What evidence did you measure in the investigation that indicated how much growth had taken place? *(Answers will vary. Most groups will have counted the number of fronds to measure growth. Other acceptable measurements may include the size of fronds, number of living plants, and/or length of roots.)*

Preparation for 2.3

- **Science Content:** What is pH a measure of? *(pH is a measure of the acidity of a substance.)*
- **Science Process:** What are two ways that the cabbage-juice indicator identifies pH during an investigation? *(pH is identified by color and by number.)*

NOTES

....

....

....

....

....

....

....

....

....

....

....

....

....

6 to 10 class periods* ▶

SECTION 2.2 IMPLEMENTATION

2.2 Investigate

Plant Growth as an Indicator of Water Quality

5 min.

Try to engage students' interest before they look at their textbooks. Catch students' attention with warm-up questions that connect to everyday experiences before the lesson begins.

2.2 Investigate

Plant Growth as an Indicator of Water Quality

Your class has been asked to think about what is a good indicator of high water quality. You may think that the way the water looks is an important indicator. You may also have considered other indicators. Some of you may have discussed the use of chemical tests to find the quality of water. In the next few sections, your class will investigate some useful water-quality indicators. These indicators can help communities determine the quality of water in their watershed.

Fertilizers used in agriculture can affect plant growth in nearby bodies of water.

People grow plants as food. They may also grow plants to make yards and parks look attractive. Fertilizers are often used to help plants grow. These fertilizers can be carried away in the runoff from fields. The fertilizer then ends up in the watershed. There it can affect plant growth in the water. In this section, you will design an experiment that shows the effects of fertilizer on plant growth.

Design Your Experiment

Your class will work together to design an experiment about the effects of different amounts of fertilizer on the growth of a water plant, duckweed. You will make observations and record data about how the plants grow. Your observations will take place over five to ten days. You will then draw some conclusions by examining your results. Your experiment should answer the following question:
• How does fertilizer concentration affect the growth of duckweed?

LT 56

Project-Based Inquiry Science

○ **Engage**

Assess students' basic understanding of the needs of living things with a series of questions. You might ask:

1. How many people ate breakfast this morning?

2. Did anyone take a food supplement? Why do people take supplements?

Continue the questions, working into the topic of plants.

*A class period is considered to be one 40 to 50 minute class.

1. What do other organisms need to grow and stay healthy? For instance, if some of you have ever had a garden or have ever grown plants indoors, what do you give plants to keep them alive? *(Most students will say water and fertilizer.)*

2. What is fertilizer? What does it do for a plant? Is it anything like the food supplements that people take? Who uses fertilizers? What do they have to do with what we have been studying?

△ Guide

Read the opening paragraphs of *Section 2.2* to focus students' attention on water quality and how fertilizers can affect water quality in watersheds.

Design Your Experiment

Your class will work together to design an experiment about the effects of different amounts of fertilizer on the growth of a water plant, duckweed. You will make observations and record data about how the plants grow. Your observations will take place over five to ten days. You will then draw some conclusions by examining your results. Your experiment should answer the following question:

• How does fertilizer concentration affect the growth of duckweed?

LT 56

Project-Based Inquiry Science

◯ Engage

Inform the class that during this section, they will design an experiment about plants and water quality that everyone will test in the classroom. Reassure them that you will help them as they go through this process.

△ Guide

> **TEACHER TALK**
>
> ❝I'm sure that you are wondering how to go about designing this experiment. Let's read the first paragraph and analyze or think about what it says. When you analyze something, you think about what it means. I can ask some questions and we can list our thoughts about what it says. In the end, we'll have a better idea of how to perform the experiment.❞

After reading the paragraph, ask the following questions to help focus students on what they will do during the experiment.

1. What will the class do together? *(Design an experiment.)*

2. What is the experiment testing? *(The effects of different amounts of fertilizer on growth of a water plant, duckweed.)*

Design Your Experiment

5 min.

Students need to know a valid reason for doing an experiment. Ask students to recall that the town of Wamego needs your input on what will happen if their watershed changes when a big factory moves in. This experiment will help them explain what the town should do.

> **META NOTES**
>
> Most students have little or no experience in designing and planning an experiment. They need practice to develop investigative skills. However rocky their first attempts might be, each experience pushes them to think more creatively about ways to solve problems.

META NOTES

Designing an experiment can be challenging for students. Much like scientists in a real laboratory, students will develop some ideas and then share and revise plans before finding one they agree upon.

3. What will you (meaning the students) do during the experiment? *(Make observations and record data.)*

4. How long will you make observations? *(Five to ten days.)*

5. What will you do with your observations? *(Draw conclusions.)*

6. What question will the experiment answer? *(How does fertilizer concentration affect the growth of duckweed?)*

NOTES

....

....

....

....

....

....

....

....

....

....

....

....

....

Get Started

Remember that when you investigate a process or a phenomenon, you want to find out what factors influence its outcome. These factors are called variables. Scientists design experiments to find out how changing the value of one variable (called the *independent* or *manipulated* variable) affects the values of outcome variables (called *dependent* or *responding* variables). You have been asked to design an experiment to study the effect of fertilizer **concentration** on plant growth. Concentration, in this case, refers to the amount fertilizer you will use. You will therefore need to design your experiment so that you will be able to see how changing the amount of fertilizer (your *manipulated* or *independent* variable) will affect the growth of duckweed (the *responding* or *dependent* variable).

You will use the materials shown in the list in your experiment. Duckweed is a plant that grows in water rather than in soil. Like other plants, duckweed requires sunlight and nutrients to grow. The plant floats in the water. It gets its nutrients from the water. They get into the plant through its roots. When conditions are right, duckweed will grow quickly over a five to ten day period.

The fertilizer you will use is in liquid form. You can add fertilizer to a container of water and place duckweed in the container. The more fertilizer you add to the container, the higher the concentration of fertilizer. The less fertilizer in the container, the lower the concentration of fertilizer.

You will need to find a way to measure the growth of the duckweed. The most common method of measuring growth of duckweed is to count fronds. Fronds are the small leaf-like structures of duckweed.

Materials
- 4 beakers or jars
- 4 duckweed plants
- water
- liquid fertilizer

concentration: the amount of a substance mixed with another substance.

Get Started
10 min.

Students need to become acquainted with a certain amount of experimental terminology if they are to present valid experimental designs for discussion.

⃝ Engage

Try to zero in on the students' understanding of what they are to test in the experiment. Demonstrate the terms *concentration* and *variable* with the help of the students. With masking tape, mark off two squares on the floor that are each two meters by two meters. Explain that both squares are the same size. Count off three students and have them stand in one square. Then have six students step into the second square. Ask: "Is the square with six students larger than the square with only three students?" *(Both are the same size.)* "What is different about the squares?" *(The number of students is the difference. One square has three students; one has six students.)* Ask: "How does the concentration of students in the square affect them?" *(The six students feel crowded.)* Explain that scientists have a way of describing these differences that uses the terms *concentration* and *variable*. Explain that a scientist might say that the square with three students has a lower concentration of students than the square with six students.

- The size of the square is the control. It is the same for each group.
- The number of students in a square is the independent variable. This is the variable that is manipulated or changed by the experimenter.
- The reaction of the students in the square is the dependent variable. The six students felt more crowded as a result of having more people in the square.

NOTES

When counting fronds, count every visible frond. Include even the tips of small new fronds. The diagram shows an example with several fronds in different positions and stages of growth. It also shows two plants joined by what looks like a stem.

You may also want to consider recording color of fronds, size of fronds, number of living plants, number of dead or dying plants, length of roots, or anything else your class decides is important. Draw sketches of what you observe.

Planning Your Experiment

To get started, each group will plan an experiment to answer the question *How does fertilizer concentration affect the growth of duckweed?* Then, after examining the different experiment ideas, the class will agree on one experiment to run. Remember to discuss and record answers to the following questions as you are planning your experiment.

Question
- What question are you investigating and answering with this experiment?

Prediction
- What do you think the answer is, and why do you think that?

Plant Growth Experiment Planning Page

Name:_____ Date:_____

Question

What question are you investigating and answering with this experiment?

Prediction

What do you think the answer is and why do you think that?

Variable Identification

- Which variable will you be changing in your experiment?
- What conditions and procedures will you control in your experiment?
- What will you measure as evidence of the variable's effect on growth?

Procedure and Data

Write detailed instructions for how to conduct the experiment. You need to include the following:
- how you set up the duckweed samples.
- how you measure changes.
- how you record data.

Variable Identification
- Which variable will you be changing in your experiment?
- What conditions and procedures will you control in your experiment? That is, what conditions and procedures will you keep the same as you change your variable?
- What will you measure as evidence of the variable's effect on growth?

Procedure and Data Collection

Write detailed instructions for how to conduct the experiment. You need to include
- how you set up the duckweed samples,
- how you measure changes, and
- how you record data.

Also, you should explain how you determine whether or not you can trust the data.

Planning Your Experiment
15 min.

Students will be more successful in carrying out an experiment if they know what steps to take at each point along the way.

○ Engage

Ask students what it means to plan something. Ask them to mention some common situations that require steps to accomplish. List their suggestions. If they are reticent about coming up with suggestions, you might talk to them about a common situation that requires a simple plan and simple materials, such as the description of washing a car.

TEACHER TALK

"Planning is an important step toward getting any job done, even if it is a small job. Have you ever thought about how to wash a car? How would you go about getting this done? You may have thought to yourself, 'First I'll put on some old clothes, then I'll get a bucket and a sponge. I'll make sure the hose is attached, and find the soap. Next, I'll hose the car down, soap it up, hose off the soap, and wipe the car down.' This is an example of a simple plan. The person thinks about what they need (materials) and the steps to take to wash the car (the plan). Washing a car isn't exactly an experiment, but it does require a plan. What do you suppose scientists think about when they have a problem to investigate?" *(They write out a list of specific steps to take and make a list of the materials they will need.)*

△ Guide

Have students assemble into groups of four for the design and planning phases. Provide each group with a copy of the *Plant Growth Experiment Planning Page*.

Provide students with a purpose for completing this page. The purpose of the planning page is to focus students' attention on the steps needed to carry out an experiment. Ask groups to study each question carefully and decide on an answer. The more thoroughly they think through their plan, the more comfortable they will be in doing the experiment.

Inform each group that they will use their completed planning pages and a poster showing their plan design as the basis of a class discussion called a *Plan Briefing*. Advise students to come to the *Plan Briefing* with evidence to support (reasons for) the steps they want to take during the experiment.

△ Guide and Assess

Check on each groups' progress. If they are struggling with variables, be prepared to offer a hint. Ask: "What will they change *(This is the independent variable—the amount of fertilizer.)* and what will react to or respond to *(This is the dependent variable — the number of fronds that will develop as a result of that amount of fertilizer.)* the change?"

Procedure and Data Collection

25 min.

Students plan the experiment.

To get started, each group will plan an experiment to answer the question *How does fertilizer concentration affect the growth of duckweed?* Then, after examining the different experiment ideas, the class will agree on one experiment to run. Remember to discuss and record answers to the following questions as you are planning your experiment.

Question

- What question are you investigating and answering with this experiment?

Prediction

- What do you think the answer is, and why do you think that?

Variable Identification

- Which variable will you be changing in your experiment?
- What conditions and procedures will you control in your experiment? That is, what conditions and procedures will you keep the same as you change your variable?
- What will you measure as evidence of the variable's effect on growth?

Procedure and Data Collection

Write detailed instructions for how to conduct the experiment. You need to include

- how you set up the duckweed samples,
- how you measure changes, and
- how you record data.

Also, you should explain how you determine whether or not you can trust the data.

> **Plant Growth Experiment Planning Page**
>
> Name:_____ Date:_____
>
> **Question**
> What question are you investigating and answering with this experiment?
>
> **Prediction**
> What do you think the answer is and why do you think that?
>
> **Variable Identification**
> - Which variable will you be changing in your experiment?
> - What conditions and procedures will you control in your experiment?
> - What will you measure as evidence of the variable's effect on growth?
>
> **Procedure and Data**
> Write detailed instructions for how to conduct the experiment. You need to include the following:
> - how you set up the duckweed samples,
> - how you measure changes,
> - how you record data.

LT 58

Project-Based Inquiry Science

Emphasize the importance of having detailed instructions to follow. An example follows.

TEACHER TALK

“Be sure to write the steps of the experiment very carefully. Tell everyone exactly how you want something done. Tell them where to write the data they collect. In an experiment, it is easier to follow a very detailed instruction. For instance, what is the difference between these two directions: 'Count the exact number of fronds on all plants in Jar A. Write that number in column A with today's date.' and 'Find out how many leaves are on the plants in Jar A and write the number in your table.' Which direction is more exact?”

Communicate Your Plan
Plan Briefing

15 min.

Group and class discussions are powerful learning tools in which students' ideas are articulated, refined, and made available for other students to consider.

You will be given a *lant Growth Experiment lanning age*. The series of questions on this page will help you plan. With your group, discuss and fill out the sections of this page. Use the hints on the planning page as a guide. Be sure to write enough in each section so you will be able to present your experiment design to the class. The class will want to know that you've thought through all of the parts of your plan.

Communicate Your Plan

Plan Briefing

To help you as you learn to design experiments, you will share your investigation plan with the class during a *lan riefing*. As you are finishing your design plan, begin to draw a poster for presentation of your design plan to the class. Your teacher will provide you with a large sheet to use for your *lan riefing* poster and possibly a template to follow.

Others in the class have planned experiments to answer the same question you are answering. Discuss your proposed experiments with the class. Be very specific about your design plan and what evidence helped you make your design decisions. You will probably see differences and similarities in these plans. This discussion will allow you to compare ideas. You will avoid making mistakes that others can see in your plan. You will talk about what is good about each plan. You should also discuss what needs to be improved. Be very specific about your design plan and what evidence helped you make your design decisions. After you discuss all the plans, the class will decide together how to set up the experiment and what to measure.

Run Your Experiment

Once you have decided on a class experiment, it will be time to set it up and run it. Set it up as your class has decided. Be sure to place the jars in an area where the duckweed can get the light it needs to grow. You will observe the jars and collect data every day for the next 5 to 10 days.

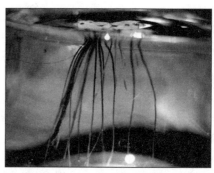

How doe duc weed re pond to fertilizer added to the water?

△ Guide

Transition students to a whole-class discussion. Remind students that the discussion is about improving their experimental plans and learning to make a useful plan.

TEACHER TALK

❝What you will be doing now is examining the plan from each group, thinking about how to improve each plan, and selecting a plan to perform as a class experiment.❞

Prepare students for the *Plan Briefing* by getting them to agree that the reason for the *Plan Briefing* is to review each groups' proposed plan and to decide what plan the class as a whole will use to test the question: *How does fertilizer concentration affect the growth of duckweed?*

Ask the groups to check that they are prepared for the *Plan Briefing*. Make sure each group has a *Planning Page*, a *Plan Briefing* poster, and that they are ready with evidence or reasons for the steps of their plan.

△ Guide Presentations and Discussions

Give students examples of what to look for in the presentations and what questions they might ask. List some questions for them to focus on, such as, "Do you think this plan will answer the experiment question? Could you carry out the procedures as they are written? How does the plan measure the dependent variable?"

TEACHER TALK

❝As you listen to each presentation, remember what to listen for. How well does the plan answer the question? Could you run this procedure? Why or why not? What changes might you make in the plan steps? What will data consist of? Is a table needed to record data? Will results be reliable? What evidence will show that results are reliable?❞

Begin a discussion after each presentation. You may need to model asking a question or two for students. Then ask a student to ask a question. Make sure the presenters respond to the student who asks the question and not to you, the teacher.

After all groups have presented, move on to how bringing together the best of the plans to use for the class wide single experiment. Project a blank *Plant Growth Experiment Planning Page* to record students' conclusions as to how the one experiment is to be conducted so that the class now has one plan to focus on.

NOTES

....

....

....

....

META NOTES

Students need practice in asking questions of other students. They need to think ahead to make sure their questions require more than a one-word answer. The question should focus on a critical part of the investigation. By modeling questions about content, you will also assist students in asking better questions.

As students begin to ask more of the questions on their own, you might have to act as a facilitator, refocusing them on the topic and periodically asking a question about something they have forgotten to ask about.

META NOTES

During a *Plan Briefing*, students share and develop new ideas when they listen to other students' ideas. They may disagree with some of what they hear, but disagreements can help students to rethink their ideas. This may be uncomfortable for some students who are not used to having to justify their ideas.

Run Your Experiment

6 to 10 class periods* ▶

Even though you are preparing the experiment, students should see connections with the experiment they proposed.

measure.

Run Your Experiment

Once you have decided on a class experiment, it will be time to set it up and run it. Set it up as your class has decided. Be sure to place the jars in an area where the duckweed can get the light it needs to grow. You will observe the jars and collect data every day for the next 5 to 10 days.

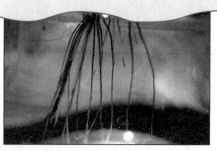

How does duckweed respond to fertilizer added to the water?

⬡ Get Going

Disposable gloves, a lab apron, and goggles should be worn when working with fertilizers. Read again the safety cautions on the liquid fertilizer product you will be using. If you are very familiar with safety rules for the materials being used, you may want to have students assist you. Set up the experiment in front of the class.

Remind students that they will make observations every day. Have them explain, from the write-up in their textbook, how counts of duckweed fronds are to be made each day. Correct any misinterpretation of the counting procedure before any data collection begins.

Inform students that if any evaporation takes place, that solution from the stock bottles that you set up earlier (*see* Section 2.2) will be added to the jars. Be sure to add the correct solution to the jar with the matching percentage of fertilizer.

META NOTES

During a discussion, students will need support to ask questions that matter to the topic, but that the rest of the class understands. Guide them to not ask the same question of each group. They also need support to ask questions as inquiries and not as accusations.

META NOTES

Listen for ideas about variables and procedures. Listen for students to demonstrate skills in sharing ideas, asking questions, and responding to peers. Compliment all participation that enhances these skills.

NOTES

....

....

....

....

....

....

....

*A class period is considered to be one 40 to 50 minute class.

Record Your Data

You will be recording data about the duckweed's growth. Record your data on a *Plant-Growth Data and Observations* page. There is space on it for recording the number of fronds, their color, the number of dead or dying plants, and anything else your class decided is important to observe and record. There is also space for you to sketch what you observe.

Plant-Growth Data and Observations

Name: _____ Class: _____

Use the following pages to record the plant growth in the four jars over a period of 5-10 days.

Observations may include the number of fronds, color of fronds, size of fronds, number of living plants, number of dead or dying plants, length of roots, or anything else your class decides is important.

Draw sketches.

Date: _____

Observations

- You will need 5 to 10 days to gather the data from this experiment. During that time, you will look at two other indicators of water quality in the next sections. Then you will return to analyze the data you have collected here.

Under certain conditions, duckweed will grow until it covers a pond's surface.

Recording Your Data

10 min. a class period

Students will begin to experiment and start to record their observations of the duckweed's growth.

Have students look back at what was agreed upon for data collection. Provide each student with a copy of the *Plant-Growth Data and Observations* page.

Once the experiment has begun, data should be collected each day at the same time.

Remind students to use their *Plant-Growth Data and Observations* page. All observations should be dated and may be in the form of drawings and descriptions. If possible, all numeric data (*number of plants to start/ numbers of living plants/counts of fronds on plants/sizes/number of dead plants*) should be in table form with the data for each day dated.

Fertilizers Contain Nitrates and Phosphates

10 min.

If students start to make connections between what they are reading about fertilizers and what is taking place in their investigation, list their ideas. Return to this list when they begin to analyze investigation data.

Fertilizers Contain Nitrates and Phosphates

Fertilizer is high in chemicals called nitrates and phosphates. A small concentration of these chemicals is good for the health of a river, lake, or stream. To grow, plants need nitrates and phosphates. However, high levels of nitrates and phosphates in the water can cause problems.

Algae and other plants that grow in the water use phosphates. However, they need only a small amount of phosphate to grow. Too much phosphate can cause large amounts of these plants to grow. Large growths of algae are called algal blooms. Algal blooms can also result from too many nitrates in the water. Algal blooms can cause problems for many reasons. They can keep light from getting to the roots of plants in the water. Also, when large growths of algae die, bacteria break them down. This makes the oxygen levels in the water decrease. In turn, fish and other animals might die.

Fertilizers and some laundry detergents are major sources of phosphates and nitrates. Excess nitrates, especially near cities or farms, can also come from sewage.

The table shows the phosphates and nitrates that can be found in water of different qualities. The "ppm" stands for *parts per million*. This term is used when only extremely small amounts of the substance are found in another substance. It means that one particle of the substance is found for every 999,999 other particles. That's like adding a drop of food coloring to 150 L (40 gallons) of water. But, as you can see from the table, even 3 or 4 parts per million of phosphate is bad for water quality. This table can help you determine the quality of any samples of water that you may want to investigate.

SureThing FERTILIZER 5 - 10 - 5
— nitrate
— phosphate
— potash

algae (singular, alga): simple organisms that live in water. Some can be as small as one cell. Some are made up of many cells. Algae made of many cells may be called "seaweed."

Water Quality and Concentration of Phosphates and Nitrates

Water Quality	Phosphate	Nitrate
excellent	1 ppm	0 ppm
good	2 ppm	0 ppm
fair	3 ppm	1-20 ppm
poor	4 ppm	30 ppm or more

ppm measures concentration in parts per million

LT 61

LIVING TOGETHER

○ Engage

Engage students by telling them a story based on information in the paragraphs on page 61 in the student text about what fertilizers contain and how these substances help plant growth. Don't focus on the vocabulary but rather, try to get students to see the big picture—that fertilizers contain substances that promote plant growth, yet sometimes this plant growth can have ill effects on watersheds.

△ Guide

Ask leading questions to help students begin to connect the presence of substances in fertilizers with water quality changes in a watershed. These substances (*phosphates especially*) come from a variety of sources (*fertilizers and laundry detergents*). Larger amounts of phosphates from farms and other sources might cause more pronounced problems in the watershed.

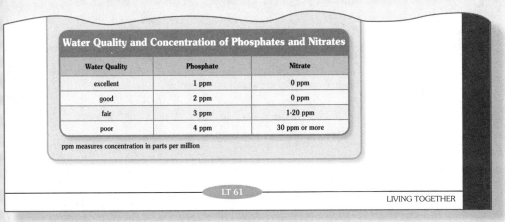

Water Quality and Concentration of Phosphates and Nitrates

Water Quality	Phosphate	Nitrate
excellent	1 ppm	0 ppm
good	2 ppm	0 ppm
fair	3 ppm	1-20 ppm
poor	4 ppm	30 ppm or more

ppm measures concentration in parts per million

LT 61

LIVING TOGETHER

◇ Evaluate

Have students study the table that compares water quality and the concentration of phosphates and nitrates. Have them summarize what they think happens to water quality.

Also, ask them if they see any trends in the data from the experiment that is similar to what is in the table.

NOTES

Stop and Think

5 min.

Students answer questions to generate class discussion.

Stop and Think

1. What are three sources of phosphates and nitrates in water?

2. Why are too many nitrates and phosphates unhealthy for a body of water?

Analyze Your Data

To help you analyze the data from this investigation, answer the following questions. Be prepared to discuss your answers with the class.

1. What effects did the fertilizer concentration have on the growth of the plants? Support your answer with evidence from the investigation.

2. Describe the relationship between fertilizer concentration and duckweed growth. (Summarize the trends you found in each of the different solutions.)

What's the Point?

Your class has been recording data about the growth of duckweed. You set up an investigation to see how fertilizer concentration affected the growth of duckweed (a water plant). Fertilizers contain nitrates and phosphates. Nitrates and phosphates are needed for plant growth.

However, when large quantities of these chemicals enter the water through runoff or groundwater, they can cause problems. They increase the growth of plants that live in the water. This can lead to the death of fish and other animals. It may even lead to the death of the plants themselves. Plant growth can be used as an indicator of the amount of nitrates and phosphates in water. If you see a thick plant growth in water, you know that water contains large amounts of nitrates and phosphates.

The amount of fertilizer in the water also can have a large impact on dissolved oxygen. Bacteria break down plants when they die. This results in reduced oxygen levels in the water. In turn, the fish and other animals that live in the water may die because there is not enough oxygen for them to survive.

Fertilizer concentration can affect the growth of duckweed.

LT 62

Ask student groups to write their answers to the questions. This will cause everyone to have to think about the information. Then use the questions as the basis of a class discussion. Listen for responses that contain some of the following:

1. Sources of phosphates and nitrates: fertilizers, laundry detergents, and sewage.

2. Too much nitrate and phosphate is unhealthy for water because it results in depleting oxygen needed by organisms that live in the water. Increased nitrates and phosphates increase plant growth in the water. This results in an increase in bacteria that decompose the plants. In doing so, the bacteria use up the available oxygen needed by aquatic animal life.

Analyze Your Data

To help you analyze the data from this investigation, answer the following questions. Be prepared to discuss your answers with the class.

1. What effects did the fertilizer concentration have on the growth of the plants? Support your answer with evidence from the investigation.

2. Describe the relationship between fertilizer concentration and duckweed growth. (Summarize the trends you found in each of the different solutions.)

Analyze Your Data

15 min.

○ Engage

Stimulate a class discussion by having students use the data from the class investigation to answer the *Analyze Your Data* questions on page 62 in the student text. Students' responses should include some of the ideas listed below and evidence from their experimental data:

1. Plant growth was different for each jar.

2. Fertilizer has an effect on plant growth. This is supported by the evidence that as the concentration of fertilizer increased, plant growth also increased.

Teacher Reflection Questions

- Students needed to design an investigation. Were there parts of the design that many students had difficulty with? How could students be better prepared for these challenges?

- How can students be guided to make more of a connection between the information they put on the *Project Board* from *Learning Set 1* and the *Big Question?*

- Did the experiment run smoothly? What could make this investigation or any other investigation run more smoothly?

NOTES

....

....

....

....

....

....

NOTES

SECTION 2.3 INTRODUCTION

2.3 Investigate

pH as an Indicator of Water Quality

Overview

◀ $1\frac{1}{2}$ *class periods* *

*A class period is considered to be one 40 to 50 minute class.

Students thought about how they might determine water quality in *Section 2.1* and they began an experiment. In this section, they will learn about their first water quality indicator, *pH*. Students observe a demonstration in which the pH of water from *Learning Set 1* is compared to the pH of distilled water. Students learn that pH is a measure of the acidity of a solution and learn that, for many living things, pH cannot be too high or too low. Students then test the pH of various liquids and compare their results with the results of those for the liquid from *Learning Set 1*. Students also mix solutions to see what happens to the acidity of the solution and consider how to advise Wamego how the acidity of water in the Crystal River might affect trout populations and their annual Trout Festival.

Targeted Concepts, Skills, and Nature of Science	Performance Expectations
pH is a measure of how acidic or basic a solution is and is one indicator of water quality.	Students should be able to say that a particular pH number will tell whether the water or solution being tested is acidic or basic. Very acidic water affects life forms adversely and is an indicator of poor water quality.
Mixing less acidic (*alkaline*) solutions with acidic solutions changes the overall pH (acidity) of a solution.	Students should be able to relate that mixing alkaline solutions with acidic solutions will raise the pH of the overall solution and therefore there is less acidity in the final solution.
Some aquatic organisms are very sensitive to changes in pH.	Students should become aware that pH affects the viability of aquatic organisms. Some organisms, such as trout, can survive in only a very narrow pH range, otherwise, they die out.

Follow all cautions
for liquids used in
this experiment.

Vinegar, isopropyl
alcohol, and baking
soda solutions can
be disposed of
by washing them
down the drain.
However, they
can be harmful
if swallowed or
come in contact
with eyes, nose, or
throat. Review all
safety procedures
with students
before beginning
the investigations.

Materials

1 per student	Safety goggles
1 per student	Apron
1 per classroom	Roll of paper towels
1 set per classroom	Jars 3 and 4 from Unit Introduction (page 4 in the student text) (or new solutions of vinegar mixed with distilled water and baking soda mixed into distilled water)
1 per classroom	Projection of pH scale (from the student text)
1 per class	Projection of pH table (from the student text) and marker
1 per classroom	Distilled water supply
1 per classroom	500 mL deionized water
1 per preparation	Large pot with lid
1 per preparation	Large pitcher (at least quart size)
1 per preparation	Method to heat pH indicator solution
1 per preparation	$\frac{1}{3}$ of a red cabbage, diced
1 per preparation	Filter paper or coffee filter
1 per preparation	Isopropyl alcohol
1 per classroom	White vinegar, 2 pts
1 per classroom	Flat, colorless soft drink, 2 L
1 per classroom	Box of baking soda
1 per classroom	Clear window cleaner (no tint or dyes added), 3-4 L
1 per group	Test-tube racks
5 per group	Test-tubes
1 per group	Test-tube stopper

Materials	
1 per group	Wax marking pencils
1 per group	Micro spatula or plastic tablespoon
1 per classroom	Funnel set
1 per group	Plastic beaker, 400-mL
2 per group	Beaker, 250-mL
30 per class	Plastic pipettes, 3-mL
1 per group	Graduated cylinder, 10-mL

Activity Setup and Preparation

The cabbage juice indicator must be prepared 1 to 2 days in advance, refrigerated until the day of use, and be at room temperature for use.

Test all of these activities ahead of time to insure that they work smoothly and that all the chemicals required react as needed. Clean glassware is important for testing solutions. In chemistry, clean glassware is glassware that has been washed thoroughly and rinsed with distilled water, then allowed to dry.

Demonstration Preparation

- You will demonstrate the use of pH as an indicator of water quality by testing the acidity of liquids in sample Jars 3 and 4 from the Unit Introduction activity on page 4 in the student text. If algae have begun to grow in these jars, you will have to make new samples using the directions in the Unit Introduction. The pH indicator that you will use is made from red cabbage.

- Jar 3 contains a solution of vinegar and water. The solution is *acidic* with a pH of about 5.

- Jar 4 is made up of about ½ tablespoon of baking soda in water. The solution is *basic* with a pH of about 8 to 9.

- Ahead of time, test the liquids in Jars 3 and 4 to make sure that each sample turns a different color. If the color changes are not significant, add a few drops of vinegar to Jar 3 and a teaspoon of baking soda to Jar 4.

To prepare cabbage-juice indicator

- Wear an apron, because cabbage juice stains.

- Prepare the cabbage juice indicator 1 or 2 days in advance. Refrigerate the mixture in an airtight container so it does not spoil. Before using, take it out of the refrigerator and let the liquid come to room temperature.

1. Dice about $\frac{1}{3}$ of a medium size head of red cabbage and put it in a large pot.

2. Fill the pot with distilled water so that the water just covers the diced cabbage.

3. Bring the water-cabbage mixture to a boil, then simmer on low heat for about 10 minutes until the water turns purple. Remove from heat and let cool.

4. While the liquid is cooling, label a clean jar and lid *cabbage-juice indicator*. Mark the date on it also and set it aside for later.

5. Over a sink, pour the cooled juice mixture into a large pitcher (*at least a quart-size pitcher*).

6. Place a coffee filter into a funnel and filter the liquid from the cabbage-juice mixture into a large measuring cup or large beaker. *Do not* fill to the top as you will be adding isopropyl alcohol to it.

7. Measure the volume of the purple cabbage-juice liquid.

8. Measure an amount of isopropyl alcohol equal to $\frac{1}{8}$ th the volume of the cabbage juice into a separate clean container. (*If you have 1 cup of filtered purple liquid, you will need $\frac{1}{8}$ th cup of isopropyl alcohol.*)

9. Add the measured isopropyl alcohol to the filtered purple liquid. Pour the mixture into an airtight container and store it in the refrigerator.

10. On the day it is to be used, remove the indicator from the refrigerator several hours before the demonstration so that it comes to room temperature.

Class Investigation Preparation
Part A: Identify Acidity

Students must wear goggles and apron. Prepare test tubes of solutions the day before. Have instructions in front of each solution and let students prepare the test tubes.

If students are preparing their own test tubes, provide marking pencils.

Have students mark tubes 1 through 5. Provide labeled beakers and a 10 mL graduated cylinder with white vinegar, colorless soft drink, baking soda, distilled water, and colorless window cleaner. Provide a clean pipette for each solution.

If the teacher is preparing the test tubes:

1. Label a set of test tubes 1 through 5 for each group.

2. Test tube 1: pour 3 to 4 mL of white vinegar into the test tube marked 1. Place in the test-tube rack.

3. Test tube 2: pour 3 to 4 mL clear soft drink (*that has been allowed to go flat*) into the test-tube marked 2; place the tube in the rack.

4. Test tube 3: pour 3 to 4 mL distilled water into the test tube marked 3; place the tube in the rack.

5. Test tube 4: using a micro spatula or small spoon, scoop a pinch of baking soda (*less than* $\frac{1}{8}$ *teaspoon*) into an empty test tube. Pour 3 to 4 mL distilled water into the test tube so that it is about ¼ full. Stopper the tube (*Do not cover the opening with your thumb.*) and shake the tube gently to dissolve the baking soda. Remove the stopper. Place the tube in the rack.

6. Test tube 5: pour 3 to 4 mL, colorless window cleaner into test tube 5; place it in the rack.

Class Investigation Preparation
Part B: Mixing Solutions Together

Set up a separate station in the classroom for Part B materials.

In this investigation, you will prepare the solutions. Students will not know what the solutions are. Students will add a baking-soda solution to a vinegar solution to see if the indicator changes colors as the pH changes.

NOTE: This investigation has students pouring a weak basic solution into a weak acidic solution. The outcome is that the overall solution comes into the neutral range. *Never*, under any circumstances, *pour water into a strongly acidic solution*. Always pour acid into water.

For each group, prepare the following:

1. Label a 250-mL graduated beaker as Solution A. Measure 10 mL of white vinegar in a graduated cylinder and pour it into the beaker. Add distilled water to the vinegar solution to fill to the 100 mL mark.

2. Label a 250-mL beaker as Solution B. Add 1 g baking soda (*about ¼ teaspoon*). Add distilled water to the baking soda to the 100 mL mark.

3. Each of the solutions should appear relatively clear. Students should not know what these solutions are.

Homework Option

Reflection

- **Science Content:** How do you think different land uses affect the pH of water in a watershed? *(Answers will vary. Most will include that some land uses will change pH and probably lower it in most cases. Students may conclude that factories and farms may affect pH the most.)*

- **Science Content:** Think about whether a garbage dump (or landfill) might affect a watershed. Pick a substance such as juice or a liquid cleanser and describe its journey through a watershed. Suggest how it might affect water quality in the watershed that it is part of. *(Answers will vary but might include that water or substances like juices in a garbage dump seep into the groundwater and eventually seep into a river. It might have more effect on the pH of a stream if the pH of the juice is very low and if it were poured directly into the water. It might have less effect if it first becomes part of the groundwater system. By the time it reaches the stream, it might be diluted.)*

Preparation for 2.4

- **Science Process:** What other water quality indicator can you think of? Think about what animals need to live. *(Answers will vary. Animals need oxygen. If animals that live in water do not have enough oxygen, then the animals throughout the watershed might begin to die.)*

- **Nature of Science:** A scientist wants to reduce the acidity of a solution. Will she add more of the same kind of acid to the solution or will she add a more basic solution to the original acid? *(The scientist must add a more basic solution to the original solution to reduce the acidity.)*

NOTES

$1\frac{1}{2}$ *class periods* ▶ *

2.3 Investigate

pH as an Indicator of Water Quality

In Section 2.1, students thought about a variety of things that might change water quality. In this section, students learn about pH, one of the indicators of water quality. They will observe a demonstration and test the pH of some unknown substances.

Demonstration

10 min.

In this demonstration, students are introduced to testing solutions with a pH indicator made from cabbage juice. They may be surprised to see differences between distilled water and solutions from Jars 3 and 4, both of which look like drinkable water. This provides motivation for the activity.

---

*A class period is considered to be one 40 to 50 minute class.

2.3 Investigate

pH as an Indicator of Water Quality

In this *Learning Set*, you have been thinking about the quality of water. You were asked to consider what is a good indicator of water quality. You may have decided that the way water looks is an important indicator. However, some pollutants are invisible. You also have begun looking at indicators of water quality that allow you to identify pollutants that are too small to see. In this section, you will investigate **pH** as an indicator. The pH of a substance measures whether it is an **acid** or not. When substances have a pH that is either very high or very low, these substances can be harmful to many living things, including you.

pH: a measure of how acidic a substance is.

acid: a solution with a pH less than 7.

pH indicator: a chemical that can be added to a solution to determine pH.

Demonstration

At the beginning of this Unit, you looked at five different samples of water. At first you may have thought that the water in Jars 3 and 4 was of a high quality. Perhaps you thought that you could drink the water in these jars.

Your teacher will add a special chemical to the water in Jars 3 and 4. This substance is called a **pH indicator**. It will tell you something about the quality of the water. This substance will also be added to a sample of water in a third jar. The water in this jar is distilled water. You can think of it as "pure" water.

LT 63

LIVING TOGETHER

◯ Engage

Have students look at the three jars on the table. All three jars should look alike except for their labels. All three should appear clear, meaning they look as if they have nothing but clear water in them. One jar is marked "distilled water" and contains just that. Jars marked "3" and "4" are from the set of five jars from the Unit Introduction.

Explain that different liquids have different acidities. Acidity is a property that is measured by testing the pH of the liquid.

Ask students what they think they can tell about the liquids in the jars by looking at them (*Answers will vary. Some may think that the water is drinkable because it is clear.*)

❝This is a useful moment to explain that the appearance of a liquid can be deceiving. Never drink something just because it might look like water—whether in a river or in a laboratory.❞

△ Guide

At this point, it might be beneficial to introduce the importance of being careful when doing procedures in the laboratory.

❝We're about to perform an investigation to see how certain liquids respond to a pH indicator. The test solution is made out of cabbage juice. We're going to put a small amount of each solution from the jars marked "3" and "4" and the one marked "distilled water" into test tubes labeled the same way. Then we will test just that small portion. When we add the cabbage-juice indicator solution we will use a kind of dropper called a pipette. Do you think we should stir the pipette into each tube? Why or why not? Where can we safely hold the pipette?❞

❝I want all of you to pay close attention to what I am demonstrating because you will use the same methods to test some other liquids yourself. If you follow the steps I am taking, you will be able to rely on your test results.❞

To make certain that all students will be able to see every part of the demonstration, make sure you plan ahead. Rearrange students' desks or raise the jars and test-tube rack if necessary. You want students to be able to see what you are doing but you don't want them crowding around and touching the demonstration materials. Model safe laboratory practices by putting on goggles and an apron before doing the demonstration. Have students wear goggles as well and then proceed with the demonstration.

1. Label three test tubes to match the labels on the three jars of liquid.

2. Pour each test tube about ¼ full with liquid from its matching jar.

3. Demonstrate testing the pH of each solution by adding 5 to 10 drops of the *cabbage-juice indicator* to the liquid in test tube 3 and then test tube 4.

Give students time to see the reaction in each test tube.

4. Involve students in the demonstration by asking them to predict what will happen when you add indicator to the distilled water sample.

5. All three test liquids will show a different color reaction depending on their pH. Ask students what they might conclude about whether all three liquids have the same pH. Ask them to give reasons for their conclusions.

Good laboratory methods aren't just a matter of safety. It is also a matter of handling materials carefully so there is no cross contamination. Cross contamination or mixing of chemicals will skew results and make the outcome of a test invalid.

During any demonstration, model good laboratory techniques for students. Practice ahead of time adding indicator solution to the individual test tubes using a pipette. Do not plunge the pipette into the unknown liquids or stir the liquids with the pipette, as that will invalidate the test. Tell students that you are holding the pipette about an inch or so above each test tube as you introduce the indicator solution.

Draw students' attention to what you are going to demonstrate.

Observe

5 min.

Students think about what an indicator reveals. They still do not know how to interpret the color changes, but they clearly can understand that looks can be deceiving; that just because several liquids look the same (before the test) does not mean that they are the same.

Observe

Answer the following questions while observing the demonstration. Be prepared to discuss your answers with your group and the class.

1. When looking at the three jars of water your teacher used in the demonstration, could you tell them apart? Why or why not?

2. After the indicator was added to each jar, was there a difference in the water in each jar? Describe what you saw.

3. Here is a picture of water from a clean swimming pool.

    a) How does the water look? Compare it to the water in the jars in the demonstration.

    b) Would you swim in this water? Explain your answer.

    c) Would you drink this water? Think about the way water in a pool smells when you explain your answer.

pH scale:
a scale used by scientists to measure the acidity of a solution.

neutral:
a solution with a pH of 7.

Acids and pH

Acids

You have probably heard about acids. You might have seen acids in bottles that have poison or caution labels on them. Acids may seem dangerous. However, many acids are very important to life. For example, the liquid in your stomach is acidic. It makes it possible for you to digest food. Vitamin C is acetic acid. It is important to maintaining a healthy body.

Too much acid can have negative effects. Sometimes your stomach makes more acid than it needs to digest food. This can cause indigestion. Outside your body, acids react with metals and cause them to corrode. Some plants need acidic soil. However, too much acid in water can have negative effects on the plant and animal life in an aquatic ecosystem. In the atmosphere, acids react with other chemicals to create acidic water. This water can then fall as "acid rain." Acid rain causes environmental damage by changing the pH of bodies of water, corroding metals, and mixing with substances it contacts to create other pollutants.

The pH scale

Scientists use the **pH scale** to measure how acidic a solution is. The scale goes from 0 to 14. Solutions with a pH less than 0 are acidic.

Project-Based Inquiry Science

△ Guide

Give students time to answer the *Observe* questions. Ask them to share their answers.

1. You could not tell the three liquids apart before the *cabbage-juice indicator* was added, but you could after it was added because the liquids changed color.

2. The color of liquids from jars changed. Students should be able to describe the color change in each.

To get a clear idea of whether students grasped the concept in the demonstration, transition students to a class discussion. Move the discussion to a new situation, such as testing the pH of swimming pool water, which is depicted in the photograph on page 64 in the student text. Rephrase questions. You might ask, "How does the color of the tested pool water relate to any of the colors of the liquids tested during the demonstration?" If students have trouble coming up with reasons for their conclusions, redirect questions to other students by saying something such as "Steven, what are some reasons why you think it might be all right to swim in this pool water?"

3. Assess students' answer to the three questions about the swimming pool:

    a) The water looks clean and clear.

    b) Students' answers and reasons will vary.

    c) Students' answers and reasons will vary.

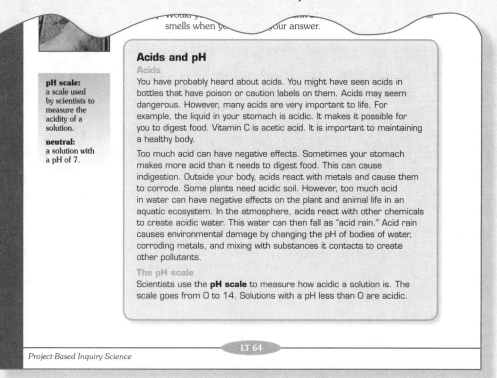

... would ... smells when yo... ...our answer.

Acids and pH

Acids

You have probably heard about acids. You might have seen acids in bottles that have poison or caution labels on them. Acids may seem dangerous. However, many acids are very important to life. For example, the liquid in your stomach is acidic. It makes it possible for you to digest food. Vitamin C is acetic acid. It is important to maintaining a healthy body.

Too much acid can have negative effects. Sometimes your stomach makes more acid than it needs to digest food. This can cause indigestion. Outside your body, acids react with metals and cause them to corrode. Some plants need acidic soil. However, too much acid in water can have negative effects on the plant and animal life in an aquatic ecosystem. In the atmosphere, acids react with other chemicals to create acidic water. This water can then fall as "acid rain." Acid rain causes environmental damage by changing the pH of bodies of water, corroding metals, and mixing with substances it contacts to create other pollutants.

The pH scale

Scientists use the **pH scale** to measure how acidic a solution is. The scale goes from 0 to 14. Solutions with a pH less than 0 are acidic.

pH scale: a scale used by scientists to measure the acidity of a solution.

neutral: a solution with a pH of 7.

LT 64

Project-Based Inquiry Science

Acids and pH
10 min.

Students read background information on what pH is. They begin to understand that pH is a useful indicator of water quality.

△ Guide

Have students read each section of information (*Acids, The pH Scale,* and *pH Indicators*) on pages 64 and 65 in the student text. After each reading, use the suggested questions under each head below to start a class discussion.

Acids

Ask students what they think an example of an acid might be. List their suggested examples.

After students have read the section on acids, ask them to identify a positive use or effect of acids in nature. Have them explain some negative effects of acids in nature.

LIVING TOGETHER

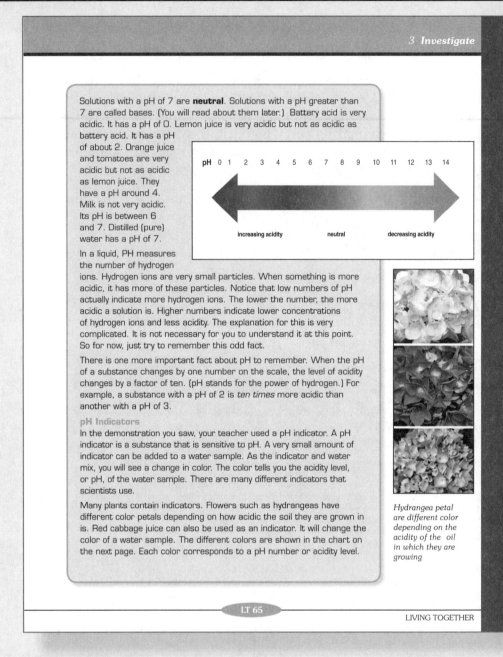

Solutions with a pH of 7 are **neutral**. Solutions with a pH greater than 7 are called bases. (You will read about them later.) Battery acid is very acidic. It has a pH of 0. Lemon juice is very acidic but not as acidic as battery acid. It has a pH of about 2. Orange juice and tomatoes are very acidic but not as acidic as lemon juice. They have a pH around 4. Milk is not very acidic. Its pH is between 6 and 7. Distilled (pure) water has a pH of 7.

In a liquid, PH measures the number of hydrogen ions. Hydrogen ions are very small particles. When something is more acidic, it has more of these particles. Notice that low numbers of pH actually indicate more hydrogen ions. The lower the number, the more acidic a solution is. Higher numbers indicate lower concentrations of hydrogen ions and less acidity. The explanation for this is very complicated. It is not necessary for you to understand it at this point. So for now, just try to remember this odd fact.

There is one more important fact about pH to remember. When the pH of a substance changes by one number on the scale, the level of acidity changes by a factor of ten. (pH stands for the power of hydrogen.) For example, a substance with a pH of 2 is *ten times* more acidic than another with a pH of 3.

pH Indicators

In the demonstration you saw, your teacher used a pH indicator. A pH indicator is a substance that is sensitive to pH. A very small amount of indicator can be added to a water sample. As the indicator and water mix, you will see a change in color. The color tells you the acidity level, or pH, of the water sample. There are many different indicators that scientists use.

Many plants contain indicators. Flowers such as hydrangeas have different color petals depending on how acidic the soil they are grown in is. Red cabbage juice can also be used as an indicator. It will change the color of a water sample. The different colors are shown in the chart on the next page. Each color corresponds to a pH number or acidity level.

Hydrangea petal are different color depending on the acidity of the oil in which they are growing

The pH Scale

After reading about the pH scale, project a copy of the scale as shown on page 65 in the student text. Ask one student to use the projection to explain the parts of the pH scale. Have a second student identify the role of the numbers on the scale that indicate increasing or decreasing acidity and that 7 is "neutral." Have a third student explain how acidity changes by factors of 10. Ask students, "What is the relationship between hydrogen ions and acidity?" Finally, ask students to think about the way the colors on the pH scale arrow resemble the colors in the demonstration liquids.

TEACHER TALK

❝After looking at the colors in the arrow, how might you now describe the three demonstration solutions? (Jar 3 was acidic with a pH down in the 2 to 4 range; Jar 4 was much less acidic, with a pH in the 9 to 14 range; distilled water was neutral with a pH in the 7 range.) How might you describe the pH of the pool water? Explain your answer choice. (The pool water had a pH in the neutral range because it tested in the color range shown for neutral on the pH scale.)**❞**

pH Indicators

⃝ Engage

Again display Jars 3 and 4 and the jar with distilled water. Remind students of the demonstration and inform them that the pH indicator you used showed that each liquid was a different pH.

TEACHER TALK

❝Look at the photographs of the hydrangea flowers in your textbook. What else besides the water from the roots might cause a change in the color of the flowers? Think about where groundwater moves. Is there anything in the soil that might affect the water? How is the color of these flowers an indicator of their chemical environment?**❞**

NOTES

....

....

....

....

....

....

....

Procedure

20 min.

Students begin to learn how acids and bases interact, even though they do not know these terminologies yet.

Part A: Identify Acidity

10 min.

Students begin to explore, observe, and identify the characteristics of acids

Procedure

In Part A, you and your group will investigate and determine the pH of five solutions with an indicator provided by your teacher. In Part B, you will attempt to change the pH of another sample by mixing it with a sample with a different pH. Follow the directions below to conduct your investigation. You will work as a group, but each of you will need to complete the data chart.

Part A: Identify Acidity

1. Place four drops of indicator in each of the samples in the test tubes you receive from your teacher. Very gently, swirl each test tube to mix the indicator. Place the test tube back in the rack before you make any observations. If the color is too faint, add one to two more drops of indicator.

2. Record the color of each test tube in your data chart.

3. Match the color of each sample with one of the colors in the chart provided here. Record the pH paired with the matching indicator color on a data table similar to the one shown.

> ⚠ You will be working with very special substances and equipment. It is important that you are very careful in handling these items to insure that you and your classmates remain safe. Your teacher will provide further instructions about science safety during your investigation.

pH	0–2	3–4	5–6	7–8	9–10	11–12	13–14
cabbage juice color	red	red-pink pink	pink- purple	pale blue	yellow green	pale green	green

	Sample 1	Sample 2	Sample 3	Sample 4	Sample 5
color					
pH					

△ Guide

Remind students again about red cabbage juice as an indicator. Inform students that they will be doing two investigations in which they use indicators to help them determine pH.

Tell them that in Part A, they will work in groups to test the acidity of five different solutions. Provide each student with a pH color scale chart on which to record their group's investigation results. In Part B, students will find out what happens when they mix two solutions of different pH values. They will need to use their color chart to interpret the result of Part B also.

"Who can tell me why you are doing this investigation? Remember how I handled the materials during the classroom demonstration? Let's think about how I added indicator to the solution in each test tube. Remember that I did not use the pipette as a mixing tool and I did not touch the test tubes with the pipette. Let's read the steps of Part A. What safety steps do you want to take? Where will you record your data? Once you have put your goggles and apron on, then you can collect the materials for your group."

☐ Assess

Suggest that each student be assigned one of the five test tubes. This way, each student has a chance to add the indicator to a solution and they each practice a laboratory procedure. Try to visit each group at least once to assess their techniques in adding indicator to the test tubes. Walk them through the procedure if necessary. Compliment students on techniques well carried out.

As students test each solution in Part A, remind them to record the resulting color and corresponding pH on their individual charts. Each student should record all five solutions tested by their group, then put away all Part A materials before getting materials for Part B.

NOTES

Part B: Mixing Solutions Together

10 min.

As students have become familiar with the nature of acids through investigation, they are ready to observe further qualities through experimentation.

Part B: Mixing Solutions Together

1. Add five to six drops of indicator to Solution A in the beaker provided by your teacher. Gently swirl the beaker. What color does the solution turn? Using the chart, determine the pH of this solution.
2. Draw a full pipette of Solution B, also provided by your teacher. Squeeze the full pipette of Solution B into Solution A. Repeat this three times. Observe what happens to the color of Solution A each time you add more of Solution B.
3. Continue to add Solution B to Solution A with the pipette until you have added 50 mL. What is the color of Solution A at this point? Use the pH color chart to determine its pH.
4. Very slowly and gently, pour the remaining 50 mL of Solution B into Solution A. What is the color of Solution A at this point? Determine the pH according to the chart on the previous page.

Analyze Your Data

1. How were the five samples you had in front of you similar to Jars 3 and 4?
2. Which of the sample or samples are most acidic?
3. Which of the sample or samples are least acidic?
4. Which is most like Jar 3 from the demonstration? What do you think is the pH of Jar 3?
5. Which is most like Jar 4 from the demonstration? What do you think is the pH of Jar 4?
6. As you added Solution B to Solution A, what was happening to the pH of Solution A? What evidence do you have of this?
7. What substances do you think you were mixing in Part B of the pH investigation?
8. How could you return the pH of Solution A to its original condition?
9. Solution A became _____ times less acidic during the investigation.

> **More about pH**
> The pH of liquid substances can be measured and reported as a number on the pH scale. The pH scale on the next page runs from 1 to 14. The diagram on the next page shows you the pH of several common substances.
>
> At the left side of the scale you see lemon juice and vinegar. You may have tasted these substances before, and if so, you know they are very

Review the procedures for Part B so that students will understand that what they are to do. They should become aware that what they are doing is different from what they did in Part A of the investigation.

Remind students that they will need to refer to the color chart from Part A to interpret the outcome from Part B.

Analyze Your Data

1. How were the five samples you had in front of you similar to Jars 3 and 4?
2. Which of the sample or samples are most acidic?
3. Which of the sample or samples are least acidic?
4. Which is most like Jar 3 from the demonstration? What do you think is the pH of Jar 3?
5. Which is most like Jar 4 from the demonstration? What do you think is the pH of Jar 4?
6. As you added Solution B to Solution A, what was happening to the pH of Solution A? What evidence do you have of this?
7. What substances do you think you were mixing in Part B of the pH investigation?
8. How could you return the pH of Solution A to its original condition?
9. Solution A became _____ times less acidic during the investigation.

Analyze Your Data

10 min.

These questions are designed to help students make sense of what they have observed. From this, they can begin to make inferences.

△ Guide

For these investigations, the class should all agree on their observations. If one group observed something different, you may want them to repeat that part of the investigation or you might repeat it for the whole class by doing a demonstration.

△ Guide and Assess

If one group observed something different from the rest of the class results, try to avoid telling them they are wrong. Instead, ask that group why they think their results differed from others. If they repeat the experiment and come up with the same observations as the class, let the group suggest reasons why they may have differed on the first try. Let students conclude that it may have been something in their technique that skewed the results. Let the class know how easy it is for solutions to change when they come in contact with another solution. Let them know that even scientists learn by practicing proper lab techniques.

Consider having each group present the answer to a question. Record these answers on the board. If one group disagrees, ask the class to think about reasons for the different answer. Students' responses should include the following on the next page.

> **META NOTES**
>
> It is important to review the observations students have recorded to affirm their observations and conclusions. Reviewing answers also helps you to spot places where errors in technique might be the reason for differing data.

NOTES

....

....

....

Answers for Part A: Identify Acidity

1. All five liquids were similar to the liquids in Jars 3 and 4 because they were clear. Their appearance was the same before adding the indicator. Some of them changed color after adding the indicator.

2. The most acidic samples were in test tubes 1 and 2.

3. The least acidic samples were in test tubes 4 and 5.

4. Sample 1 was most like Jar 3.

5. Sample 4 was most like Jar 4.

Approximate pH values of test substances: white vinegar— pH 3; clear, flat soft drink— pH 5; distilled water— pH 6.5 to 7; baking soda solution— 8.0 to 8.5; colorless window cleaner— 9 to 10.

Answers for Part B: Mixing Solutions Together

6. As Solution B was added to Solution A, the acidity of the combined solution decreased. Evidence for this: the color changed to a color that indicates a less acidity on the color chart.

7. Answers may include that they were mixing a solution with a low pH with a solution that had a high pH.

8. You could return the pH of Solution A to nearly its original condition by adding more of the acidic solution (*Solution A*) to the mixture.

9. Answers will vary with degree of color change. Students can approximate the amount of change by multiplying the change in value by 10.

More about pH

10 min.

Students read more in depth about pH. They are introduced to the terms base, basicity, *and* alkaline. *A class discussion helps students make connections between water quality and its potential effects on organisms that live in the water.*

More about pH
The pH of liquid substances can be measured and reported as a number on the pH scale. The pH scale on the next page runs from 1 to 14. The diagram on the next page shows you the pH of several common substances.

At the left side of the scale you see lemon juice and vinegar. You may have tasted these substances before, and if so, you know they are very

○ Engage

Have students read the first two paragraphs and study the enhanced pH scale on page 68 in the student text. They will see a number of familiar substances of the chart. Some students will find that they are more familiar with acids and bases than they expected so that they may not be intimidated by the new information and added terms.

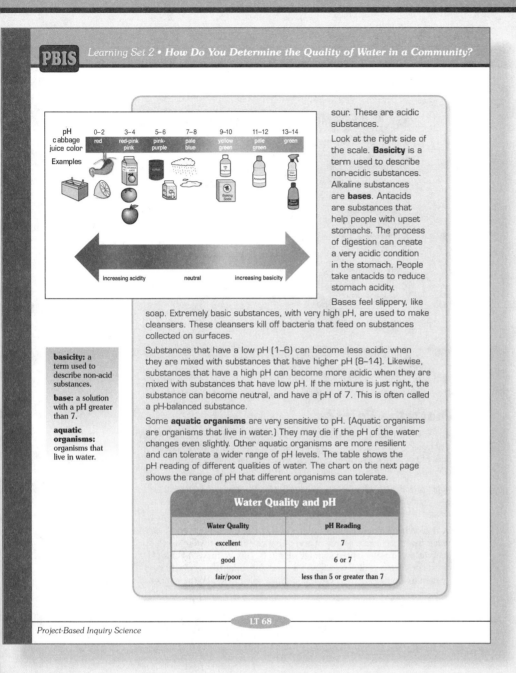

sour. These are acidic substances.

Look at the right side of the scale. **Basicity** is a term used to describe non-acidic substances. Alkaline substances are **bases**. Antacids are substances that help people with upset stomachs. The process of digestion can create a very acidic condition in the stomach. People take antacids to reduce stomach acidity.

Bases feel slippery, like soap. Extremely basic substances, with very high pH, are used to make cleansers. These cleansers kill off bacteria that feed on substances collected on surfaces.

Substances that have a low pH (1–6) can become less acidic when they are mixed with substances that have higher pH (8–14). Likewise, substances that have a high pH can become more acidic when they are mixed with substances that have low pH. If the mixture is just right, the substance can become neutral, and have a pH of 7. This is often called a pH-balanced substance.

Some **aquatic organisms** are very sensitive to pH. (Aquatic organisms are organisms that live in water.) They may die if the pH of the water changes even slightly. Other aquatic organisms are more resilient and can tolerate a wider range of pH levels. The table shows the pH reading of different qualities of water. The chart on the next page shows the range of pH that different organisms can tolerate.

basicity: a term used to describe non-acid substances.

base: a solution with a pH greater than 7.

aquatic organisms: organisms that live in water.

Water Quality and pH

Water Quality	pH Reading
excellent	7
good	6 or 7
fair/poor	less than 5 or greater than 7

LT 68

Project-Based Inquiry Science

△ Guide

Have students read the remaining information. After reading, ask each of the class groups to write two or three questions about the information in the paragraph to help students become more familiar with the new terminology. Then, have each group ask their questions of the class as a whole.

Draw students' attention to the *Range of pH that Different Organisms Can Tolerate* chart on page 69 in the student text. Have the groups describe the effect changes in water quality might have on each organism in the table. Students might describe the tolerance of bacteria as very wide, from acidic to basic, whereas trout have a very narrow range of pH that they can tolerate, any changes would affect them adversely.

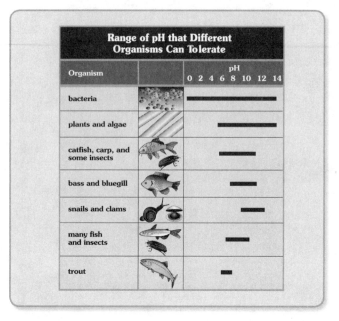

Range of pH that Different Organisms Can Tolerate		
Organism		pH 0 2 4 6 8 10 12 14
bacteria		
plants and algae		
catfish, carp, and some insects		
bass and bluegill		
snails and clams		
many fish and insects		
trout		

What's the Point?

You now know that pH is the level of acidity of a substance or solution. The pH of water is very important. Living things can survive only within certain ranges of pH. When the water changes pH, especially when it happens suddenly, it can kill the living things in and around the water.

The pH of water can change when the water is contaminated by industrial waste, agricultural and urban runoff, or drainage from mining operations. The type of contamination depends on the way the land is used.

Why do you need to know the pH of the water? The pH can tell you whether there might be a source of acid or basicity in your stream. Most organisms prefer water with mid-range pHs (6.0–7.5). Water within this pH range is considered of high quality and is close to neutral. Trout, which are important to Wamego and St. George, can live only in a very narrow range of pH.

LT 69

LIVING TOGETHER

After students practice assessing the range of tolerances, draw students back to the original problems faced by the town of Wamego starting on page 8 in the student text. Let them work in their groups to consider a relationship between what they have learned about pH and Wamego's challenge.

TEACHER TALK

❝So now let's think about an important reason to have studied pH. Go back and look at the challenge presented at the very beginning on pages 8 through 11—where the town of Wamego has to consider making some major changes to stay alive. Think about some of the things you have learned about how pH affects water quality and relate that to Wamego's problem.❞ (If students have difficulty making a connection, draw them to the paragraphs on page 9 in the student text, and ask, How would a change in pH affect Wamego's annual Trout Festival?)

Assessment Options

Targeted Concepts, Skills, and Nature of Science	How do I know if students got it?
pH is a measure of how acidic or basic a solution is and is one indicator of water quality.	**ASK:** What is one way to determine the water quality in a community? **LISTEN:** Students should be able to say that testing the pH of water will give an indicator of water quality. **ASK:** What pH reading would tell that a community's water has an acidic pH? **LISTEN:** Students' responses should include that an acidic pH would be in the range of 0 to 6.
Mixing less acidic (*alkaline*) solutions with acidic solutions changes the overall pH (*acidity*) of a solution.	**ASK:** What happens when a base is added to an acid? **LISTEN:** Students should relate that when a base is added to an acid, the overall acidity of the solution is changed and the pH of the acid (*and the base*) changes. The acid becomes less acidic and the base becomes more acidic.
Some aquatic organisms are sensitive to changes in pH.	**ASK:** Using the *Range of pH* chart, tell what organisms have the greatest range of tolerance and which has the least range of tolerance. **LISTEN:** Students should be able to interpret the chart to say that bacteria have the greatest pH range of tolerance and trout have the least range of tolerance. **ASK:** What might the pH of a stream be if it suddenly becomes crowded with snail and freshwater clams and trout disappear? **LISTEN:** Students should be able to say that the pH of the stream has become more alkaline (*or basic*) and is up around 8 to 12.

Teacher Reflection Questions

- In this section, students are introduced to pH. Did students have trouble understanding that pH is related to acidity? Did they come away with a sense of what acidity is?

- Were students able to make connections to the driving question of the *Learning Set*? Where were they able to do this?

- What improvements could be made to the investigations to make them go smoothly and efficiently?

NOTES

SECTION 2.4 INTRODUCTION

2.4 Investigate

Dissolved Oxygen as an Indicator of Water Quality

◄ *1 class period**

*A class period is considered to be one 40 to 50 minute class.

Overview

Students have studied the effects of one water quality indicator, pH. In this section, students read about dissolved oxygen and learn that aquatic organisms need it to sustain life. The teacher uses a probe to demonstrate how factors such as temperature and turbulence affect the level of dissolved oxygen in water. After completing this section, students return to *Section 2.2*, the fertilizer investigation, to learn how increased plant growth can cause a chain of events that reduces the level of dissolved oxygen in a body of water.

Targeted Concepts, Skills, and Nature of Science	Performance Expectations
Most aquatic organisms use dissolved oxygen for respiration.	When asked what fish, clams, snails, and other aquatic organisms rely on for breathing, students should be able to identify dissolved oxygen in water as their source of oxygen.
The amount of dissolved oxygen in water increases as temperature decreases and as turbulence increases.	Students should be able to say that the amount of dissolved oxygen in water increases as temperature decreases and as turbulence increases.
Dissolved oxygen is an indicator of water quality.	Students should be able to relate that the amount of dissolved oxygen in water is an indicator of the quality of the water.
Tools can be used to collect and display data.	Students should be able to tell the function of certain types of tools used for collecting and displaying data.

LIVING TOGETHER

Materials	
1 per classroom	Safety goggles
1 per classroom	Apron
1 per classroom	Probeware with dissolved oxygen testing kit
1 per classroom	Temperature probe
1 per classroom	Laptop or probeware display
1 per classroom (optional)	LCD projector or TV monitor
4 per classroom	Water samples at 4 different temperatures (very cold, cold, room temperature, and hot)
1 per class	Method to heat and cool water for dissolved oxygen testing
4 per classroom	500-mL glass beakers or equivalent-size jars
1 per classroom	Plastic spoon
1 per classroom	Supply of distilled water
1 per classroom	Roll of paper towels
1 per classroom (optional)	Projection of data tables from student text

Clean up spills. Be careful when using hot water.

Activity Setup and Preparation
Demonstration

- Test all demonstrations ahead of time to ensure that they work smoothly and that you have all of the equipment required. Make sure you understand how to work the probeware kit.

- Set up and run the dissolved oxygen and temperature probes on a computer along with a method for displaying results (*an LCD projector or a TV monitor*). If you cannot find a way to project information on a computer screen, have an alternate plan in place to have student record results on the board using a transparency of the students' data tables.

Homework Options

Reflection

- **Science Content:** Describe what might happen to water quality during a heat wave. *(Answers will vary but might include that a heat wave would increase the temperature of water. This would result in a decrease in dissolved oxygen and therefore, a decrease in its ability to support animal life. Increased temperature results in decreased oxygen levels.)*

- **Science Content:** Describe what might be the differences in levels of dissolved oxygen in a white-water river (*one with lots of turbulence in some parts*) and a pond where water does not move appreciably. *(Answers will vary but should include that dissolved oxygen would be high in the river, especially in areas where there is a lot of turbulence. Large amounts of atmospheric oxygen gets mixed in a river that moves and rolls a lot. A pond is still and quiet. It warms up in the sunlight and does not move. Water that is quiet and warm will have less dissolved oxygen.)*

Preparation for 2.5

- **Science Process:** Think of some ways in which human activities contribute to a decrease in water quality. *(Answers will vary. Many will include that human activities such as businesses, farming, and recreation can decrease water quality.)*

- **Science Content:** What are two ways that natural occurrences contribute to pollution of streams and lakes or ponds? *(Answers will vary but may include that heat from the Sun can warm up some bodies of water to the point where they might lose dissolved oxygen. Rushing water after a heavy rain might cloud water in a river.)*

SECTION 2.4 IMPLEMENTATION

1 class period* ▶

2.4 Investigate

Dissolved Oxygen as an Indicator of Water Quality

10 min.

Students read a summary of information about the importance of and sources of dissolved oxygen. They learn how natural processes affect the amount of dissolved oxygen in water. It is also a way for you to assess what your students have absorbed thus far about things to test for in water.

2.4 Investigate

Dissolved Oxygen as an Indicator of Water Quality

O_2

So far you have looked at plant growth and pH as indicators of water quality. Dissolved oxygen (DO) is another important indicator of water quality. All organisms that live in the water need oxygen to survive. Testing for DO can help you determine the health of a body of water.

The oxygen molecule is two oxygen atoms bonded together. The formula for an oxygen molecule is O_2 (O-two). Dissolved oxygen is oxygen molecules surrounded by many water molecules. You cannot see these molecules because they are too small. However, aquatic animals use DO for respiration (breathing).

Most of the DO in water comes from the atmosphere. As water moves along, oxygen is mixed into it. Another source of DO is aquatic plants. Oxygen is formed by plants in the process of photosynthesis. Excellent water has high levels of dissolved oxygen. Natural changes can affect the amount of DO. Humans can also affect the "health" of the water.

Demonstration

Your teacher is going to demonstrate how to measure the amount of DO in several water samples. You will investigate the effect of two different factors, temperature and **turbulence**, on the amount of dissolved oxygen.

Most of the dissolved oxygen in water bodies comes from the atmosphere, but some comes from plants in the water.

Dissolved oxygen is best measured in a unit called *percent saturation*. This indicates how much dissolved oxygen is in a sample compared to what it can hold. For example, water at 28°C (82°F) can hold 8 ppm of dissolved oxygen. When it has 8 ppm of oxygen in it, it is 100% saturated. If that same sample of water had only 4 ppm of dissolved oxygen, it would be 50% saturated.

LT 70

Project-Based Inquiry Science

◯ Engage

Have students read the opening paragraphs to review the indicators they have already worked with and to identify dissolved oxygen as a new indicator. Talk about how aquatic animals obtain oxygen in a way that is different from the way air-breathing animals obtain oxygen. Prepare students to learn about how turbulence gets oxygen from the air into water where it available for animals that depend on dissolved oxygen.

*A class period is considered to be one 40 to 50 minute class.

"Oxygen is not a new term for most of you. Let's review what you might know about oxygen. First of all, where does oxygen come from? How important is oxygen to you? Could you live without it? You really depend on oxygen, don't you? Other animals, such as fish, are just as dependent on oxygen, but they all live in water. Fish, clams, snails and other animals that live in water can't take the oxygen from the air as you do. How do they obtain oxygen? Water particles (H_2O molecules) contain oxygen. But fish and other aquatic animals can't take the oxygen that is part of these water particles. Instead, these animals use oxygen that is mixed into the water they live in. Today, you are going to talk about oxygen in a new way. You're going to talk about oxygen that gets mixed into moving water. This oxygen is called dissolved oxygen. Your textbook refers to it as DO."

△ Guide

Tell students that you are going to do two demonstrations concerning dissolved oxygen in water. The first demonstration will be about how temperature affects the amount of oxygen that can dissolve in water. The second demonstration will test how turbulence, or stirring, affects how much oxygen can be dissolved in water at different temperatures.

Make certain that all students can see each demonstration. Rearrange desks if necessary, especially if you do not have access to probeware that is set up to record on an LCD. Test whether the sample and the probe can be put on an overhead projector so that all students can see the procedure.

Demonstration

Your teacher is going to demonstrate how to measure the amount of DO in several water samples. You will investigate the effect of two different factors, temperature and **turbulence**, on the amount of dissolved oxygen.

Most of the dissolved oxygen in water bodies comes from the atmosphere, but some comes from plants in the water.

Dissolved oxygen is best measured in a unit called *percent saturation*. This indicates how much dissolved oxygen is in a sample compared to what it can hold. For example, water at 28°C (82°F) can hold 8 ppm of dissolved oxygen. When it has 8 ppm of oxygen in it, it is 100% saturated. If that same sample of water had only 4 ppm of dissolved oxygen, it would be 50% saturated.

LT 70

Project-Based Inquiry Science

Demonstration

5 min.

Students are introduced to the use of probes to test the effects of temperature and turbulence on dissolved oxygen.

○ Engage

Have students read the paragraph that explains *percent saturation* so that they can begin to understand how water at different temperatures can contain different amounts of oxygen. You may wish to remind students about their initial introduction to *ppm* (*parts per million*) on page 61 in the student text and revisit that discussion. If students need some reinforcement to understand the scale of *ppm*, use an analogy involving numbers of cars.

TEACHER TALK

"To get a better understanding of how small eight parts per million is, picture a bumper-to-bumper line of cars stretching all the way from Ohio to the Pacific Ocean. The length of an average car is about 20 feet. The total distance is about 2480 miles. The total number of cars, bumper-to-bumper, would number about 650,000 cars. Eight ppm would amount to eight of those cars out of all the remaining cars."

NOTES

Temperature and Dissolved Oxygen

Your teacher will set up four samples of water at different temperatures and use a temperature probe to measure the temperature of each sample. An oxygen probe will be used to measure the amount of oxygen in each sample.

Make your observations. Record them in a table similar to the one shown.

Dissolved Oxygen and Temperature				
	Sample 1	Sample 2	Sample 3	Sample 4
temperature				
dissolved oxygen				

Turbulence and Dissolved Oxygen

Your teacher will use the same samples as in the previous demonstration. The dissolved oxygen in each sample will be measured before and after stirring.

Predict what will happen to the dissolved oxygen when the samples are stirred.

Make your observations. Record them in a table similar to the one shown.

turbulence: the violent disruption, agitation, or stirring up of something (in this case, of water).

Dissolved Oxygen and Stirring					
		Sample 1	Sample 2	Sample 3	Sample 4
	temperature				
dissolved oxygen	before stirring				
	after stirring				

Prepare students for the demonstration about the relationship between temperature and dissolved oxygen. Ask if they think that water can hold different amounts of oxygen at different temperatures. How much oxygen do they predict water can hold at high and low temperatures?

Temperature and Dissolved Oxygen
10 min.

Students observe that temperature directly impacts levels of dissolved oxygen and think about how this affects the ability of aquatic organisms to stay alive.

META NOTES

The warmer a solution, the faster its molecules move and the further apart they move.

As temperatures increase, there is more opportunity for dissolved oxygen particles to escape from the surface of the liquid.

META NOTES

As the temperature of a solution decreases, molecules move more slowly and they move closer together. In cold water, there is less opportunity for dissolved oxygen molecules to escape. Therefore, cold water holds onto more of the dissolved oxygen.

"When water gets warm, it is because energy moves from a heat source (such as the Sun) to the water. This causes the particles (molecules) of water to move faster. They bump into each other and move apart from each other. Other particles in the water, such as any dissolved oxygen particles (molecules) that are there, also get bumped around. Eventually, if the temperature is high enough, some of the water particles and some of the dissolved oxygen particles escape from the surface of the pond into the air. Knowing that, what relationship can you make on a summer day between water in a pond that gets heated by the Sun and the amount of dissolved oxygen in the pond?" (The warmer the pond becomes, the less dissolved oxygen there will be because the energy will cause more and more particles of water and dissolved oxygen to leave the pond.)

Then ask what might happen if the temperature of the pond water were to decrease.

"What do you think might happen to the particles in the pond when temperatures become lower (decrease)? Do the particles move faster or do they move more slowly? Are they closer together or farther apart? Does a cooler pond hold more dissolved oxygen than a warmer pond? Why do you think water can hold more dissolved oxygen as it gets colder? (There is less movement among the particles and less dissolved oxygen will get bumped from the surface of the pond.) During what time of year do you think the pond has more dissolved oxygen? What are some reasons for your choice?"

○ Get Going

Ask students, "What are some benefits from dissolved oxygen in the water?" During this investigation, draw students' attention to think about how temperature affects the ability of water to hold dissolved oxygen. Ask, "Give a reason why it is important for water to be able to hold the greatest amount of dissolved oxygen that it can hold." *(Because organisms living in the water need dissolved oxygen to survive.)* Ask students why it might be important for someone to know this information. Ask them to think how this might relate to Wamego's annual Trout Festival.

Describe the four samples of water—very cold, cold, room temperature, and hot. These are not temperatures, but relative descriptions. Explain that you will use a temperature probe to take the temperature of each sample and students will be able to read the temperature on the LCD monitor.

Demonstrate how the temperature probe is used. Let students read the temperature from the LCD if possible. If that is not set up, then, as much as possible, let students take turns reading the temperature probe and recording it on a transparency for the whole class. Remind students to record the temperatures on their individual data charts.

Before you use the dissolved oxygen probe, ask students to think about each temperature reading and predict whether organisms will be able to survive in it. Ask why.

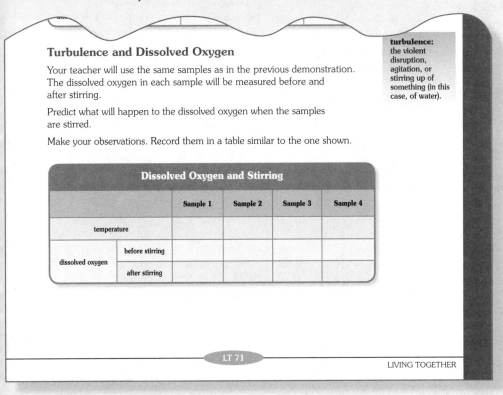

Turbulence and Dissolved Oxygen

Your teacher will use the same samples as in the previous demonstration. The dissolved oxygen in each sample will be measured before and after stirring.

Predict what will happen to the dissolved oxygen when the samples are stirred.

Make your observations. Record them in a table similar to the one shown.

turbulence: the violent disruption, agitation, or stirring up of something (in this case, of water).

Dissolved Oxygen and Stirring		Sample 1	Sample 2	Sample 3	Sample 4
temperature					
dissolved oxygen	before stirring				
	after stirring				

LT 71

LIVING TOGETHER

△ Guide

Explain that during the second demonstration, you will check the temperatures and dissolved oxygen, but this time, you will stir each sample vigorously, creating turbulence. Then you will use the *DO* probe to recheck the sample's *DO* reading at that temperature.

While you are replacing the four samples, ask students to predict what they think will happen to each sample as it is stirred vigorously. Ask them to give reasons for their predictions. Have a student volunteer to record predictions.

Demonstrate how turbulence affects the dissolved oxygen levels by first taking the temperature of the water, then the *DO* level, and then vigorously stirring the sample for about 10 seconds. Then immediately measure the *DO* level of that sample again. Ask students to record the results.

Turbulence and Dissolved Oxygen
10 min.

Students observe that dissolved oxygen in water increases when the water is stirred vigorously.

META NOTES

Because time will have elapsed and temperatures of the four water samples will tend to rise or fall to room temperature, replace all four samples and begin the turbulence demonstration by quickly measuring the temperature of each new sample. Explain to students why you are replacing the water samples.

Analyze Your Data

5 min.

Demonstrating how to measure dissolved oxygen, gives students the chance to make observations that will generate and stimulate a class discussion.

Analyze Your Data

1. Describe the trends you see in these results.

2. What appears to be the relationship between temperature and dissolved oxygen found in water?

3. What appears to be the relationship between turbulence (stirring) and dissolved oxygen found in water?

4. When would a river be stirred up like the water in the jar?

LT 72

Project-Based Inquiry Science

Ask students to answer each analysis question, basing their answers on the evidence they observed in the demonstrations. Tell students that they will use their answers in a class discussion.

If necessary, guide students as they answer question 4 by asking them, "How does the structure of the river (*its bottom and sides*) or what is inside it, affect how the water flows?"

Students' responses to the questions should include the following:

1. Trends: As the temperature of the water decreased, the levels of dissolved oxygen increased. As the temperature increased, the levels of dissolved oxygen decreased. As the water was stirred vigorously,

its dissolved oxygen increased. But the amount of increase depended on the temperature of the water. Warmer water had a smaller increase in dissolved oxygen than cold or very cold water.

2. The higher the temperature, the lower the dissolved oxygen levels.

3. Dissolved oxygen increases with turbulence.

4. Answers will vary. A river might be more turbulent when it is moving fast and dropping in elevation. It might become more turbulent after a storm when there is more water or if it is narrow and has a rocky bottom.

NOTES

....

....

....

....

....

....

....

....

....

....

....

....

....

Dissolved Oxygen

10 min.

Students learn more details about dissolved oxygen and its relationship with water quality.

Dissolved Oxygen
Temperature, turbulence, and amount of plant growth all affect the amount of dissolved oxygen in water.

Temperature
Colder water can hold more dissolved oxygen than warmer water. In fact, water at 8°C (46°F) can hold 12 ppm dissolved oxygen. Compare this to only 8 ppm in 28°C (82°F) water. You should have seen a clear trend in your results suggesting this relationship.

Turbulence
The more turbulent the water, the higher the amount of dissolved oxygen it will have. As water is mixed and splashed, it traps oxygen from the air. This oxygen dissolves into the water. Water that is moving rapidly and mixes with air has more DO than water that is still.

Rivers should be saturated with dissolved oxygen. That is because rivers are always flowing. Rivers that fall below 90% saturation are considered to be less than desirable. The table on the next page shows the amount of DO in different qualities of water.

◯ Engage

Ask students to read the background information on pages 73 and 74 in the student text to help them pull together ideas about the effects of turbulence, temperature and plant growth on water quality.

◇ Evaluate

Evaluate students' understanding of the importance of dissolved oxygen to maintaining life in water by reading aloud the closing statement under *Plant Growth* on page 74, "Thus, a water body with a large amount of plant growth around it will eventually have very low levels of dissolved oxygen." Ask students to work in their groups to use information they have learned thus far to discuss how to keep this from happening.

Plant Growth

Large numbers of plants in and around a body of water can rob water of its dissolved oxygen. When plants and other organic material (anything that is or was once living) falls into water, the material begins to break down (decompose). Bacteria decompose the material. In the process of doing this, bacteria use up dissolved oxygen. Thus, a water body with a large amount of plant growth around it will eventually have very low levels of dissolved oxygen.

Water Quality and Dissolved Oxygen	
Water Quality	**Dissolved Oxygen (% of saturation)**
excellent	91 - 100%
good	71 - 90%
fair	51 - 70%
poor	50% or less

What's the Point?

The amount of oxygen (O_2) that can dissolve in water depends on two factors. One is the temperature of the water. Cool waters can dissolve more oxygen than warm waters. The amount of dissolved oxygen (DO) also depends on turbulence. Quickly flowing, turbulent areas, like waterfalls, have high levels of DO. By comparison, slowly moving or still water, like lakes and ponds, will have lower levels of DO.

Aquatic organisms need DO to survive. Fish use their gills to obtain DO from the water. They need oxygen to breathe and live. The organisms that live in a body of water depend on the amount of DO in the water. Generally, the more DO, the greater the number of organisms that can live in the water.

Large amounts of plant growth in water or very near water can rob the water of its DO. This is because when the plants die, bacteria use DO as they decompose the dead plants.

NOTES

Assessment Options

Targeted Concepts, Skills, and Nature of Science	How do I know if students got it?
Aquatic organisms use dissolved oxygen for respiration.	**ASK:** What do fish, clams, and snails use to breathe in water? **LISTEN:** Fish, snails, clams, and many other organisms depend upon dissolved oxygen in water to breathe. **ASK:** A student discovers dead fish and other aquatic animals in a stream. What might be one explanation? **LISTEN:** One possible explanation students might give is that there isn't enough dissolved oxygen in the water to support the animals.
The amount of dissolved oxygen in water increases as temperature decreases and as turbulence increases.	**ASK:** How are temperature, dissolved oxygen, and turbulence in water related? **LISTEN:** You should hear students relate that as the temperature of a body of water decreases or becomes colder, the amount of dissolved oxygen increases. Also, as turbulence increases, the amount of dissolved oxygen in a body of water increases. **ASK:** Which body of water would you expect to have a lot of healthy animal life—a small lake with lots of sunlight and little movement or a shaded, rushing river? Why? **LISTEN:** Students should relate that the shaded, rushing river would have more dissolved oxygen to support life because it is turbulent and the shade would make the temperature of the water cooler.

Targeted Concepts, Skills, and Nature of Science	How do I know if students got it?
Dissolved oxygen is an indicator of water quality.	**ASK:** Describe the ability of pond water that is warm and still to support life that depends upon oxygen. **LISTEN:** Students should be able to say that warm, still, pond water probably has a low amount of dissolved oxygen and therefore will not support life that depends on oxygen.
Tools can be used to collect, display, and analyze data.	**ASK:** What are some ways that water temperature and dissolved oxygen can be detected during an investigation? **LISTEN:** Students should be able to describe that probes can be used in an investigation to report temperature and dissolved oxygen levels. Students may also relate that these probes can be connected to a computer so that data can be reported visually and recorded.

Teacher Reflection Questions

- In this section, students are introduced to dissolved oxygen. Did students have trouble understanding that dissolved oxygen is a measure of how many oxygen molecules are in a volume of water?

- Were students able to make connections to the driving question of the *Learning Set?* Where were they able to do this?

- What improvements could be made to the investigations to make them go smoothly and efficiently? Did the demonstrations go smoothly? How can this be improved?

NOTES

....

....

....

....

....

2.5 Read

1 class period* ▶

More Water-Quality Indicators

Overview

Students have looked at three water-quality indicators: phosphate and nitrate levels, pH, and dissolved oxygen. Students learned that temperature and turbulence can affect levels of dissolved oxygen in a body of water. In this section, students learn more about the influence of temperature. They read about the effects of thermal pollution on photosynthesis in water plants and learn that excessive growth of these plants can diminish dissolved oxygen. Students learn that *turbidity* is a measure of water clarity and that fecal coliform bacteria in a body of water is an indicator of poor water quality. This information adds to science knowledge that students can use when they diagnose possible water-quality problems that might be faced by the town of Wamego if its community and its ecology are changed.

**A class period is considered to be one 40 to 50 minute class.

Targeted Concepts, Skills, and Nature of Science	Performance Expectations
Some common water-quality indicators that can be measured are pH, temperature, turbidity, levels of dissolved oxygen, nitrates, phosphates, and fecal coliform bacteria.	When asked for indicators of water quality, students should be able to identify temperature, turbidity, and fecal coliform bacteria as some of the factors that affect water quality.
Thermal pollution, the result of human activity, causes water temperatures to increase, reduces water quality, and harms the ecosystem.	Students should be able to say that thermal pollution and the rise in water temperature harms the ecosystem because it decreases available oxygen and therefore, harms aquatic organisms.
Turbidity is a measure of how opaque water is and may increase due to disturbances in land structure or in the river bed as a result of human activity.	Students should be able to relate that turbidity increases because of changes in land use and in rivers.
Fecal coliform bacteria can cause disease for humans and other animals when the bacteria enter a watershed in contaminated runoff or untreated sewage.	Students should be able to say that the presence of fecal coliform bacteria in a water source indicates poor quality water.

Homework Options

Reflection

- **Science Process:** How would a scientist out in the field measure the turbidity of a body of water? *(The scientist would lower a Secchi disk into the water and then raise it. A reading would be taken at the point where the disk disappears from view as it is lowered and reappears as it is raised in the water.)*

- **Science Content:** Why do scientists have to think about dissolved oxygen as well as thermal pollution when fish, such as trout, begin to die in a stream? *(Answers will vary, but students should include that some organisms, such as trout, survive in a narrow temperature range. If the water temperature rises, the fish cannot survive because the increased temperature lessens the amount of dissolved oxygen.)*

Preparation

- **Science Content:** You and your group are scientists preparing to do some tests in the field. You want to test for changes in temperature in the local river. You are not certain where this is happening. Is it near the big bakery? Is it further down stream near the factory that manufactures cars? Explain how your group would test for temperature differences in the river. *(Answers will vary but students' answers should include that temperatures should first be checked further up the river as close to its source as possible, then again near the factories in question.)*

- **Science Content:** Your group consists of the committee in Wamego that is preparing for the annual Trout Festival. Some people on the committee want to allow motor boats on the river this year during the festival and some want to ban the use of motor boats. What water-quality indicator might be more common if motor boats were allowed? How might this affect the trout and the festival? *(Answers will vary but turbidity would probably increase if motor boats are allowed. This could lead to an increase in the water temperature (thermal pollution) as particles in turbid water absorb more heat from sunlight. This might reduce the number of trout available because it would change the temperature range in which they normally live.)*

1 class period* ▶

2.5 Read

More Water-Quality Indicators

In this section, students complete their introduction to major water-quality indicators that are used to test water throughout a watershed.

Temperature

10 min.

Students begin to appreciate the complexity of the ways in which temperature can affect an ecosystem.

META NOTES

It has been found that trout feed best when the temperature of the water is rising into their optimum feeding range (10°C to 20°C) rather than when it is falling into this range from a higher temperature.

*A class period is considered to be one 40 to 50 minute class.

SECTION 2.5 IMPLEMENTATION

2.5 Read

More Water-Quality Indicators

You have investigated nitrates and phosphates, pH, and dissolved oxygen as indicators of water quality. Now you will read about three more indicators of water quality. You will read about and discuss the indicators, temperature, turbidity, and fecal coliform.

Temperature

Temperature is very important to water quality. As you saw in the previous section, temperature affects the amount of DO in the water. It also affects the rate of photosynthesis. Under certain conditions, photosynthesis increases at higher temperatures. Therefore, the amount of plant growth increases. This can be a good thing. Photosynthesis adds DO to the water. However, plant growth can increase too much. The plants will eventually die. Then, decomposing bacteria will use up the DO.

Science Connection

Thermoregulation is the ability of an organism to maintain a fairly steady body temperature. Some organisms do this by living in environments with appropriate temperatures. Other organisms rely on their bodies to produce or dispel heat.

he e carp need water temperature of 15° – 34°C (59° – 93°F) to li e

Temperature also affects the balance of an ecosystem. Most organisms are suited to live in a given range of temperature. Some organisms, such as trout, prefer cooler water. Others, such as carp, need warmer conditions. As the temperature of a river or lake increases, cool-water animals will leave or die. Warm-water animals may replace them. Most organisms cannot survive in temperatures of extreme heat or cold. Temperatures that are too high or too low can stress an organism. This stress makes organisms more prone to disease. It also makes it more difficult for them to react to pollution.

LT 75

LIVING TOGETHER

△ Guide

Have students read the opening paragraph to refresh their knowledge of the water-quality indicators they have already studied and to introduce them to three additional indicators: thermal pollution, turbidity, and fecal coliform.

○ Engage

Ask a student to summarize what the class has already learned about temperature and its effect on water quality.

Explain that sometimes when an industry is built near a river, it is because the factory needs water to cool down parts in the manufacturing process. The water is drawn from the river and later returned to the river. But the temperature of the returning water is higher. This can have an effect on the plants and animal life in the river and change the quality of the river water. Refer students' attention to the table on page 76 in the student text that shows original temperatures and the temperature of changed water (*in parentheses*).

TEACHER TALK

"In the table named Water Quality and Temperature, the first temperature is the temperature of the original water source, such as a river or a pond. The temperature in parentheses is the temperature of the water after it has been used and returned to the river or pond. What do you notice about these temperature differences? (As the difference between the original water temperature and the temperature of the returning water becomes greater, the quality of the water becomes less.) Scientists may use these differences as an indicator of water quality."

△ Guide and Assess

Explain that the range of water temperature that trout can live in ranges from 2°C to 23°C, and they feed best in water that ranges between 10°C and 20°C. List this feeding temperature range and also list the temperature information about carp from the caption on page 81 in the student text. Ask the class to compare the temperature ranges and discuss the dissolved oxygen levels they might find at these two temperature ranges. Have students discuss whether trout and carp would be found living in the same stream. Listen for students to begin to make a connection between stream temperatures and what might happen if the temperature of the Crystal River in Wamego were to rise.

TEACHER TALK

"Think about what you know about how temperature and dissolved oxygen are related. Which fish, the trout or the carp, probably lives in a stream with higher dissolved oxygen? If a factory along a river caused the temperature of the water to rise, which fish would most fishers be catching in the river? If the temperature of the Crystal River rises, will the town of Wamego have to change the name of its festival? Why or why not?"

Turbidity

10 min.

Students should become aware that water quality is not merely a question of whether water is safe to drink, but also whether it is of high enough quality to support plant and animal life in and around it.

---

thermal pollution: a change in temperature in a natural body of water that is caused by humans.

turbidity: how cloudy, murky, or opaque something is.

A Secchi disk can be used to measure turbidity. The disk is lowered and raised in the water. A depth reading is taken at the point where the disk disappears and reappears.

Humans cause many changes in the temperature of a body of water. This is called **thermal pollution**. One source is warm water added by industries. Scientists test temperature by measuring the temperature of the river, lake, or stream at its source. Scientists then move to another point in the river or lake. They take a second measure of the temperature. If the difference in temperatures is very large, there is a problem. The table shows the water quality for various changes in temperature.

Water Quality and Temperature

Water Quality	Temperature Change Between Samples
excellent	0–2°C (0–4°F)
good	3–5°C (5–9°F)
fair	6–10°C (10–18°F)
poor	10°C (18°F) or greater

Turbidity

How clear water looks can also help you determine the quality of the water. **Turbidity** is the measure of how clear water is. Materials such as clay, silt, organic and inorganic matter, and microscopic organisms can be suspended in water. These cause turbid, or murky, water. The murkier the water, the greater the turbidity.

Not as many types of organisms can live in turbid water as can live in clear water. This is because there is less sunlight that can get through the water. Also, the suspended particles absorb sunlight. Therefore, water temperature increases. This causes oxygen levels to fall. (Remember, warm water holds less oxygen than colder water.)

Turbid water may be the result of soil erosion, urban runoff, and/or bottom sediment disturbances. Bottom sediment disturbances can be caused by boat traffic. They can also result from large numbers of bottom feeders. (Bottom feeders are organisms that stay near the bottom of the water when feeding.)

LT 76

Project-Based Inquiry Science

◯ Engage

Turbidity is a term that students may not be familiar with. Ask them what they think the word *turbidity* means and list their ideas.

To enable students to get an idea of what turbidity is, you might demonstrate with a jar of water with some soil. Initially show the jar after its contents have been standing for an hour or more. Mud will be settled at the bottom and the liquid should be fairly clear. Then stir the contents of the jar. The contents should cloud up or become turbid with particles of soil. The particles will soon begin to settle out.

△ Guide

Explain that various types of materials make water turbid or difficult to see through. Much of the particulate matter is made up of fine particles of soil. If the water becomes turbulent, then the fine particles can cloud the water.

Encourage students who might want to make a model Secchi disk to test various levels of turbidity. This can be made of simple materials on a small scale with water of various clarities in the classroom.

META NOTES

Bacteria, viruses, and other pathogens can be harder to detect in turbid water. Turbid waters can absorb and retain more heat and contribute to the growth of pathogens in a contaminated water source.

NOTES

Fecal Coliform

10 min.

Fecal Coliform

Fecal coliforms are bacteria. They occur naturally in animals' digestive tracts. They aid in the digestion of food. Fecal coliform bacteria are found in feces. They appear in the feces of humans and other warm-blooded animals such as cattle and birds. Sometimes there is too much of this type of bacteria present. Then there is the possibility that harmful microbes are also present. These microscopic organisms or viruses can cause disease.

Even though there may be very few harmful microbes, it only takes a small amount to make a person sick. In water, if fecal coliform counts are high, there is a greater chance of harmful microbes being present. Swimming in waters that have high fecal coliform counts can increase a person's risk of getting sick. Microbes can enter the body through the skin, cuts, nose, ears, or mouth. Fecal coliform bacteria can enter a river through runoff or sewage discharge.

There is a very simple determination of water quality when it comes to fecal coliform. Water that tests negative (no fecal coliform) is good quality. Water that tests positive for fecal coliform is considered poor quality. It is that simple.

Sewage discharge that is not treated in some way can introduce harmful bacteria into waterways.

LT 77

LIVING TOGETHER

△ Guide

Inform students that water is treated before it enters their homes and after it leaves their homes. Water that has been treated and enters their homes is safe to drink and is called *potable water*. Some of this water is pumped from aquifers, reservoirs, lakes, or wells. Water that leaves their homes is made up of all the water from kitchens, bathrooms, dishwashers and washing machines and is called *wastewater*. Water quality in the United States is protected by the Safe Drinking Water Act, amended in 1996. This act is made up of rules that regulate the nation's drinking water supplies. Tell students how water is treated in most towns and cities.

TEACHER TALK

"Wastewater from homes, schools, hospitals, and businesses is collected in sanitary sewers and carried to sewage-treatment plants. At a sewage-treatment plant, water goes through a series of screens or filters that remove debris. Then, it is mixed with microorganisms (bacteria) that feed on the very small materials that might still be in it. Eventually, this mixture has oxygen stirred into it. The oxygen causes the microorganisms to expand and sink to the bottom where they can be removed.

The water is then disinfected with chlorine. This means that any remaining harmful pathogens are killed. Chlorine is very strong so other chemicals are added to neutralize the chlorine. Finally, some of the treated wastewater is released into streams and in some cases, an ocean. Some of this water is sprayed onto large grassy areas such as golf courses. Some is stored.

Sewers are part of the infrastructure of a city. When more businesses are built, more homes, schools, and shopping facilities are also built and infrastructures have to also be expanded."

Explain that much of the water that people can drink is pumped from groundwater sources such as aquifers and surface-water resources, such as rivers, lakes, and streams. A smaller percentage of people have their own wells. Water is treated before it is allowed into homes to be consumed.

With the above information in mind, ask students why it would be important for Wamego to be able to treat increased waste materials if the new FabCo factory is built. List their ideas.

META NOTES

The Safe Drinking Water Act of 1996 lists maximum allowable levels of nearly 100 contaminants that can exist in drinking water.

NOTES

....

....

....

....

....

....

....

Stop and Think

20 min.

During the class discussion, students should be encouraged to ask each other questions that assist in refining and articulating their answers.

Stop and Think

1. In the previous section, you observed how temperature affects the amount of dissolved oxygen. What trend did you observe?

2. How can increased plant growth both increase and decrease dissolved oxygen in water?

3. An ecosystem includes the organisms that live in it. How can changes in temperature in an ecosystem change the organisms that live there?

4. What is thermal pollution?

5. How do scientists measure the quality of water using temperature?

6. Why is it important to measure the turbidity of the water?

7. What causes turbid water?

8. Why is fecal coliform measured to determine the health of a river?

9. How do fecal coliform bacteria end up in a river?

10. What role do humans have in altering temperature, turbidity, and amount of fecal coliform in water?

What's the Point?

These three indicators, temperature, turbidity, and fecal coliform, are critical to understanding the quality of a body of water. Each can easily be affected by the different land uses you and your classmates have been learning about.

Now that you have reviewed six indicators, you understand much better how scientists can evaluate the quality of water in a community. Next you will be applying your knowledge of these indicators to diagnose possible water-quality problems in a watershed.

LT 78

Project-Based Inquiry Science

○ Get Going

Let students spend ten minutes answering the questions on their own. Then, transition them to a class discussion. If, during the class discussion, students disagree, ask students with opposing viewpoints to defend their stands. Most student responses should contain at least the following:

1. As temperature increases, dissolved oxygen decreases.

2. Aquatic plants release oxygen into the water through photosynthesis. This increases the level of dissolved oxygen. However, increased plant growth means that more plants are available to decay in the water. Bacteria that decompose dead vegetation use the dissolved oxygen during this process.

3. Changes in temperature change the level of dissolved oxygen in the water. This affects what can or cannot survive in the water.

4. Thermal pollution occurs when the temperature of the water rises, often because of human activity.

5. Scientists measure temperature in rivers, lakes, or streams at the source and again at another point farther down the river, lake, or stream. If the temperature difference is greater than 10°C, the water quality is poor; if the temperature difference is small (0° to 2°C), water quality is classified as excellent.

6. The more particles in water, the more sunlight it absorbs. This increases the temperature of the water and decreases the levels of dissolved oxygen.

7. Turbid water is caused by soil erosion, urban runoff, and disturbances in the sediments in the bottom of the river.

8. When fecal coliform is measurable in a stream, it means that there is a greater chance that pathogens are present in numbers to cause health problems.

9. Fecal coliform bacteria end up in a river when untreated sewage is dumped or leaks into the river.

10. One way humans affect water quality is through changes in land use or increases in systems that stress an existing system.

◇ Evaluate

Listen for students' understanding of how these important indicators of water quality can affect their health and the health of the environment in which they live.

NOTES

....

....

....

....

....

Assessment Options

Targeted Concepts, Skills, and Nature of Science	How do I know if students got it?
Some common water-quality indicators that can be measured are pH, temperature, turbidity, levels of dissolved oxygen, nitrates, phosphates, and fecal coliform bacteria.	**ASK:** Select one of these three indicators and write a paragraph that explains how a scientist might use the indicator to check on water quality. **LISTEN:** Answers will vary based on the selected indicator but should include the following: For temperature, thermal pollution is checked in a stream by comparing the measured temperature of the water at its source and again at a second point along the stream. For turbidity, a scientist might use a Secchi disk to check the clarity of the water in the same place on two different days. For fecal coliform, a scientist would check for the presence of the bacteria at places where water enters a stream.
Thermal pollution, the result of human activity, causes water temperatures to increase, reduces water quality, and can harm organisms in an aquatic ecosystem.	**ASK:** During a very hot, bright summer day, you notice an unpleasant scum and smell coming from the lake in the town park. What is one explanation for these changes? **LISTEN:** Answers will vary but you should hear students say that increased temperatures caused any plants in the lake to grow rapidly and eventually die off. Bacteria in the water decomposed the dead plants and used up dissolved oxygen.

NOTES

....

....

....

Targeted Concepts, Skills, and Nature of Science	How do I know if students got it?
Turbidity is a measure of how opaque water is and may increase due to disturbances in land structure or in the river bed as a result of human activity.	**ASK:** Land has been cleared of trees and plants for a development of new homes. Shortly after, there is a heavy rainstorm. The local stream, which is normally clear, becomes muddy with sediment. Describe the turbidity of the stream and give one explanation for the change in the stream's turbidity. **LISTEN:** Answers will vary but students should be saying that the stream is muddy because the land has been cleared of vegetation and a lot of soil was carried away in runoff because there are no plants to hold the soil in place.
Fecal coliform bacteria can cause disease for humans and other animals when the bacteria enter a watershed in contaminated runoff or untreated sewage.	**ASK:** There is a break in a sewer line in your town and people are told to boil any water used for drinking, cooking, and brushing teeth. Why are they told to do this? **LISTEN:** Answers will vary but students should be able to say that these precautions are being taken to prevent people from consuming or using water that may be contaminated with fecal coliform.

Teacher Reflection Questions

- In this section, students learned about contaminants in water. Do they understand that in real life, all drinking water contains a number of contaminants at allowable levels?

- Were students able to make a connection between the water-quality indicators they learned about and the information they learned about in the watershed discussion in *Learning Set 1?*

- Did most students participate in class discussions? How can this be further encouraged?

Learning Set 2

1 class period* ▶

Back to the Big Question and Challenge

Overview

Students have learned about several water-quality indicators in this *Learning Set* and they learned about how land use can affect the water in a watershed in *Learning Set 1*. In this section, students apply what they have learned about water-quality indicators in *Learning Set 2* to the land use that they explored in *Learning Set 1*. In their groups, students discuss and determine what type of water indicators should be tested for in their assigned land use. They use these discussions to answer the question: *How do you determine the quality of water in a community?* For a context, students will refer to the features of their land use as shown in the Rouge River photographs they studied earlier in the Unit.

*A class period is considered to be one 40 to 50 minute class.

Targeted Concepts, Skills, and Nature of Science	Performance Expectations
Explanations are claims supported by evidence, accepted ideas, and facts.	Students should be able to write an explanation and recommendations based on science knowledge and evidence from their investigations that back their claims.
Scientists make claims (*conclusions*) based on evidence (*trends in data*) obtained from reliable investigations.	Students should be able to make claims based on trends in their data and in the class's data.
Human activity can affect the ecology of a community. Humans use rivers for residential, commercial, industrial, and agricultural purposes. These activities affect water quality along a river.	Students should be able to infer that residential, commercial, industrial, and agricultural activities are human activities that can affect water quality.
Some common water-quality indicators that can be measured are pH, temperature, turbidity, levels of dissolved oxygen, nitrates, phosphates, and fecal coliform bacteria.	Students should be able to say that tests for pH, dissolved oxygen, temperature, turbidity, and fecal coliform bacteria levels are tests that would be performed to determine water quality.

Materials	
1 per group	*Idea-Briefing Notes* page
1 per group	Colored markers set
1 per group	Presentation materials
1 per group	Class *Project Board*
1 per student	Class list of criteria and constraints
1 per group	*Applying Indicators to My Land* page

Homework Options

Reflection

- **Science Content and Process:** Edit and add to the rough draft of recommendations for the Wamego town council that you started at the end of *Learning Set 1*. Modify any recommendations or comments you made at that point by incorporating new information you acquired in *Learning Set 2*. *(Updated statements should contain expanded recommendations supported by evidence collected in investigations conducted during* Learning Set 2*.)*

- **Science Process:** While out in the field, you take two samples of water from two different locations along the same stream. Back in the lab, you decide to test them for pH. Sample A tests green and Sample B tests in the pink-purple range. Explain how you would test the water samples for pH and what ranges these color reactions mean. Is Sample B more acidic or more basic than Sample A? *(You would use a pH-indicator solution or pH-indicator paper to test each sample. The green sample has a pH reading of 13 to 14. The pink-purple sample has a pH reading of 5 to 6. Sample B is more acidic than Sample A.)*

BACK TO THE BIG QUESTION AND CHALLENGE IMPLEMENTATION

1 class period* ▶

Learning Set 2

Back to the Big Question and Challenge

10 min.

Students begin to think about how to connect what they know about water quality indicators to land use.

Learning Set 2

Back to the Big Question and Big Challenge

How does Water Quality Affect the Ecology of a Community?

Over the past few days, your class has examined several water-quality indicators:

- pH
- dissolved oxygen
- nitrates and phosphates
- temperature
- turbidity
- fecal coliform

You also examined the factors that can lead to changes in these indicators.

During previous activities, you looked at how land use can affect the waters in a watershed. Your group has spent some time looking at one of four land uses: residential, commercial, industrial, and agricultural. As a wrap-up to this *Learning Set*, you will apply what you have read about indicators to your assigned land use.

In the beginning of this *Learning Set* you were asked the question *How do you determine the quality of water in a community?*

You will now answer this question, but not in the way you might expect. During this activity, your group will determine which water-quality tests would be best for the land use you were assigned in *Learning Set 1*. You will review the photos of your land use. Your group will discuss and determine the water-quality tests you should conduct. You will base your decisions on features of land use found in the photos.

LT 79

LIVING TOGETHER

○ **Engage**

Review with the class what they have accomplished in this *Learning Set*. They have investigated how plant growth can be an indicator of water quality. They have tested solutions for pH and learned that certain aquatic organisms can tolerate only certain ranges of pH. They have learned about other water quality indicators such as temperature, dissolved oxygen, turbidity, and fecal coliform bacteria.

△ **Guide**

Now ask students to recall their investigation about land use from *Learning Set 1*. Have them summarize what happened to water in their particular investigation. Explain that they will be using this information again as they

*A class period is considered to be one 40 to 50 minute class.

come to the end of this *Learning Set.* Explain that connecting what they have learned in both *Learning Sets* will enable them to begin to answer the question: *How does water quality affect the ecology of a community?* To focus students' attention on the *Big Challenge,* you might rephrase the question and change the word *community* to *Wamego.*

TEACHER TALK

"The information you learned about and investigated in Learning Set 1 isn't separate from what you have been talking about in Learning Set 2. Now it is time to make a connection between the two. Think about the land use investigation you did in Section 1.6. You're going to look at the land use again and add some new information—the information about water-quality indicators. To do this, you will use photographs from the Rouge River. This activity will help you to build a more realistic picture of how land use might be affecting water quality because now you have a better idea of how to test the water. It will also help you to answer the question: How does water quality affect the ecology of Wamego?**"**

Explain to the students that you will all review the example of the activity in the text so that everyone understands how they are to use the photographs. Tell students that the example of the car wash in the text is just one scene. Then use the bulleted statements on page 80 in the student text to illustrate how evidence from investigations and science knowledge about water-quality indicators has been used to apply information to the car wash scene.

Ask students what questions they have. Even if they appear to understand what to do now, they will probably have more questions once they begin to work with their own land use photographs.

Divide the class into their original eight groups so that, once again, there are two groups working on each type of land use—namely residential, commercial, industrial, and agricultural land uses. Explain that they will work in their groups and create a poster that summarizes the groups' decisions.

Explain that after completing their discussion, the class will hold an *Idea Briefing* during which time, each group will present their findings and recommendations. Other students will be able to ask questions and groups will be able to improve their posters.

Each test you decide to conduct needs to have an explanation justifying the use of the test. You will need to

- use evidence from the photos to suggest why the test is appropriate.
- describe what issue or problem your test is intended to identify.

For example, suppose that one of your pictures features a car wash that dumps used water into a creek that runs next to the car wash. Your group might decide that the detergent in the car wash water would do something to disrupt the water quality of the creek, specifically pH. You have several pieces of information and some knowledge from your pH investigation and research.

- You know from your investigation of pH that the water should not have a pH outside the 6.0–7.5 range.

- Your group also read that detergents can have a pH around 10–11.

- Your investigation showed how the pH of a solution can change as it mixes with a different solution of lower or higher pH.

- Your research shows that many living organisms cannot easily survive changes in pH.

Thus, when soapy water mixes with the creek, the creek's pH could move above desired levels. So, your group might decide to do a pH test to see if the pH of the creek is a problem.

Your teacher will provide you with an *Applying Indicators to My Land Use* page. Review how the example on the next page would be recorded on this sheet.

Applying Indicators to My Land Use

What water quality indicator do you want to test?

- *pH*

What feature or aspect in the photos suggests you should test this?

- *The car wash is dumping soapy water into the creek.*

What do you predict would be the outcome of the test? Provide a specific number or value you expect to see from the test.

- *The pH might be as high as 8 or 9.*

What facts or useful information do you have from your research and investigations that suggest this is a good test and an important one for good water quality? Provide at least two.

- *Water should not have a pH outside the 6–7.5 range.*
- *pH of a solution can change as it mixes with a different solution of lower or higher pH.*
- *Many living organisms cannot survive quick changes in pH.*

What process or condition that occurs in the land use could cause possible problems with this indicator?

- *Detergents can have a pH around 10–11, so the dumping of the soapy water might be changing the pH of the creek to unsafe levels.*

LT 81

LIVING TOGETHER

Applying Indicators to My Land Use
10 min.

Students work in groups to create presentations that connect their classroom observations to their real-life experiences.

◯ Get Going

Distribute a copy of the *Applying Indicators to My Land Use* page to each student and the appropriate Rouge River land use photograph to each group. Let students know where poster materials are available. Let students know that they have about 10 minutes to complete their group discussion and plan their poster.

☐ Assess

Circulate through the room listening to group discussions. Some groups may have difficulty identifying possible problems in the photographs. Assist these groups by asking questions such as: "What type of land use is this?", "How many trees are in the photograph?", "Does this scene look like a place where you would picnic?", "Do you think the houses affect the pond because they are so close to it?", "How do you think the smoke from that smokestack might affect the water?"

Some students might look at the photo and think of actions or situations that are plausible for the scene, but not explicitly shown in the photo. This is acceptable so long as students justify their conclusions with evidence from the photo.

Tell students that you will give them a two-minute warning to prepare them for the start of the *Idea Briefing*. Use a timer if you have one.

NOTES

Communicate Your Ideas

Idea Briefing

An *Idea Briefing* is like a *Solution Briefing*. It is a presentation that allows presenters and audiences to communicate effectively about an idea or solution. The *Idea Briefing* provides an opportunity to share what people have tried and what they have learned from their attempts at solving a challenge.

Your *Idea Briefing* should focus on

- the factors that exist in your land-use photos that suggest there might be a problem
- the water-quality tests you want to perform
- the possible problems you might expect in the indicator's test results.

Your poster should address these three points. Use the information you recorded on the *Applying Indicators to My Land Use* page to make your poster. Your teacher will give more guidance on how to lay out your poster and include important information.

Be sure to fill out an *Idea Briefing Notes* page as you listen to everyone's presentation. Record the evidence people cite for justifying the test of an indicator for each land use.

Update the *Project Board*

The question for this *Learning Set* was *How do you determine the quality of water in a community?* The *Idea Briefing* should have helped you think about the water quality tests and how they might help you determine water quality. Discuss what you've learned with your class. You might formulate and discuss several new ideas that should be recorded on the *Project Board*. Record your new understanding in the *What are we learning?* column of the *Project Board*. Be sure to include evidence from your investigations.

Now, return to the *Big Question* and the *Big Challenge*. Think about the connection between what you have determined about water quality and the *Big Question*. Describe the connections between your new knowledge and the *Big Question* in the last column of the *Project Board*.

LT 82

Project-Based Inquiry Science

Communicate Your Ideas *Idea Briefing*

20 min.

Having two groups present each land use provides an opportunity for more in-depth analysis and application of information about that land use.

△ Guide Presentations and Posters

Explain to students what they will accomplish during an *Idea Briefing*. Remind students that their *Idea Briefing* needs to focus on the three points listed on page 82 in the student text. These three points must also be shown on the group's poster.

Remind students that their posters need to indicate the tests they wish to conduct and reasons for using these particular tests. Also, tell them that posters can be modified as a result of the *Idea Briefing* when other students have given them new ideas to think about.

"The purpose of an Idea Briefing is to share your conclusions with the class, hear advice from the class, and help other groups improve on their ideas. In many ways, sharing information is similar to how scientists often work when they solve a problem. Sharing information about how land use affects ecology will help you to develop recommendations for the challenges faced by Wamego."

Remind students again of the three points to include in their presentations. Tell them that you and the rest of the class will be listening for these factors during the presentation.

"Your presentations should clearly describe the question you are answering. Include the kind of land use in your opening statement. You might say something like: 'We investigated how agricultural land use might affect the water quality in a community.' Then be clear about the water-quality indicators that you recommend using and explain why these indicators are the ones to use."

Inform groups that they will have five to seven minutes for their presentation. Tell them that you will let them know when they have two minutes left, then one minute so they can sum up their discussion.

△ Guide and Assess

Let the rest of the class know that during the presentations, they are to look at the photo used by the group, listen to the types of indicators the group thinks they would use and their reasons for doing so. They should be listening for an explanation supported by evidence that is in the photograph. Remind them that they can ask questions if they do not understand something or do not agree with the presenting group. Point out that these comments and questions are to be made politely.

At the end of the presentations and discussions, give students time to revise their conclusions and their posters.

☐ Assess

During each presentation, listen for students to make connections between land uses, sources of pollution, and water quality indicators. Students should be able to describe all three for their particular photograph and justify their choices.

The answer table summarizes problems, tests, and possible causes that students might identify in their assigned land use photos. Refer to the answer table as each group presents. If a group is unable to think of more

---

META NOTES

When students listen to their classmates giving a presentation on something they are familiar with, they may learn what to say or how to say something and refine their own presentations.

than one item for their photograph, use the table as a resource to suggest an indicator the students might have overlooked.

NOTE: See the back of the book for the answer table.

After each land use has been presented, give the audience a few minutes to ask questions and make observations. Make sure the student asking the question directs the question to the group presenting and not to you.

Assess students' ability to construct and pick out claims, evidence, and science knowledge and construct explanations.

Give groups time to edit their posters. Collect posters for future use in the Unit.

...or each land use.

Update the *Project Board*

The question for this *Learning Set* was *How do you determine the quality of water in a community?* The *Idea Briefing* should have helped you think about the water quality tests and how they might help you determine water quality. Discuss what you've learned with your class. You might formulate and discuss several new ideas that should be recorded on the *Project Board*. Record your new understanding in the *What are we learning?* column of the *Project Board*. Be sure to include evidence from your investigations.

Now, return to the *Big Question* and the *Big Challenge*. Think about the connection between what you have determined about water quality and the *Big Question*. Describe the connections between your new knowledge and the *Big Question* in the last column of the *Project Board*.

Update the Project Board

10 min.

△ **Guide**

Let students know that after their discussions, they will be ready to answer the question: *How does water quality affect the ecology of a community?* The answer to this question will come from information that they have entered in Columns 3 and 4 of their *Project Board*. Let students know that after they have answered this question, they will be ready to write a draft of part of their recommendations for the Wamego town council.

Transition students to recommendations by letting them know that their explanations go into the fifth column of the *Project Board*. Their recommendations will be about how to address the challenge. Their recommendations will be based on the evidence they have entered in Columns 3 and 4.

○ **Get Going**

Have students work in groups to compose their best group recommendation. Remind them to review their explanations to make sure it reads logically.

△ Guide

At this point, you should lead students to a class discussion. Ask one group to share their recommendation and allow other groups to comment on the recommendation. Display each statement where the class can edit it before it is transferred to the *Project Board*. Write the recommendations the class makes in Column 5. Let students know that when the final letter to the Wamego town council is put together, it will contain all the recommendations.

Assessment Options

Targeted Concepts, Skills, and Nature of Science	How do I know if students got it?
Explanations are claims supported by evidence, accepted ideas, and facts.	**ASK:** What might explain the disappearance of a certain species of fish from a stream near your home? **LISTEN:** Listen for explanations that include that the stream has become less turbulent and as a result, the water temperature is higher. Less turbulence means that there is less dissolved oxygen. If the fish rely on these qualities, they will not be able to stay alive if the stream has changed in these respects.
Scientists make claims (conclusions) based on evidence (trends in data) obtained from reliable investigations.	**ASK:** Scientists have measured a large algae growth in a stream near a farm. Why might the scientists want to talk with the farmer about how she manages her use of fertilizer? **LISTEN:** Listen for students to say that the increased algae growth may indicate increased use of nitrates and phosphates by the farmer. She may not be managing her use of fertilizer and it is affecting the stream quality.

Targeted Concepts, Skills, and Nature of Science	How do I know if students got it?
Human activity can affect the ecology of a community. Humans use rivers for residential, commercial, industrial, and agricultural purposes. These activities affect water quality along a river.	**ASK:** A builder wants to build a new complex of 400 apartments, a gas station, and a car wash on the edge of a small river. Why might some people not want this project to be built? **LISTEN:** Listen for students to include that there would be increased runoff from the parking areas, and that there might be problems from gasoline and soap seeping into the groundwater.
Some common water-quality indicators that can be measured are pH, temperature, turbidity, levels of dissolved oxygen, nitrates, phosphates, and fecal coliform bacteria.	**ASK:** You are measuring two samples of water from a river. One sample has a pH of 9.5. The second sample has a pH of 7.5. Which sample is more acidic? Explain your answer. **LISTEN:** Listen for students to say that the sample with the lower pH is more acidic than the sample with the pH of 7.5. On the pH scale, the lower the number, the greater the pH.

Teacher Reflection Questions

- What difficulties did students have in integrating information from *Learning Sets 1* and *2*? How successful were they in making justifications for their use of particular water-quality indicators and what can be done to improve their comfort in doing so in future presentations requiring justification of choices?

- How comfortable were students in using the photographs as a basis for their decisions? Do the photos contain sufficient information for students to base conclusions? How many groups also referred to their investigations with their land use models?

- During the *Idea Briefing*, did everyone participate in the presentations? What can be done to improve participation by everyone in each group?

NOTES

LEARNING SET 3 INTRODUCTION

Learning Set 3

How Can Changes in Water Quality Affect the Living Things in an Ecosystem?

◄ $8\frac{1}{2}$ *class periods**

Throughout Learning Sets 1 and 2, students concentrated on the abiotic features of watersheds. Learning Set 3 introduces the role and importance of the biotic portion of these ecosystems. Students classify organisms and learn that living things can serve as an indicator of water quality. They investigate the affects of photosynthesis on water quality. They are introduced to and model feeding relationships in food webs.

Overview

In *Learning Set 3*, students learn that water quality can affect the number and variety of organisms that live in and around it. In doing so, students' understanding of the importance of changes to water in a watershed expands from merely looking at changes in the water to how these changes affect living things trying to survive there. With the use of dichotomous keys, students classify common aquatic macroinvertebrates based on their physical characteristics. Through a case study investigation, students also learn that some species of aquatic organisms can be monitored, and therefore act as biotic indicators of water quality. Using elodea, students learn the rudiments of photosynthesis and observe that some organisms are more tolerant of polluted water conditions than others. Students learn about food chains. This leads them to study the characteristics of populations and communities. Using the computer program *NetLogo*, students simulate population changes in a model community. *NetLogo* can be used with *MAC, Windows, Linus,* and *Unix*. The Unit culminates with students preparing a presentation to the Wamego town council that explains scientifically, the ecological impacts that are likely to occur if the town takes on a new industry.

> **LOOKING AHEAD**
>
> In Section 3.4, students will use elodea, a live aquatic plant, for an investigation. Order the materials either Monday or Tuesday for delivery on Wednesday or Thursday. Elodea is relatively easy to care for but is sensitive to low temperatures, so avoid cold conditions.
>
> Section 3.6 requires access to a computer lab for the *NetLogo* program or simulations can be used in a classroom with a single computer and an LCD projector.

*A class period is considered to be one 40 to 50 minute class.

Targeted Concepts, Skills, and Nature of Science	Sections
Human activity can affect the ecology of a community. Humans use rivers for residential, commercial, industrial, and agricultural purposes. These activities affect water quality along a river.	3.3
Scientists classify organisms based on criteria including physical appearance, feeding relationships, and relationships to the environment.	3.2
The growth and survival of organisms depends on the physical conditions of its environment (*ecosystem*).	3.3, 3.4
Scientists use tools such as the dichotomous key to classify and identify different organisms.	3.2
Biotic and abiotic components interact in an ecosystem.	3.1, 3.3, 3.4
Scientists can determine water quality using biotic indicators.	3.3
Living organisms are made of cells, get energy from the environment, grow and develop, reproduce, and respond to changes in their environment.	3.2
Organisms may interact in several ways as producer and consumer, predator and prey.	3.2, 3.4, 3.5, 3.6, 3.7
Living things that consume other organisms are called consumers.	3.5, 3.6, 3.7
Bacteria act as decomposers.	3.7

- Tracing food chains in various environments helps students gain a deeper understanding of how interdependent organisms are on each other and their ecosystems. By middle school, students have some understanding of this interdependence. However, this understanding is limited to relationships between two different organisms. Often, students think animals are dependent on humans for food and shelter. Some students have misconceptions about adaptations. They may think that all animals can change their food preference according to what is available when their normal food source disappears. They also may think that animals can change their body structure to accommodate changes in the environment. (Jungwirth, 1975; Clough and Wood-Robinson, 1985a.)

- Aquatic macroinvertebrates are common in river systems. They are linked to each other and to the abiotic features of their environment. When students make food webs, they begin to understand the complexity of life in an ecosystem. Many students may think that only vertebrates are animals (Mintzes et al., 1991). Some students have difficulty understanding hierarchical classification schemes and may not, for example, think of birds as animals (Bell, 1981). Some students may not understand that grass, vegetables, and some trees are plants. (Osbourne & Freyberg, 1985.)

- Many students think that plants get food from outside of themselves, rather than making their own food during photosynthesis. (Anderson et al., 1990.)

- Computer-based technology can extend learning by helping students perform cognitive tasks they might otherwise not be able to do in the classroom. (Salmon and Perkins, 1991.) Using software helps students to make predictions and explanations. (Linn and His, 2000.) Using technology in the classroom enables the teacher to become more of a guide and collaborator rather than the expert resource of all information. (Dwyer, 1994.)

Understanding for Teachers

Concepts in this *Learning Set* bring students closer to understanding the complex relationships that exist in an ecosystem. Students learn about some real organisms that might be present in the Crystal River near the fictional town of Wamego. This gives them a context for seeing the effects of human intervention in the case study of the Marry Martans River Mystery. Students also begin to understand that in any ecosystem, matter and energy are being transferred as one organism feeds on another. Through the use of the *NetLogo* software, they also begin to appreciate that when a food chain or web is interrupted because a specific organism has not survived, the ecosystem becomes unbalanced.

$8\frac{1}{2}$ *class periods* * ▶

LEARNING SET 3 IMPLEMENTATION

Learning Set 3

How Can Changes in Water Quality Affect the Living Things in an Ecosystem?

10 min.

Students begin to think more specifically about what interacts within an ecosystem. They focus more on the role and importance of living things within a watershed.

Learning Set 3

How Can Changes in Water Quality Affect the Living Things in an Ecosystem?

ecosystem: all the living things in a given place, along with the nonliving environment.

interaction: a kind of action in which two or more organisms have an effect on each other.

You have studied watersheds and water quality. You modeled how water can affect the land over which it flows. You also modeled how a river is affected by changes in land use. Changes in land use affect how water flows in the watershed. They also affect the quality of the water.

Factories might use water from a river. They may return it to the river at a higher temperature. Runoff from farms and lawns can increase the amount of phosphates and nitrates in the river. You read about how this affects the quality of the water. The quality of water, in turn, can affect the organisms that live in and around it.

A watershed, the land, the water flowing through it, and the plants and animals living in it are all part of what scientists call an **ecosystem**. An ecosystem is made up of both nonliving and living components. In an ecosystem, the living things have **interactions** with one another and with the nonliving things.

Human activities, water quality, and the types of organisms that live in the watershed are all connected. When you understand the interactions among living and nonliving things in an ecosystem, you can better measure the effects your actions might have on the ecology of the community.

In the first two *Learning Sets*, you investigated the nonliving parts of an ecosystem—watersheds and how they change. In this *Learning Set*, you will look more closely at the living things in an ecosystem. You will answer the question *How can changes in water quality affect the living things in an ecosystem?*

What do you think this otter depends on to live?

LT 83

LIVING TOGETHER

○ **Engage**

To begin to build a more complete picture of the concept of an ecosystem, ask students to tell you what they have interacted with thus far today. Record their ideas. Then, ask students to classify each of the listed items as living (L) or nonliving (NL). At the end, explain that everyone exists within a system made up of living and nonliving things. Explain that this system is called an *ecosystem.* Make sure students understand that ecosystems are not limited to an idyllic woods or a local stream. Get students to identify themselves as part of many different ecosystems.

3.0

TEACHER TALK

"Each day, no matter where you are or what you are doing, you are an active and important part of many ecosystems. Ecology isn't just about being out in the woods or at a lake. What are some of the settings you find yourself in each day? (Whether you are riding on the school bus, walking down the street, sitting in a classroom, or sleeping in your bed, you are a part of an ecosystem.) What are some of the living and non-living parts of each of these settings?"

△ Guide and Assess

As students describe various possible ecosystems, ask them why their example qualifies as an ecosystem. Make sure you hear them include both living and nonliving parts. If they mention only animals, ask them what forms of plant life there might be for the nonliving parts. At first, students often forget that air and soil are important nonliving parts of ecosystems. A few students may mention bacteria and fungi, which are important living parts of almost every ecosystem, but not always easily recognized.

△ Guide

Introduce students to the *Big Question* for this *Learning Set: How can changes in water quality affect the living things in an ecosystem?* Refocus students' attention to water as an important nonliving factor in ecosystems and listen for them to connect what they have learned in *Learning Sets 1* and *2* with this discussion on interactions between living and non-living things.

NOTES

....

....

....

....

....

....

....

NOTES

SECTION 3.1 INTRODUCTION

3.1 Understand the Question

Thinking about Ecosystems

◀ *1 class period**

Overview

Students are introduced to the terms *abiotic* and *biotic* in place of "nonliving" and "living" parts of an ecosystem. They also use the term *habitat* in place of "environment" and begin to think about organisms living together and interacting in a "community." The class updates the *Project Board*, focusing on what they would like to learn about the biotic portion of an aquatic ecosystem.

Targeted Concepts, Skills, and Nature of Science	Performance Expectations
Biotic and abiotic components interact in an ecosystem.	Students should be able to say that living organisms *(biotic components)* interact with or depend upon items such as air, water, soil, oxygen, light, temperature, and nutrients *(abiotic components)* in an ecosystem.

Materials	
1 per class	Class *Project Board*

Homework Options

Reflection

- **Science Content:** Describe an interaction that might occur between living and nonliving components in your environment. What evidence do you have that the interaction had taken place? *(Answers will vary but should include a living organism and a nonliving factor. Examples might include a bird drinking from a bird bath; a dog gulping water from a water dish; a bird or insect flying through the air; a fish swimming in a stream; a tree bending in the wind; or a plant turning in the direction of sunlight. Be sure that students describe evidence of the interaction.)*

*A class period is considered to be one 40 to 50 minute class.

- **Science Content:** Describe an interaction between two or more living components of an ecosystem. What evidence tells you that the interaction has taken place? *(Answers will vary but should include two living organisms interacting in some way. Examples might include one bird feeding another, a dog or cat moving her offspring from one place to another, an insect sipping nectar from a flower, an insect biting a human or annoying a dog, an insect larva feeding on the leaves of a plant, or a wild animal feeding on the flesh of another animal. In each instance, ask students to describe evidence that the interaction took place.)*

Preparation for 3.2

- **Science Content:** Describe the relationships among water plants, fish, and insects that live near the water. *(Answers will vary but may include that some fish live among certain water plants; that the plants provide some of the dissolved oxygen on which the fish depend, and that the insects may be a source of food for some of the fish.)*

- **Science Content:** What are three organisms that live near water and would have some sort of relationship that depends on water? *(Answers will vary but may include fish, water plants, frogs, insects, trees, raccoons, and others. Accept all reasonable responses.)*

NOTES

....

....

....

....

....

....

....

....

SECTION 3.1 IMPLEMENTATION

◀ *1 class period**

3.1 Understand the Question

Thinking about Ecosystems

abiotic: nonliving parts of an ecosystem.

biotic: living parts of an ecosystem.

habitat: the place where an organism lives and grows naturally.

community: groups of organisms living together in a certain area. The organisms interact and depend on one another for survival.

biome: a community of plants and animals living together in a certain kind of climate.

Science Connection

A group of ecosystems that have the same climate and similar communities is called a **biome.** You can learn more about biomes if you turn to the end of this *Learning Set.*

An ecosystem is made up of both nonliving and living parts. The nonliving parts are called **abiotic** components. They include things like water, soil, oxygen, temperature, light, and chemicals. The living parts are called **biotic** components. They include organisms like plants and animals.

The environment around a river can change a great deal from stream to stream. It can also change in different locations along the same river. As the environment, or **habitat**, changes, the types of organisms that live there also change. Scientists call a group of organisms living together a **community**. To understand how organisms interact with their environment and one another, scientists collect and organize lots of information about the organisms in a community and the environment in which they live.

Update the *Project Board*

Your class started a *Project Board* to help you keep track of your investigations and questions regarding water quality in a community. At the end of *Learning Set 1,* you updated the *Project Board* with information about water moving in a watershed. After you completed *Learning Set 2,* you added the results of your investigations on water quality.

Consider what you might like to know about the biotic parts of the aquatic ecosystem you have been talking about. What are some ideas you have about the types of organisms that live in the aquatic ecosystem? Discuss

LT 84

Project-Based Inquiry Science

3.1 Understand the Question

Thinking about Ecosystems

10 min.

Students make a connection in the relationship between living and nonliving components of an ecosystem.

○ **Engage**

As an example of what students already know about the interactions of biotic and abiotic factors in the environment, have students recall the relationship described in *Sections 2.4* and *2.5* between water temperature and the narrow range in which trout thrive. To begin to get students to think about their own habitats, ask questions that connect things, such as air, soil, and water, to their lives. Students may cite examples of needing to breathe air that contains oxygen, using soil to grow food, and that living organisms need water to survive.

*A class period is considered to be one 40 to 50 minute class.

❝I want you to think about the importance of relationships between biotic and abiotic factors that you already know about. For instance, what do you recall about water temperature and the trout in the Crystal River flowing by Wamego? Let's list some abiotic factors that you depend upon. What are two abiotic components that you need each day? *(Oxygen and water.)* What abiotic factors enable food to grow?**❞** *(Soil and water.)*

△ Guide

Guide students to think about various types of habitats. They might first think about a house or apartment. To get them to think beyond the obvious, remind them that a *habitat* is a place or environment where interactions take place between biotic and abiotic components. Ask them again to think of other places where interactions take place. Record their suggestions. In each case, ask them to justify or provide reasons for their choices. *(Their reasons should include that there are some abiotic factors on which the biotic factor depends. Some examples of habitats might be a squirrel's nest in a tree, a bird's nest, a bee colony, a leaf eaten by insect larvae, fish in a stream, a hole in the ground where a mole lives, wood in which termites live, a rotting vegetable or fruit where fungi or bacteria are living, a garden, a lawn, a wool sweater eaten by moth larvae, rotting food in a refrigerator, or a rotting log.)*

Update the Project Board

10 min.

Lead students in updating the Project Board *with information they have learned about how living and nonliving components of an ecosystem interact.*

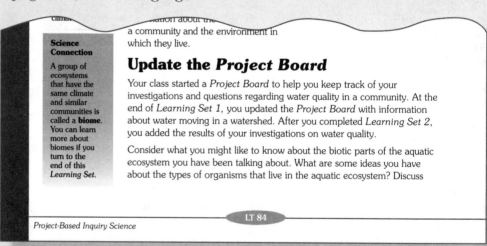

Science Connection

A group of ecosystems that have the same climate and similar communities is called a **biome**. You can learn more about biomes if you turn to the end of this *Learning Set.*

...climate...mation about the...a community and the environment in which they live.

Update the *Project Board*

Your class started a *Project Board* to help you keep track of your investigations and questions regarding water quality in a community. At the end of *Learning Set 1,* you updated the *Project Board* with information about water moving in a watershed. After you completed *Learning Set 2,* you added the results of your investigations on water quality.

Consider what you might like to know about the biotic parts of the aquatic ecosystem you have been talking about. What are some ideas you have about the types of organisms that live in the aquatic ecosystem? Discuss

LT 84

Project-Based Inquiry Science

Review the kinds of materials to be included in the first two columns of the *Project Board.* Column 1 is what students think they know. Column 2 is reserved for questions students think they would like to investigate. Although students updated the *Project Board* as they completed *Learning Set 2,* new information about biotic and abiotic factors, and habitats, may have sparked new ideas they would like to include and investigate.

what you think you know about how living things interact in nature. Maybe you have some ideas already about how living things can be impacted by water quality and how those impacts could affect other organisms in an ecosystem. Your teacher will help your class to discuss your ideas and questions and then record them on the *Project Board*.

What's the Point?

In this section, you thought about ecosystems. An ecosystem includes all the living and nonliving parts of an environment. You also considered what you might think you know about the types of living things in an aquatic environment. You have updated the *Project Board* to record the questions you have and would like to investigate. Your class might have a lot of ideas and questions about how water quality can affect living things. Once all the different ideas are recorded, your class can pursue investigations that focus on these ideas and questions.

As students prepare to make suggestions for the *Project Board*, remind students that their goal for this *Learning Set* is to answer the question: *How can changes in water quality affect the living things in an ecosystem?*

Record what students say they think they know in the first column.

Guide students' ideas about questions to investigate. You are preparing students to work on a case study in *Section 3.2* that involves looking at the effects of fertilizer on the numbers and variety of aquatic organisms at different points along a river. This is in preparation for answering the *Big Question* for the town of Wamego.

"What are some things that you could investigate about interactions that occur in nature between organisms and their environments? You have talked about the reaction that some fish have to water temperatures— how do other organisms that live in the water react to changes in water quality? What might you have to observe that would provide evidence that an interaction occurred among the organisms in the water?**"**

Record suggestions that students make in Column 2 of the *Project Board*.

Assessment Options

Targeted Concepts, Skills, and Nature of Science	How do I know if students got it?
Biotic and abiotic components interact in an ecosystem.	**ASK:** What are some abiotic resources in an ecosystem upon which living things depend?
	LISTEN: Students should be able to say that in most ecosystems, biotic parts of an ecosystem depend upon having water, oxygen, and food resources.

Teacher Reflection Questions

- What evidence do I have after this lesson that students understand that an ecosystem is made up of living and nonliving things that interact with each other and depend on each other?

- How well did I use the *Project Board* to focus students on the *Big Question?* What other questions can I ask to guide students in filling in Column 2 of the *Project Board?*

- What can I do the next time to involve more students in a class discussion?

SECTION 3.2 INTRODUCTION

3.2 Investigate

Classifying Macroinvertebrates

◄ *1 class period* *

Overview

Students are introduced to the concept of classification. Through classification, students come to understand something about the diversity and abundance of aquatic organisms. Students watch a video about how aquatic organisms are collected. The video demonstrates the processes and tools that scientists use to collect organisms from a river setting. The collected samples are grouped according to common physical characteristics and then counted. These actions determine the diversity and abundance of the sampled organisms. Students will then classify select macroinvertebrates with the help of a simple dichotomous key. Using another key provided by their teacher, students will be able to name their organisms. Students read about the importance of cells and how scientists classify organisms into kingdoms, phyla, orders, families, genus, and species.

*A class period is considered to be one 40 to 50 minute class.

Targeted Concepts, Skills, and Nature of Science	Performance Expectations
Scientists classify organisms based on criteria including physical appearance, feeding relationships, and relationships to the environment.	Students should be able to say that organisms are often classified by their physical characteristics.
Scientists collect, organize, and identify organisms from the field using a variety of tools.	Students should be able to relate that when collecting organisms in the field, scientists might use a variety of tools including nets and buckets.
Scientists use tools such as the dichotomous key to classify and identify different organisms.	When asked what a scientist would use to identify a specific organism, students should say that a dichotomous key would be used.
Living things are made of cells, obtain energy from their environment, grow and develop, reproduce, and respond to changes in their environment.	When asked what the characteristics of living things are, students should be able to reply that living things are made of cells, obtain energy from their environment, grow and develop, reproduce, and respond to changes in their environment.

Materials	
1 per classroom	*Far Mill River* sampling DVD
1 per classroom	Video player
1 set per group	Macroinvertebrate classification card set
1 per classroom	Giant water bug

Activity Setup and Preparation

Practice identifying each type of specimen before giving them to students. This will help when students begin to wonder what structures they are looking at in the organism to make an identification because you will have already encountered similar problems and have been able to solve them.

Homework Options

Reflection

- **Science Process:** How does a dichotomous key help to identify organisms? *(Answers will vary but should include that a dichotomous key assists in identifying organisms by grouping organisms according to common physical characteristics and eliminating some choices because their physical characteristics differ from one another.)*

- **Nature of Science:** A scientist in Italy and a scientist in China are both working to find a cure for a deadly disease caused by a certain kind of bacteria. Why would both scientists have someone check the identification of the bacteria being used for the research in their individual laboratories? *(Answers will vary but should include that if they are to discuss their work on the disease, they must be certain of the identity of the organism and that they are both working with the same organism. Other answers might include that the two scientists might eventually put their work together and they would not have to repeat tests already done.)*

Preparation for 3.3

- **Science Content:** Scientists want to check on the effect of fertilizer runoff from two farms on the abundance and diversity of organisms in a local river. What two factors will they have to check on in their research? *(Answers might vary but should include that the scientists should check on the amount of fertilizer running into the river and the kinds and numbers of organisms found in the river at each sampling point.)*

- **Science Process:** What kinds of tools will scientists checking a river for macroinvertebrates have to use? *(Answers will probably include buckets, sieves, nets, and dichotomous keys.)*

NOTES

1 class period * ▶

3.2 Investigate

Classifying Macro-invertebrates

10 min.

Set the scene by introducing common examples of the diversity and abundance of organisms so that students can deduce the meanings of the new vocabulary terms when they are introduced.

SECTION 3.2 IMPLEMENTATION

3.2 Investigate

Classifying Macroinvertebrates

The presence of different kinds of animals can indicate a healthy ecosystem.

aquatic ecosystem: an ecosystem located in a body of water.

diversity: difference (in this case, the different types of animals).

abundance: the amount (in this case, the number of a type of animal).

classify: arrange or sort by categories.

macroinvertebrate: an organism that does not have a backbone and can be seen with the naked eye.

When scientists investigate the characteristics of an ecosystem, they look at both the living and nonliving things. Ecosystems that are located in a body of water are called **aquatic ecosystems**. Sometimes, scientists specifically study the different types of animals that live in the water. Consider some of the animals you know that live in an aquatic ecosystem.

When you have finished watching the video, discuss the video with your group. What different tools and techniques did the people use to collect the animals? What process did they use to classify the animals? Often, scientists are interested in finding out exactly what kind of animals live in streams. They want to identify how many different types of animals live in the water. This is called **diversity**. They also want to know how many of each type of animal is in the stream. This is called **abundance**.

To understand the diversity and abundance of different animals in the water, scientists have to sample the water. They pick a spot in the water and collect samples of water with the animals in it.

Then they pick all the animals they can see out of the water. They sort them, or **classify** them, based on how they look. Next, the scientists take each group of animals and count how many of each one they have found. They count the abundance of the animals.

One group of animals that live in the water is called the **macroinvertebrates.** Macro means that you can see it with your eyes. You don't need any tools like a microscope or hand lens to help you see it. Invertebrate means that the animal does not have a backbone. Common macroinvertebrates that live in the water include all kinds of insects, snails, and crustaceans such as crayfish.

LT 86

Project-Based Inquiry Science

○ Engage

Ask students to name types of organisms (*aquatic organisms*) that they might find living in or around a pond, lake, or river. Record their suggestions. *(Answers will vary but will probably include fish, frogs, birds, many different kinds of insects, raccoons, muskrats, turtles, salamanders, groundhogs, and snakes. The variety of organisms represents a diversity of organisms.)*

Then, ask students to estimate how many of each aquatic organism they might find in these places. Explain that an estimate might not be exactly the same number one gets if one counts every organism separately. *(An estimate of the number of aquatic organisms often varies with the size of*

---

*A class period is considered to be one 40 to 50 minute class.

the organism, so that one would estimate finding one muskrat in the area, and two or three raccoons or turtles, but one might estimate that there are a dozen fish, a dozen frogs, and hundreds of insects.)

Once students have thought about the numbers of possible members of each group, explain that these numbers tell you the abundance of each animal in the area.

△ Guide

Now that students have a reasonably clear idea of the terms *diversity* and *abundance*, you can begin to use these terms.

Go on to explain that another important factor is to classify and identify each organism by its specific name. Use the following explanation and example to focus attention on the importance of classification.

TEACHER TALK

"Classification of organisms is a very important part of any research. For instance, there are many kinds of mosquitoes in the world. You can use the word mosquito to describe any one of them, but that is just the insect's common name. Malaria is a disease that causes more than one million deaths each year worldwide. While there are few cases of malaria in the United States, it has occurred here and is carried by a mosquito named Anopheles quadrimaculatus. Another mosquito in the United States is called Aedes albopictus. This kind of mosquito carries the West Nile virus. Scientists are working on ways to cure both diseases. If you were a scientist working to cure malaria, why would it be important for you to know that the mosquitoes in your research project were all Anopheles quadrimaculatus and not a mixture of other types? How helpful would your research results be if you had a mixture of types of mosquitoes in your laboratory?"

NOTES

....

....

....

....

....

Procedure

30 min.

The goal of this procedure is for students to recognize that scientists group specimens according to how they look (physical characteristics) with the help of paired questions in a dichotomous key, rather than merely matching a specimen to an image of the organism.

Observe

You are going to watch a video of scientists. They are collecting river-water samples to classify and count the animals that live in the water. As you watch the video, think about the types of animals they are trying to collect. Look at the kinds of tools they are using. Think about why they use those tools to collect the animals.

Procedure

1. In this investigation, you will be classifying macroinvertebrates by using pictures of the animals. The macroinvertebrates shown in the pictures are similar to those collected by the scientists in the video. Just like the scientists in the video, you will use a **dichotomous key** to help you classify the animals based on how they look. Your goal is to put animals into groups by similar characteristics. These groups are usually very small, containing only one or two animals in a group. To use the key accurately, you will have to observe the pictures carefully and answer questions about several characteristics of the animal.

2. Begin with one macroinvertebrate picture. In your group, observe the characteristics of this animal. Look carefully at its body parts. Compare the body parts to those listed on the key. For example, look to see if the animal

Materials
• macroinvertebrate classification cards

dichotomous key: a key used to identify living things.

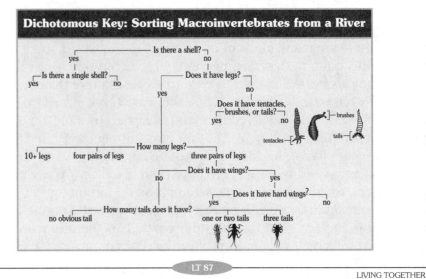

Dichotomous Key: Sorting Macroinvertebrates from a River

LT 87

LIVING TOGETHER

○ Engage

Provide students with the big picture of what they are going to do during this investigation:

- Step 1: A video will show students how aquatic specimens are collected and the tools scientists use to collect them. Stop the video at 4 minutes, 53 seconds before they show the classification chart.

- Step 2: Students will be given specimens to sort using the *Sorting Macroinvertebrates from a River* key in their text (*page 87*). Students track information they observe about each specimen with an *Identifying Macroinvertebrates* chart (*page 88*) for their specimens.

- Step 3: Use a photo and name key to name each specimen and write the name on the *Identifying Macroinvertebrates* chart (*page 88*).
- Steps 4 and 5: Students will be given another organism to classify and name, using keys they have already practiced with.

⬡ Get Going

Get students into groups of three and distribute tangrams, one assembled and one unassembled, to each group. Assign half of the groups in the class to solve the puzzle without words and half of the groups in the class to solve the puzzle with words. Make sure students understand which set of directions they should use, and make sure each student in each group understands what role he or she has in the group.

Once groups are ready, with tangrams in place and pencils and paper in the observers' hands, have them get started. Remind observers to record the time when the puzzle solvers in their group finish.

TEACHER TALK

❝To start this investigation, we will watch a short video of scientists collecting *aquatic macroinvertebrates.* What kinds of tools would you use if you were scooping things from a pond or river? Watch carefully for the kinds of tools the scientists use to collect their specimens. **❞**

Start the video but stop it at 4 minutes, 53 seconds, before the classification chart is shown.

△ Guide and Assess

Facilitate a class discussion by asking "What tools did you observe the scientists using to collect specimens?"

Focus students' attention on the tools the scientists used. Ask why collecting specimens from water would require these kinds of tools.

△ Guide

To get students thinking about what a scientist might do next and to connect with the concepts of abundance and variety, ask students to put themselves in the place of the scientist.

TEACHER TALK

❝If you had just collected organisms from a pond or river, what do you think you would do next? What would you do when you got back to your laboratory?**❞** *(Answers might vary but students might say that they would sort through what they had collected to see what variety of organisms they have (diversity) and how much (abundance) of each they have. Then, they would begin to classify the specimens.)*

LIVING TOGETHER

Explain that students are going to get the opportunity to classify similar specimens. To introduce the activity, first demonstrate the procedure to the class. Give each group a card with a picture of a macroinvertebrate. Make sure all groups get the same picture. Guide students, in a class demonstration, to look at the parts of the body parts of specimens. Introduce students to the *Dichotomous Key: Sorting Macroinvertebrates from a River* table in the student text by calling attention to the questions and the choices on the key. Then, explain what students are to do at each set of questions.

Model the classification of one organism by asking each question and showing students how to work through the *Dichotomous Key: Sorting Macroinvertebrates from a River* key and how to organize and construct a chart that classifies and identifies characteristics of macroinvertebrates. You may want the students to record a picture or a drawing of their macroinvertebrate. The chart needs to contain a space for students to record the macroinvertebrate's characteristics and name.

TEACHER TALK

"Let's pretend that you have just come back from a collecting trip like the scientists in the video and you want to begin to classify the specimens in your collecting jars. You can do this by using the questions in the *Dichotomous Key: Sorting Macroinvertebrates from a River* in your textbook. This series of questions is called a key. Why do you think it is called a key?" *(Responses will vary but some students may equate it with a key that unlocks or opens a door to reveal something or solves a puzzle.)*

Always begin by answering the question at the top first. The first question is "Is there a shell?" Look at your specimen. Does it have a shell? If so, your answer to the first question is "yes." If it does not have a shell, then your answer to the first question is "no" and you proceed to the question that asks, "Does it have legs?" Observe your specimen and answer that question. Then proceed to the next question. Don't skip any questions unless one of the choices tells you to do so. Record the characteristic you find at each question on your *Identifying Macroinvertebrates* chart."

⬡ Get Going

After guiding students, let each group work at tracking a different specimen through the *Dichotomous Key*. Walk around the room, observing whether individual students are able to use the key effectively and identify characteristics on their specimens. If anyone is having trouble, recall any difficulties you had during your preparation for this exercise when you checked out each specimen. Some students may have trouble understanding the difference between tentacles and legs or counting the numbers of legs.

has a shell or tentacles, or count the number of legs. See which body parts you can identify.

3. Begin at the first question of the key, "Is there a shell?" Answer that question yes or no, and follow the correct line to the next question. For example, if your animal doesn't have a shell, you need to decide next if it has legs. Continue reading and answering the questions in order until you have come to the end of the questions for your animal.

4. Make a table for the first animal you classified. Record the answer to each question on the key. Make a space on your table for each characteristic of the animal and a space for your answer. Also make a space for the name of the organism; you will record that later.

5. Practice classifying organisms by using the key for two or three more animals. If you are confused about the animals' characteristics, discuss your ideas in your group. Using a key like this is not easy. You might have debates about where an animal belongs. If you cannot decide on the answer to one of the questions, record that in your table.

Procedure

You have not yet named the macroinvertebrates you were classifying. You will now use another key to help you classify and name your organisms. This key is similar to the first one, but you will end up with a name for each macroinvertebrate rather than just a classification. For each organism, follow the key until you are able to name the organism. On your table, record the name of each macroinvertebrate.

Reflect

Classifying animals using pictures can be difficult. What did you find to be difficult? How did your group make decisions when you ran into difficulties?

Communicate

Select one of the macroinvertebrates you have classified and named. Share this macroinvertebrate with your class. As you describe your animal, make sure you identify how you determined the answer to each of the questions on the key. As you listen to other's descriptions, look carefully at the characteristics and determine if you would have answered each question the same way they did. Be sure you understand why each group classified their macroinvertebrate the way they did. If you do not understand something or think they classified their organism incorrectly, be sure to ask questions and make suggestions. Focus on the characteristics of the organism and what you see in each picture to support your ideas.

LT 88

Project-Based Inquiry Science

Field guides are different from dichotomous keys. A *field guide* shows photographs or drawings of individual organisms, often on a regional basis. *(Insects of Eastern United States; Birds of Western United States.)* Field guides usually describe where an organism lives, what its habitat is like, what it feeds on, and the basics of its life stages and its scientific *(Latin)* name. *Dichotomous or analytical keys* contain highly detailed illustrations of the smallest parts of organisms and a series of pairs of numbered choices that eventually lead to the scientific name of organisms. Field guides rarely contain keys.

Guide

Now that students have their samples to work on, make sure that they are using the *Dichotomous Key* properly, as demonstrated by you earlier. Walk around the room as they work to ensure that students are following the flow of *Dichotomous Key* and that students are using the pictures and drawings to correctly identify their specimen.

Remind students to fill in their charts as they identify and classify their specimen.

Scientific names consist of, for the most part, two Latinized words. Use of the Latin name enables the organism to be universally recognized by all scientists, no matter what language they speak. The two parts of the scientific name, genus and specific epithet, often stand for a characteristic of the organism, the person who is recognized as having discovered the organism, or the place where the organism was first discovered.

☐ **Assess**

Monitor students' group work and listen for progress with the naming procedures. Have students describe specific questions within the key that they find difficult to answer.

Ask individual groups to demonstrate how they classified a specific specimen.

Have each group share how they identified and named their organism. Ask students why the organism is different from other specimens. Students should base their reasons on divergences in the key.

◇ **Evaluate**

Steps 4 and 5: Now that students have had experience using a key to identify organisms, provide each group with an unknown specimen to classify and name using the methods they learned earlier.

NOTES

....

....

....

....

....

....

....

....

....

....

Investigate

Why Classify Living Organisms?

Scientists have determined that all living things on Earth, past and present, share the following characteristics:

- They are made of units called cells.
- They get energy from the environment.
- They grow and develop.
- They reproduce.
- They respond to changes in the environment.

Living things are also different from one another. The **cell** is the basic structure and function of living things. Some living things are made up of only one cell. That single cell can do everything it needs to stay alive. Some living things are made up of many cells. You are made up of about 100 trillion cells!

Living things are different in other ways as well. Some living things, such as plants, can make their own food. Plants can also be divided into many smaller groups. For example, some plants have flowers.

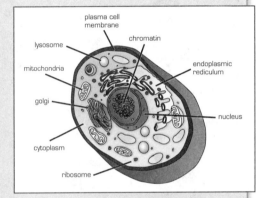

cell: the structural and functional unit of all living organisms. It is sometimes called the "building block" of life.

taxonomist: a scientist who classifies organisms by characteristics.

species: a group of organisms that look alike and can breed with other members of the group and produce fertile offspring.

Classifying living things takes a lot of time and hard work. Scientists called **taxonomists** create systems for classification. They separate the millions of **species** of living things on Earth based on their differences and put them into groups. The systems they create help scientists communicate with one another.

Why Classify Living Organisms?

10 min.

Reading technical text is a very different skill than reading narrative text. Have students read aloud and stop after each paragraph to summarize what they have read. The focus should be on developing an understanding of the content.

○ Engage

Before having students look at the content on page 89 in the student text, explain that the content on the page will expand their science knowledge about living organisms as it explains the basic structure common to all living organisms, namely, the cell and the characteristics of living organisms. Emphasize that the characteristics listed (*made of cells, energy from the environment, grow and develop, reproduce in some way, and respond to changes*) are shared in some form by all living things.

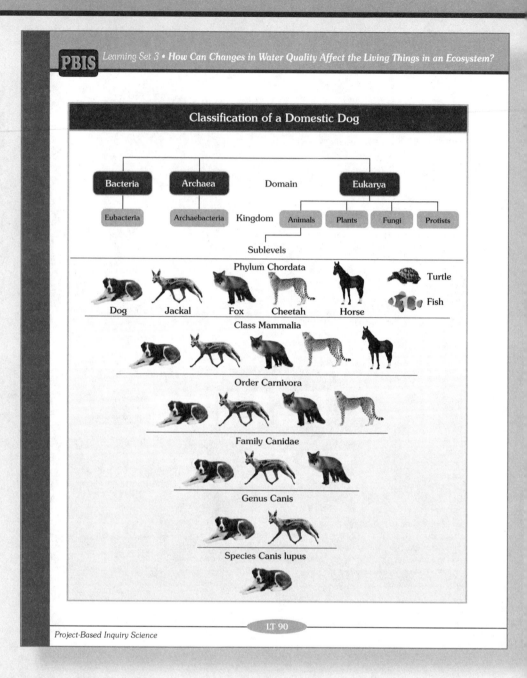

Classification of a Domestic Dog

△ Guide

Draw students' attention to the *Classification of a Domestic Dog* chart on page 90 in the student text. Point out the major groups at the top of the page. Explain that this chart represents, in abbreviated form, the classification of all known living things. The classification of the domestic dog from Domain through Species is featured, but explain that every known living organism can be classified in the categories they see along the left margin—Domain to Kingdom to Phylum through Species.

What's the Point?

0 to 5 min.

The chart on the previous page helps you see how animals are classified. At the top of the chart are large categories. For example, all animals are in the Kingdom Animalia. There are about 1.5 million species of animals on Earth and all have some characteristics in common. The characteristics of animals make them different from plants or bacteria.

These 1.5 million animal species are then divided further based on having a backbone (chordata). There are only about 50,000 species of chordates—most animals do not have a backbone. You will see that each category has fewer animal species than the one above it. The animals in the last rows are very similar to one another. At the bottom of the chart, only the dog is in the Species category, *Canis lupus*.

The number of Kingdoms into which organisms are classified has varied over time. Scientists are constantly finding new living organisms that they didn't know existed. Because of this, the classification methods they are using to order the groups of animals are changing.

What's the Point?

Scientists attempt to bring order to the many different organisms that live on Earth. To do this, they sort, or classify, living things based on their characteristics. Often, physical characteristics are used to classify animals. In the activity in this section, you observed animals and sorted them based on their physical characteristics. Scientists often use tools like a dichotomous key to assist them in classifying different organisms. Taxonomists continue to update these classification systems because new species are being found all the time. Even humans fit into the classification system of living things.

What characteristics might you use to classify this dragonfly?

Assessment Options

Targeted Concepts, Skills, and Nature of Science	How do I know if students got it?
Scientists classify organisms based on criteria including physical appearance, feeding relationships, and relationships to the environment.	**ASK:** What is the most common system scientists use to classify organisms? **LISTEN:** Students should be able to answer that scientists classify organisms by their physical characteristics.
Scientists collect, organize, and identify organisms from the field using a variety of tools.	**ASK:** What are some of the tools that scientists might use to collect aquatic invertebrates from a pool or stream? **LISTEN:** Students should recall the tools that they saw used in the video, which may have included but were not limited to buckets, sieves, and collecting jars with lids, labeled by location where collected.
Scientists use tools such as the dichotomous key to classify and identify different organisms.	**ASK:** What does a dichotomous key require you to do in order to identify an organism? **LISTEN:** Students should be able to say that in a dichotomous key, there are a series of questions. Choices are made in how to answer each question based on the physical characteristics of the organism.
Living things are made of cells, obtain energy from their environment, grow and develop, reproduce, and respond to changes in their environment.	**ASK:** What characteristics are shared by all living things on Earth? **LISTEN:** Students should be able to relate that living things are made of cells, obtain energy from their environment, grow and develop, reproduce, and respond to changes in their environment.

Teacher Reflection Questions

- One goal of this section was to have students learn the importance of classification. Do students understand the importance of having a system that lets scientists in separate facilities know they can be doing research with the same species of organism?

- Did students understand how the dichotomous key worked, having to make choices between two aspects of the same characteristic?

- In this section, students used preserved specimens. How did you work with students to help them make their observations and work like scientists?

NOTES

....

....

....

....

....

....

....

....

....

....

....

3.3 Explore

1 class period* ▶

The Marry Martans River Mystery: Macroinvertebrates in an Ecosystem

Overview

Students use the skills and science knowledge that they learned about collecting and classifying macroinvertebrates in *Section 3.2* to solve a mystery case study. They explore why different parts of a river have different types of macroinvertebrates and learn that the presence or absence of these organisms can indicate something about water quality. Students analyze data to find out which of three farms may be a source of excessive fertilizer pollution for a local river. They compare their analysis with those of an ecologist and an EPA (*Environmental Protection Agency*) expert.

*A class period is considered to be one 40 to 50 minute class.

Targeted Concepts, Skills, and Nature of Science	Performance Expectations
Some species of macroinvertebrates can indicate water quality.	Students should be able to say that the presence or absence of some macroinvertebrates might indicate problems in water quality.
Scientists can determine water quality using biotic indicators.	Students should be able to say that the presence or absence of certain biotic components can be used to determine water quality.
Biotic indicators may help determine a pollution source.	When asked how a pollution source can be determined, students should be able to say that some biotic indicators may help to determine a pollution source.
Human activity can affect the ecology of a community. Humans use rivers for residential, commercial, industrial, and agricultural purposes. These activities affect water quality along a river.	When asked what factors can affect the ecology of a community, students should be able to say that human activity can affect the ecology of a community.

Materials	
1 per class	Optional: projections of images and tables from the student text.

Activity Setup and Preparation

Study the details of the Marry Martans River Mystery thoroughly so that you can tell it as a story to the class before allowing them to read it in their textbooks.

Homework Options

Reflection

- **Science Process:** What pieces of information were important in determining the source of pollution in Marry Martans River and Lake? *(Students answers might vary but should include the diversity and abundance of the macroinvertebrates collected from the water and the honest description of the amount of fertilizer used by each farm was important for determining the source of pollution.)*

- **Science Content:** In the Marry Martans River Mystery, explain how the macroinvertebrates became indicators of water quality problems. Include the terms *diversity* and *abundance* in your explanation. *(Answers will vary but students should include that some macroinvertebrates were sensitive to changes in their environment and did not tolerate changes in the water quality. When the usual diversity and abundance of macroinvertebrates changes, a source of pollution nearby or upstream is indicated.)*

Preparation for 3.4

- **Science Content:** What do plants need to photosynthesize? *(Plants need a source of light, preferably sunlight. They also need carbon dioxide, water, and nutrients.)*

- **Science Content:** What substance in the cells of plant leaves enables photosynthesis to take place? *(Students should respond that chlorophyll is the substance in the cells of plant leaves that enables photosynthesis to take place.)*

1 class period* ▶

3.3 Explore

The Marry Martans River Mystery: Macro-invertebrates in an Ecosystem

Examine a Case Study

20 min.

Telling the story of the Marry Martans River informs students of a real problem and provides them with data collected by ecologists for analysis. The Marry Martans River Mystery case study utilizes skills and information that students learned in Section 3.2 and engages students by connecting to prior experiences.

META NOTES

Projecting the image of the Marry Martans River and Lake will help students follow the details of the story as you tell it.

*A class period is considered to be one 40 to 50 minute class.

SECTION 3.3 IMPLEMENTATION

3.3 Explore

The Marry Martans River Mystery: Macroinvertebrates in an Ecosystem

ecologist: a scientist who studies the relationships between organisms and their environment.

You watched a video of scientists collecting macroinvertebrates. You should now have a good sense of how scientists organize and classify macroinvertebrates. Once scientists identify macroinvertebrates in an ecosystem, they can use this information to better understand the conditions in an ecosystem.

You also learned about diversity and abundance. Recall that diversity refers to the types of organisms found in an environment. Abundance refers to the number of each type. In this activity, you will examine the diversity of macroinvertebrates in an area. You will see how diversity can indicate water quality and ecosystem health. You will be working with some macroinvertebrate data collected by an **ecologist**. The ecologist has been asked to help the residents of a small community solve a mystery. What you learn from this case study will help you address this *Learning Set*'s question.

Examine a Case Study

A group of residents live on a small lake called Marry Martans Lake. The Marry Martans River flows into the lake at one end. The lake drains back into the river at the other end. (See the picture on the next page.) Over the past few months, the residents have noticed a lot of algae growing in the lake. The young people in the community know about water-quality indicators from their science classes. They remember that sudden algae and plant growth could be a sign of high amounts of fertilizer running off into the river.

The young people and their parents decide to investigate the case. Where might the fertilizer be coming from? They discover that there are three farms upriver. These farms are upstream from the lake and border the river. They wonder if fertilizer runoff from the farms is causing the problem. The residents discuss this with the farmers. Each of the three farmers denies that they have a fertilizer-runoff problem.

Project-Based Inquiry Science

LT 92

◯ Engage

Begin this section by projecting an image of the Marry Martans River and Lake system from page 93 in the student text and by having students close their textbooks. Then, tell the case study of the Marry Martans River Mystery as a story to the class. Make certain that as you tell the story, you include the details of the story in order.

Once you have told the story, ask students to think back to some things that they learned in *Section 3.2*. You might ask, "What skills did you learn about in *Section 3.2* that might help in this story?" *(Classification of macroinvertebrates.)* Ask them how this knowledge could be important to answering the questions raised in the Marry Martans story.

Each farmer knows that the river and the land are important to the community. They also know that the land use is needed for their business. Each farm claims to have safeguards in place that are designed to make sure harmful runoff does not enter the river.

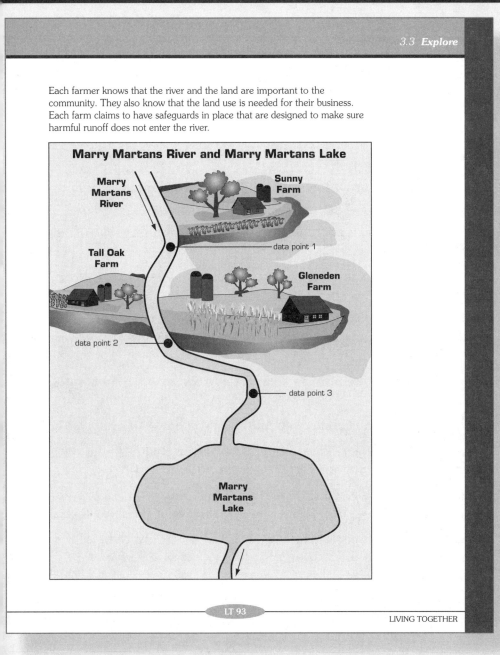

Marry Martans River and Marry Martans Lake

Marry Martans River

Sunny Farm

Tall Oak Farm

data point 1

Gleneden Farm

data point 2

data point 3

Marry Martans Lake

LT 93

LIVING TOGETHER

△ Guide

Students will need to know and use the details of the Marry Martans Mystery, so after you have told them the story, they can review it by reading the case study on their own. This will prepare students to summarize the story. Record the main points of their summary. Once students have summarized the problem, ask if anyone wants to add details that may have been missed.

△ Guide and Assess

If students have difficulty summarizing the story and picking out the important details, ask the questions on the next page to provide some structure and organization to their descriptions.

"What feeds the Marry Martans Lake? *(The river.)*

What change have people recently noticed in the lake? *(A lot of algae and plant growth.)*

What does sudden growth in algae and other plants often mean? *(Increased fertilizer is getting into the river from somewhere.)*

What is normally a source of fertilizer? *(Runoff from farm fertilizer applications.)*

Where are the farms in relation to the lake? *(The farms are upriver from the lake.)*

What did the residents discuss with the farmers? *(Whether the farmers used all the safeguards designed to make sure harmful runoff did not reach the river.)*

What safeguard do the farmers take? *(The residents have some question about this.)*

Who did the residents contact about their problem? *(An ecologist.)*

What does the ecologist know about macroinvertebrates in the river? *(There are twelve macroinvertebrates common to this river.)*

What does the ecologist decide to do? *(Collect macroinvertebrates from three locations along the river—one below each of the three farms.)*

What does the ecologist's team use to collect macroinvertebrates? *(Methods and tools [nets] similar to those used in the video that students watched.)*

What does the team use to identify the organisms they collect? *(A dichotomous key.)*"

The residents contact a local ecologist. They explain their problem. The ecologist knows that there are twelve macroinvertebrates commonly found in this river. Some of the macroinvertebrate populations can be harmed by pollution.

The ecologist's team decides to collect macroinvertebrates from the three locations in the river identified on the map on the previous page. The locations are upstream from the lake. Each data collection point is located at the end of the runoff zone for one of the farms. These points are the last places where runoff from each farm would enter the river.

The team uses nets to collect macroinvertebrates from each location. They use a method like the one you watched in the video. Then each macroinvertebrate is identified using the dichotomous key. The table shows the different macroinvertebrates the ecology team collected at each point.

Location	Macroinvertebrates Collected		
data point 1 (Sunny Farm)	• midge larva • blackfly larva • leech • lunged snail	• cranefly larvae • sow bug • crayfish • dragonfly larvae	• mayfly larvae • caddisfly larvae • gilled snail • riffle beetle
data point 2 (Tall Oak Farm)	• midge larva blackfly larva	• leech • lunged snail	
data point 3 (Gleneden Farm)	• midge larva • blackfly larva • leech • lunged snail	• cranefly larvae • sow bug • crayfish • dragonfly larvae	

Analyze the Data

See if you can figure out the mystery. Work with your group to analyze the data table and answer these questions. Later, your class will meet and discuss everyone's answers.

1. If polluted runoff could harm macroinvertebrates, which farms seem to be harming the macroinvertebrates? Support your answer with data and science knowledge.

2. Does the data support the claim by any of the farms that their pollution-control measures are working? Explain your answer with data and science knowledge.

Analyze Your Data
10 min.

Knowing that twelve macroinvertebrates are common to the Marry Martans Rivers gives students a baseline to analyze the specimens collected by the ecologist's team at the three data points.

META NOTES

When students analyze a situation, they study details to see if relationships and patterns exist among the data.

◇ **Evaluate**

Tell students that they will work in their groups to analyze the data in the table on page 94 in their text. To study the data, tell them that they are to discuss and answer the first four questions. Inform the groups that they will then discuss their answers as a class.

Evaluate students' ability to interpret the data table by listening to their responses to the first four questions. While individual groups may have different interpretations of some of the data, most responses to the first four questions should include the following:

1. If polluted water is harmful to macroinvertebrates, then it seems that there is harmful runoff from Tall Oak Farm and Gleneden Farm. There are far fewer types of macroinvertebrates at the Tall Oak Farm

META NOTES

Most of the organisms collected at the data points are immature stages of insects. Many immature insect stages need to develop in water. Therefore, if insect development is interrupted, populations of organisms common to an area begin to change. The ecology or balance of the area changes.

data collection point than at the other collection points. There are fewer macroinvertebrates at Gleneden Farm than at Sunny Farm, but this may be the result of the runoff from Tall Oak Farm.

2. The data supports the claim that Sunny Farm pollution control measures are working because the diversity of macroinvertebrates is high at the data point for this farm.

3. Macroinvertebrates common to all three data points are: Midge larvae, blackfly larvae, leech, and the lunged snail. If pollution is in the river, somehow it does not affect any of these four macroinvertebrates.

4. Macroinvertebrates found only at one data point are: Mayfly larvae, caddisfly larvae, gilled snail, and the riffle beetle. These were found only at the first data point (*Sunny Farm*). If pollution is in the river, it must occur further downstream from this point and kills these organisms because they are not present in the river water at data points 2 or 3.

NOTES

....

....

....

....

....

....

....

....

....

3.3 Explore

The substances that farmers use on their fields can end up in local bodies of water.

3. List the macroinvertebrate species that were common to all three data points. If pollution is in the river, what effect do you think the pollution has on this set of macroinvertebrates?

4. Which species of macroinvertebrates were found at only one data point? Which point was this? If pollution is in the river, what effect do you think the pollution has on this set of macroinvertebrates?

The ecologist reviewed the data and believes that there may be more pollution entering the river than the farmers know. The ecologist visits each of the three farms with an expert from the Environmental Protection Agency (EPA). The EPA is a government agency that oversees and regulates how our society interacts with the environment. The EPA expert knows a lot about farm pollution.

The ecologist and EPA expert inspect the pollution-control measures that each farm is taking. Their findings are shown in the table on the next page.

LT 95

LIVING TOGETHER

△ Guide

The analysis of the data is not yet complete. To prepare students for a class discussion of the results of the research carried out by the ecologist and other scientists, refocus students' attention on the Marry Martans River story. Ask them, "What does the ecologist do with the data collected on the macroinvertebrates?" *(The ecologist consults with an EPA scientist and checks the pollution control measures being employed by the three farms.)*

◇ Evaluate

In preparation of a class discussion, have students work in their groups to study the *Pollution Control Analysis* data *(on page 96)* obtained when the ecologist and EPA expert looked at the control measures being employed

META NOTES

The Environmental Protection Agency (EPA) functions to protect the human health and the health of the environment of the United States. The EPA employs trained scientists who provide expert advice on how to rehabilitate contaminated water resources.

Scientists often share data and call upon the expertise of other scientists who may have more specific knowledge of a particular area of study.

at each farm. Ask them, "Does the data show that runoff is affecting any of the organisms living in the river? Which farm, if any, was having trouble controlling harmful runoff that could pollute the river?"

Have students work in their groups to answer questions 5, 6, and 7 using the data in the table.

5. The data from the ecologist and the EPA expert reflects the results found by the ecologist's team when they collected macroinvertebrates at three data points. That means that scientists found problems with pollution control measures at Tall Oak Farm and Gleneden Farm.

6. The earlier macroinvertebrate data suggest that some macroinvertebrates were not collected at data point 2 because they could not live in polluted water.

7. The macroinvertebrates collected at data point 1 that were not collected at data points 2 and 3 are unable to tolerate the pollution emitted at data point 2. There appears to be a lower pollution level downstream at data point 3. It is unclear if the organisms at that data point are affected by the pollution that seems to have been contributed by Tall Oak Farm.

NOTES

....

....

....

....

....

....

....

....

Location	Macroinvertebrates Collected
Sunny Farm	Farm is adding very low levels of pollution to the river from runoff. In fact, the farm is below the limit allowed by the EPA.
Tall Oak Farm	Pollution-control measures are not working properly. Farm owner was unaware of gaps in the runoff system. Farm is contributing very high levels of pollution to the river from runoff. Runoff levels are above the limit allowed by the EPA by 200% (3 times what is allowed).
Gleneden Farm	Pollution-control measures are working for the most part. One gap in the system is a major source of runoff pollution. Runoff levels are above the limit allowed by the EPA by 15%.

5. Earlier, you decided which farms might be polluting the river. How does the data from the EPA report match up with the data collected at data points 1, 2, and 3?

6. What does the earlier macroinvertebrate data suggest about the ability of macroinvertebrates at data point 2 to live in polluted water?

7. Look at data point 1. There are several macroinvertebrates that only appear at data point 1, but not at data points 2 and 3. What does the data from the EPA suggest about the ability of those macroinvertebrates to live in polluted water?

Communicate Your Ideas

Once all of the groups have completed these questions, your teacher will lead you in a discussion. You will review the answers and ideas generated by your group. Your teacher will ask you to group the macroinvertebrates from this activity into one of three groups.

Tolerant	Macroinvertebrates that can survive in polluted water.
Mildly Tolerant	Macroinvertebrates that can sometimes survive in polluted water.
Intolerant	Macroinvertebrates that cannot survive in polluted water.

LT 96

Project-Based Inquiry Science

Communicate Your Ideas

10 min.

All students are encouraged to participate in class discussions, to ask questions and provide answers based on information from the data that has been collected.

△ Guide

Once groups have completed answering the analysis questions, ask them, as a class, to group the kinds of macroinvertebrates as *tolerant, mildly tolerant,* and *intolerant* based on descriptions in the table at the bottom of page 96. Discuss the meaning of each of these terms.

Record a public list of the students' groupings of macroinvertebrates according to tolerance. Ask students to defend their choice of the term for each group of organisms. Listen for students to use the following terms as they present justification for their grouping choices.

- **Tolerant:** Midge larvae, blackfly larvae, leech, and the lunged snail are classified as tolerant because they were observed at all three collection points and, therefore, seem to be able to survive in both polluted and non-polluted water.

META NOTES

Learning to use formal scientific terms will enhance students' presentations when they later advise the Wamego town council.

- **Mildly Tolerant:** Cranefly larvae, sow bugs, crayfish, and dragonfly larvae, are classed as mildly tolerant. They were found at the first data collection point where pollution controls were best and also at the third data collection point, which is considered to have some pollution runoff.

- **Intolerant:** Mayfly larvae, caddisfly larvae, gilled snail, and the riffle beetle are unable to tolerate pollution because they were found only at the first data collection point, before pollution entered the river.

△ Guide Discussions

Prepare students for a class discussion by telling them that all groups will listen while one group summarizes their conclusions about the study. Then, each group should be prepared to ask a question or make a comment about the data. Remind students that their questions, answers and comments should be based on the data that has been collected.

Remind students that when someone is talking, they are to listen politely. If a student wishes to ask a question, remind the student to direct his or her question to the student who has been making the presentation, not to you, the teacher.

Facilitate the class discussion by projecting all diagrams and data charts available or as handouts.

Initiate the discussion by asking guiding questions, such as, "Which farm seems to have had the least problem with pollution? What evidence do you have in the data that supports your claim?" Always listen for students to support their statements with evidence based on the data provided.

Encourage language such as, "I agree with...because..." or, "I disagree with ... because...." "Can you explain why ... because I don't understand what you mean." Explain that by utilizing polite, formal language during this discussion, students are being prepared to have their conclusions taken seriously when they present their suggestions for solutions to the *Big Challenge* faced by the Wamego town council.

Explore

Biotic Indicators

Biotic Indicators

In *3.2*, you read that one way to classify organisms is by their physical features. Those features evolved over very long periods of time. Animals have those specialized features because those features allowed previous generations of that animal to survive. These previous generations reproduced and had offspring with those features. Thus, the features keep being passed on to the next generation.

For example, certain macroinvertebrates have wings that allow them to avoid being easily captured by other organisms that try to eat them.

These macroinvertebrates pass this trait onto their offspring. The feature remains important in the organism. Other features might help animals avoid capture. They may also help them hunt food, better digest nutrients, camouflage the organism, or help them build a habitat (nest or den, for example). These features, or *survival tools*, help protect populations of organisms.

However, these protections or features are not a guarantee that an organism will avoid death. You probably are aware that, in nature, some organisms are the food source for other organisms. If you begin to see a decrease in the number of organisms in a population, then it could be a sign that the organism is being hunted and captured by other organisms in the ecosystem. You will learn more about this in a later section.

One other interpretation of a decrease in organism population is that the environment changed. In *Learning Set 2*, you read about how changes in water quality can be very harmful to the health of some organisms. Macroinvertebrates are one set of organisms that serve as a **biotic** (or living) **indicator** of water quality and health of the ecosystem.

biotic indicator: an indicator that an organism is or was alive (could be a fossil).

These animals serve as biotic in icators of the health of their aquatic ecosystem.

Biotic Indicators

5 min.

Changes in water quality can have a domino effect on organisms. If conditions are harmful to one group, other organisms that depend on that group will also be affected negatively.

◇ Evaluate

Share the main concept of this science knowledge section on page 97—that populations of organisms can act as biotic indicators because they can reflect the health of their aquatic ecosystem. Populations normally change when one group of organisms supply the needs of other organisms as when the heron in the photograph feeds on the fish. However, populations can also change if the environment changes. Ask students, "How and why did macroinvertebrates act as biotic indicators in the Marry Martans River Mystery?" (*The macroinvertebrates acted as biotic indicators because they are organisms whose populations were affected by a negative change in their environment.*)

What's the Point?

Some of the residents near Marry Martans Lake noticed a sudden increase in the growth of algae. They worried that the nearby farms had runoff problems. Macroinvertebrates are an example of living things that can serve as a biotic indicator of water quality. Explosion in growth of algae and plant life can also indicate a change in water quality due to fertilizer runoff. The residents were able to use macroinvertebrate data to support their ideas about pollution in the Marry Martans River.

In this activity, you could see that when pollution entered the river, the variety of macroinvertebrates in that area decreased. Macroinvertebrates that are tolerant of pollution can survive new pollutants being added to the water. Intolerant and mildly tolerant macroinvertebrates will not survive when water becomes polluted. This is another way scientists often classify organisms. Organisms can be grouped by how they react to changes in their environment.

Algal growth can make water unlivable for some macro- and

Assessment Options

Targeted Concepts, Skills, and Nature of Science	How do I know if students got it?
Some species of macroinvertebrates can indicate water quality.	**ASK:** How did some species of macroinvertebrates reflect the quality of the water in the Marry Martans River? **LISTEN:** Listen for students to say that at several points along the river, several species of macroinvertebrates were unable to tolerate pollution that was occurring.
Scientists can determine water quality using biotic indicators.	**ASK:** How did scientists use information from biotic indicators to determine the water quality of the Marry Martans River? **LISTEN:** Listen for students to say that the scientists determined water quality by collecting samples of macroinvertebrates at three separate locations.
Biotic indicators may help determine a pollution source.	**ASK:** How can biotic indicators help determine a pollution source. **LISTEN:** Listen for students to say that biotic indicators can be used to determine a pollution source when populations of the organisms change negatively below the source of the pollution.
Human activity can affect the ecology of a community. Humans use rivers for residential, commercial, industrial, and agricultural purposes. These activities affect water quality along a river.	**ASK:** What human activity caused harmful substances to pollute the Marry Martans River and Lake? **LISTEN:** Listen for students to respond that poorly operating pollution-control measures on two of the three farms caused a change in the macroinvertebrate populations of the river.

Teacher Reflection Questions

- What tools helped students prepare for discussing the Marry Martans River Mystery?

- How well were students able to analyze the differences in the kinds of organisms at each data collection point?

- What connections were you able to make to the *Big Challenge* in this section?

SECTION 3.4 INTRODUCTION

3.4 Investigate

1 class period* ▶

Effects of Turbidity on Living Things

Overview

Students revisit and experiment with turbidity as a water-quality indicator. They experiment with elodea in two different light conditions and learn that increased turbidity affects (*by decreasing*) the number of oxygen bubbles produced by plants in aquatic ecosystems. A teacher demonstration, using probes that measure dissolved oxygen, corroborates their elodea experiment. Students also read details about the process of photosynthesis.

Targeted Concepts, Skills, and Nature of Science	Performance Expectations
The chlorophyll in plants uses energy from sunlight to make food in a process called photosynthesis.	Students should be able to say that plants make their own food when they undergo photosynthesis.
Organisms may interact in several ways as producer and consumer, predator and prey.	Students should be able to say that green plants can make their own food and that animals need to consume other organisms to get the energy they need to stay alive.
By following how water flows in and over land in an ecosystem, ecologists can learn how water is affected by organisms and by the land in the ecosystem.	Students should be able to say that water quality affects the organisms that live in an ecosystem.
Plant growth can affect water quality.	When asked what factors can affect water quality, students should be able to say that, when excessive, plant growth can increase turbidity, and therefore, decrease water quality.

*A class period is considered to be one 40 to 50 minute class.

Materials	
1 per group	Artificial light source
2 per group	Shoe boxes, one with a lid
4 per group	Sprigs of elodea, each 2 ½-3 inches long
2 per group	Small plastic zipper bags labeled, each with ½ tsp. baking soda
per class	Source of distilled water (*Do not* use chlorinated tap water.)
1 per student	*Elodea Investigation* page
1 per class	Class *Project Board*
1 per classroom	Photosynthesis kit for data logger, for teacher demonstration (optional)

Activity Setup and Preparation

Student elodea investigation: Order materials for the elodea investigation on either Monday or Tuesday for delivery on Wednesday or Thursday. If the investigation is not to be carried out until the following Monday, elodea specimens must be cared for over the intervening days. Elodea is sensitive to temperature; therefore, cold temperatures need to be avoided. When the elodea sprigs arrive, they may be floating in water or rooted in sand or gravel and water. If floating, cover the sprigs completely with distilled water or pond water, but do not use chlorinated tap water. If rooted, cover with distilled water or pond water. Place the plant materials in a bright area, but not directly under a lamp or in direct sunlight until you use them.

Teacher demonstration: Order materials sufficiently in advance to set up a trial run using the kit materials before you try this as a class demonstration. Use fresh elodea each time you demonstrate with this equipment.

Homework Options

Reflection

- **Nature of Science and Science Content:** How can aquatic plants help improve water quality? How can aquatic plants be indicators of water quality? (*Aquatic plants help water quality by maintaining levels of dissolved oxygen in water through the process of photosynthesis. Aquatic plants can be indicators of water quality by how well they grow. An excessive growth of plants in water may* *indicate that phosphate or nitrate levels from fertilizer runoff are too high. This might lead to the depletion of dissolved oxygen needed by other organisms in the water. A decrease in aquatic plant growth may indicate that turbidity is high. It might mean that sunlight is not reaching the green leaves so that they can make food through photosynthesis.)*

- **Science Content:** What is photosynthesis? Why is photosynthesis important? *(Students should describe the process of photosynthesis as a process whereby plants utilize light energy and carbon dioxide from the atmosphere, plus materials within their own cells (water and chlorophyll in chloroplasts) to make their own food. This food is their source of energy. Students should be able to say in some way that photosynthesis is important because it distinguishes green plants, which make their own food, from animals and other organisms, which depend on sources outside of themselves for energy. Photosynthesis is also important because it is the process that provides oxygen that organisms use during respiration.)*

Preparation for 3.5

- **Science Content:** List the ingredients that make up one meal. Trace one of these ingredients back to its original energy source. *(Students choices will vary but all should eventually trace back to the Sun's energy in a plant.)*

- **Science Content:** Describe what organisms might be dependent on aquatic plants in an ecosystem such as the one you live in. *(Answers will vary but students should at least include that a variety of fish are dependent on aquatic plants in almost any ecosystem they choose. They might know of some birds that are also dependent on the fish that depend on the aquatic plants.)*

SECTION 3.4 IMPLEMENTATION

◄ *1 class period**

3.4 Investigate

Effects of Turbidity on Living Things

Macroinvertebrates can serve as an indicator of water quality in an ecosystem. That is just one example of how living things can serve as an indicator of water quality. Recall that in *Learning Set 2*, your class investigated how high levels of nitrates and phosphates can affect plant growth. Nitrates and phosphates increase the growth of plants. Plants can serve as an indicator of nitrates and phosphates in the water.

You also read about turbidity in *Learning Set 2*. Turbidity is a measure of how easily light passes through the water in a river or stream. Land development is common in many ecosystems. The soil that runs off the land often ends up in the river. The particles of soil mix in the river. This makes the water turbid. If turbidity is high, the water appears very murky and cloudy. Light cannot easily pass through highly turbid water. But what does this mean for living things in the water and the ecosystem? How might water plants be affected by highly turbid water?

Plants can serve as an indicator of water quality.

You will work with your group to investigate this issue. You are going to study the effect of sunlight on the life of an aquatic (water) plant called elodea (ih-LODE-ee-uh).

Elodea is found naturally in many streams and rivers. It is found in North America and other continents. Elodea is easy to grow. Because of this, it is a plant commonly kept in aquariums. It is also often used in the laboratory to investigate how plants function. The question you will investigate is *How do plants react to changes in the amount of sunlight they receive?*

Run the Investigation

You will work with your group to investigate how the amount of light received affects elodea, but the results of your investigation can be applied to land plants as well.

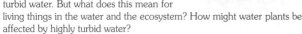
LT 99

LIVING TOGETHER

3.4 Investigate

Effects of Turbidity on Living Things
5 min.

Through investigation, students learn how turbidity affects living things. In doing so, they learn about photosynthesis and how various factors interact to affect water quality.

○ **Engage**

Ask students what they recall about the turbidity in *Learning Set 2*. If students should become stalled, begin to rebuild their recollections by asking a series of questions that connect a variety of concepts that students have learned.

*A class period is considered to be one 40 to 50 minute class.

META NOTES

During the process of scaffolding, the teacher coaches students in assembling pieces of information that they need to create a big picture.

TEACHER TALK

❝Think back to how you would define the term turbidity. *(The measure of how clear water is.)* What are some reasons water becomes less clear? *(Answers will vary but students should include that runoff washes soil into streams or ponds, and that sediments at the bottom of streams or ponds become stirred up.)* Do you think more organisms would live in turbid water or clear water? *(More live in clear water.)* Do you think that sunlight penetrates a turbid stream as well as it penetrates a clear stream? *(No.)* In addition, how is the temperature of a clear stream different from a cloudy stream? *(Increased turbidity causes the temperature of the water to rise.)* How does this affect the amount of dissolved oxygen? *(Dissolved oxygen becomes less available to organisms.)*

Knowing all of these factors, what would you predict about how turbidity affects the ecology of the stream? *(Students should be able to predict that a turbid stream affects a stream negatively; more turbid waters will decrease the ability of aquatic plants to generate oxygen, water temperatures will change and thus aquatic animals and other organisms will be negatively affected.)*❞

△ Guide

Inform students that they will work in groups to investigate the question, *How do plants react to changes in the amount of sunlight they receive?*

to investigate how plants function. The question you will investigate is *How do plants react to changes in the amount of sunlight they receive?*

Run the Investigation

You will work with your group to investigate how the amount of light received affects elodea, but the results of your investigation can be applied to land plants as well.

LT 99

LIVING TOGETHER

Run the Investigation

10 min.

In this investigation, the reactions of elodea plants to normal light conditions and to turbid light conditions are compared.

△ Guide

After clarifying the concept of turbidity with students, read through the instructions to make certain that students understand how they are to proceed.

☐ Assess

After students have read the instructions, make certain that they understand the overall focus of the investigation by asking them, "What will the plants be exposed to?" *(Two different levels of light,)* "What do the different levels of light represent?" *(Ideal light conditions for plants growing in water and poor or insufficient light conditions for plants growing in water,)* and "What is being tested?" *(How elodea reacts to changes in the amount of sunlight it receives.)*

3.4

Elodea is an aquatic plant that lives in freshwater streams. You will use four sprigs of elodea in your investigation. You will expose two sprigs to sunlight and keep the other sprigs out of the sunlight.

Keeping the sprigs out of the sunlight would be similar to the conditions in a high-turbidity stream or river where very little sunlight is reaching the elodea. You will record observations of what happens to the plants at the start of the investigation and again after thirty minutes.

The results of your investigation should help you answer the question, *How do plants react to changes in the amount of sunlight they receive?*

Procedure: Set Up Your Investigation

1. Your teacher will provide you with the materials in the list. You will have about ten minutes to set up your experiment.

2. Label two plastic zipper bags with your name.

3. Take four sprigs of elodea. Use scissors to trim the stem end of each sprig. Pull the leaves off the bottom of each stem so there are no leaves in the first 1 cm ($\frac{1}{3}$ in.). Crush the stem end of each sprig with your fingers.

4. Put two sprigs of elodea in each sealable bag. Your teacher has added a half-teaspoon of baking soda to each bag. Add enough water to each bag to cover the plant. Seal the bags. Make sure the water is not leaking.

5. Carefully observe the plants inside the bags. Record your initial observations on an *Elodea Investigation* page similar to the one shown on the next page.

Materials
- 4 sprigs of elodea
- 2 transparent plastic zipper bags (small)
- 2 shoe boxes, one with a lid
- baking soda
- distilled or pond water
- access to a lamp or a light tray
- *Elodea Investigation* page

LT 100

Procedure: Set Up Your Investigation
15 min.

Turbidity is modeled by placing elodea samples in a box with a lid, thereby cutting off the source of light needed in the process of photosynthesis.

○ Engage

Distribute the *Elodea Investigation* page to each student for recording their predictions and observations during the investigation.

△ Guide

Remind students of the importance of careful and complete observations and of keeping accurate records during any investigation. Ask, "Why are written records important to an investigator?" *(Written records are a source of evidence in an investigation.)*

Explain that you will distribute materials for the investigation and that you will also demonstrate how to treat the elodea stems for use.

Prepare students by showing them how to trim the bottom of the stem, pull leaves off about 1 cm of the bottom of the stem, and to crush the bottom of the stem.

To prevent some problems, demonstrate how to put the sprigs into each bag, how to put in the correct amount of water, and how to seal the bags.

Distribute the distilled water, resealable plastic bags, each with ½ teaspoon of baking soda, four elodea sprigs to each group, and two shoe boxes. Have students label their bags with their group name first.

☐ Assess

Circulate through the room to monitor students as they set up their two plastic bags. Be prepared with paper towels if needed.

⬡ Get Going

Remind students to write a prediction on the *Elodea Investigation* page and record their "Before" observations before they submit them to the different light conditions. *(Students should record that both bags are identical at the outset of the investigation.)*

Have each group place one bag in the shoe box with a lid and one in the shoebox without a lid and set both boxes under a lamp. Both boxes should be the same distance from the lamp. Then leave both boxes undisturbed for 30 minutes while students gain some science knowledge about photosynthesis.

△ Guide

To complete the investigation setup, talk about the importance of controlling factors during an investigation. An investigation must be a fair test. Ask, "Why is this investigation an example of a fair test?" You might describe it as having only two kinds of variables, one that you change intentionally (*the manipulated variable*) and the one that they will measure in response to the thing that is changed (*the responding variable*).

To tie this explanation to the specific investigation they are about to do, you might add:

> **TEACHER TALK**
>
> **❝**A fair test is one in which you keep all of the parts of the experiment the same except for one condition, the manipulated variable. What will you use in this test that is the same? *(Same plants, same kind of water, same size plastic bags, and the same distance from the light source.)* What is the one condition that will be different? *(The exposure to light-one will be exposed to light and one will not be exposed to light.)* How the plants respond will be your responding variable.**❞** *(The responding variable will be how well the sprigs produce oxygen bubbles depending on their exposure to light.)*

6. Put one bag with elodea *inside* a shoe box. Put the shoe box under the lamp or in the light tray, if available.

7. Put the other bag with elodea *inside* another shoe box. Place the lid on the shoe box. Put the shoe box under the lamp or in the light tray, if available.

8. Leave the shoe boxes undisturbed for thirty minutes.

9. What do you think will happen during your experiment? Record your prediction on your *Elodea Investigation* page.

● You will return to this investigation later during class.

Elodea Investigation

Name: _____ Date _____

Prediction

Describe what you think will be the effect of keep elodea out of the light. Also, what difference do you think you will see between the two plants at the end of the investigation?

Observations

Record details of how the plants look and what you see in the bags. Draw a sketch of the elodea to help you describe what you see.

With Light	Without Light
Before	Before
After	After

What did you notice in the bag of elodea that was exposed to light?

What did you observe in the elodea that was kept in the dark?

What do you think caused the changes you observed?

Understanding Photosynthesis

You may remember from previous science classes that plants require certain things in order to grow. Recall the duckweed in *Learning Set 2*. You placed the plants in water. Then you added fertilizer (nutrients) to the water and placed the plant under the growth lights. Each of these factors was important for the growth of the plant.

Photosynthesis refers to a complex process that keeps plants alive and helps them grow. It takes place in plants, algae, and some bacteria. These organisms use water, **carbon dioxide**, nutrients, and light to survive and grow. They get these things from their environment.

photosynthesis: the process in which sugar and oxygen are produced by green plants and some other organisms using water, carbon dioxide, and energy from sunlight.

carbon dioxide: a colorless, odorless gas commonly found in air; carbon dioxide is used by plants in the process of photosynthesis.

Understanding Photosynthesis

10 min.

The basic concepts of photosynthesis are explained in this science text box. Students can use this information to make connections between the process of photosynthesis and the investigation they are performing. The interdependence of plants and animals through photosynthesis and respiration is underscored.

△ Guide

Have students read the information about photosynthesis to gain background knowledge and prepare them for interpreting what they will observe in the investigation that they have set up. Tell students the details of photosynthesis, but de-emphasize the terminology and focus students' attention on the larger function of photosynthesis—to capture sunlight energy and convert it into a form the cells can use. You might want to project an image of the *Photosynthesis-Cell Respiration* diagram on page 103.

META NOTES

Reading technical text is a different skill from reading narrative text. Students engaged in project-based investigation benefit from technical text as it provides background projects to related contexts.

"There are many new terms in the description of photosynthesis on pages 101 and 102, but let's try to look at the bigger picture of what is going on here. Photosynthesis is really a story of moving energy from one place to another.

Think about what you need to survive every day *(Energy.)* Organisms *(both plants and animals)* need energy to survive. The only way you and I, as animals, can get the energy we need is by eating food. A couple of hours later, you have to eat all over again to keep going. Green plants are different from animals. Green plants can make their own food. Green plants are important living things because they can take sunlight energy and change it into food energy. That is something animals cannot do.

Green plants use carbon dioxide, water, and sunlight to make sugar (and oxygen) in their green leaves. Overall, during the process of photosynthesis, energy in sunlight is changed into food energy in green plants."

◇ **Evaluate**

Give students a chance to ask questions or ask them to summarize what they have heard thus far.

"What is being transferred during the process of photosynthesis? *(Energy)* What can green plants do that animals cannot do?" *(Green plants can make their own food.)*

NOTES

....

....

....

....

....

....

glucose: a type of sugar. It is the main source of energy for living organisms.

sugar: a chemical compound produced by plants during photosynthesis. Sugars provide a source of energy used by living organisms.

oxygen: a colorless and odorless gas produced by plants during photosynthesis and used by animals for respiration.

chlorophyll: a green substance found in plants that is used to capture energy from sunlight during photosynthesis.

chloroplast: the part (organelle) of a plant cell that specializes in photosynthesis. It contains chlorophyll.

They produce **glucose** (a **sugar**) and **oxygen**. Glucose is used as an energy source. The energy is needed to power the plant's life processes. Life processes include activities such as growing new tissue and reproduction. The oxygen produced is released into the environment.

Light is essential for photosynthesis. If plants cannot get enough light, they cannot produce the glucose they need to grow. The light can come from the Sun or from indoor lights.

In plants, photosynthesis occurs mainly in the leaves. The leaves are made up of cells that contain a green substance. The substance is called **chlorophyll**. It captures the energy of sunlight. Chlorophyll is found in the **chloroplasts** of plant cells. To function properly, plants and algae also need small quantities of mineral nutrients, such as nitrate and phosphate. Land plants get these nutrients from the soil. Aquatic plants get them from the water.

All organisms need a source of energy to function. Animals eat plants and other animals to make energy. Plants make their energy through photosynthesis.

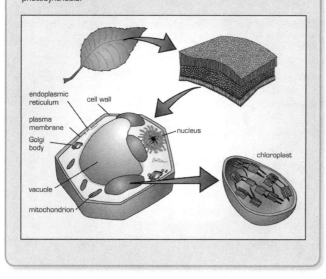

endoplasmic reticulum   cell wall

plasma membrane

Golgi body

nucleus

chloroplast

vacuole

mitochondrion

LT 102

Continue telling the story of the transfer of energy from sunlight to food energy.

TEACHER TALK

"A second thing that happens when plants change sunlight energy into food energy is that they release a waste product. This waste product is called *oxygen. Oxygen* is very important. You breathe in some oxygen every time you take a breath. Did you wonder why? Oxygen goes into your blood cells, gets carried to your body's cells, and is used to break down food in your cells. During this process, energy is released from the food. This is the energy you use to think, walk, and talk. This process is called *respiration. Respiration* is the way that your body releases energy you need from the food you have eaten."

cell respiration: the process by which food and oxygen are converted to energy, carbon dioxide, and water in a living cell.

The processes of photosynthesis and **cell respiration** are connected. The oxygen produced by plants is used by animal cells during respiration. The carbon dioxide produced during respiration is released into the environment. This carbon dioxide is taken up by plants to photosynthesize. Animals and plants are therefore tightly linked. Each needs the other, and both rely on the environment to survive.

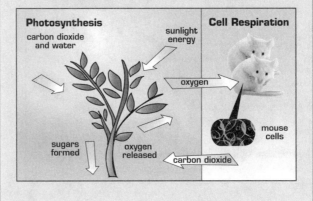

LT 103

LIVING TOGETHER

META NOTES

Once they get back to their investigation with the elodea, students will have a better understanding of photosynthesis and respiration. Connect this reading to the investigation after students have removed the covers from the shoe boxes.

Observation and Recording Data

1. After about thirty minutes, return to your samples. Gently remove the lid and compare the contents of the two bags.

   Record any differences you observe in the bags.
   - What changes occurred in the bag of elodea that was exposed to light?
   - What changes occurred in the bag of elodea that was kept in the dark?
   - What differences do you see between the two bags? (Think about what happens during photosynthesis and see if you can spot any differences in what photosynthesis produces.)

2. Record your observations on an *Elodea Investigation* page. Although you are working on the investigation as a group, each member of the team should record his or her own observations. Later, you will discuss your results among your group and compare them to those of other groups in the class.

Analyze Your Data

Once you have collected your data, discuss the results of your experiment with other members of your group. The questions below should help you organize your discussion. Take notes of what others are saying during the discussion.
- What did you notice in the bag of elodea exposed to light?
- What did you observe in the elodea that was kept in the dark?
- What do you think caused the changes you observed?

Answer the following questions:

1. What is produced during photosynthesis? Record at least two things that are the result of photosynthesis.

2. How do you think the investigation you conducted with elodea is related to photosynthesis?

3. How do the results of this investigation help you answer the investigation question *How do plants react to changes in the amount of sunlight they receive?*

4. What do the results of your investigation suggest will happen to water plants in a river that has very high turbidity? How might that affect other living things in the ecosystem?

Project-Based Inquiry Science

Observation and Recording Data

30-40 min.

Students should observe the presence of oxygen in the form of bubbles around the stems of the elodea exposed to light. The bubbles are an indicator of the formation of oxygen during photosynthesis.

△ Guide

At the end of 30 minutes, have students turn off the lights and remove the lid of the shoe box. Ask them to look at the bag that was exposed to the lamp and the one that was kept in the dark in the box. Warn them not to jostle or handle the bags unnecessarily.

Remind students that it is important to record their observations on their *Elodea Investigation* page. Encourage them to think about what they have just read about photosynthesis while they observe what is going on in the two bags.

META NOTES

The bag that was exposed to light should have more bubbles than the bag that was kept in the dark. The bag that was exposed to light will also feel (use a gentle touch with a finger) more firm than the bag in the covered box. As oxygen builds up in the bag, the pressure in the bag increases.

◇ **Evaluate**

Monitor students' observations. Focus their attention on what is happening at the end of the stem where it was crushed. Students should be able to tell you that they see bubbles coming from the end of the elodea stems.

If the elodea stems are releasing many bubbles, suggest that students count the number of the bubbles and compare bubble production in both light-exposed and non-exposed treatments.

Analyze Your Data

5 min.

Students answer questions based on their observations and newly acquired science knowledge and prepare to discuss their answers within their group.

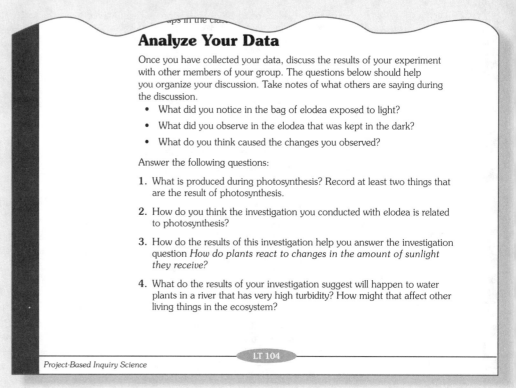

Analyze Your Data

Once you have collected your data, discuss the results of your experiment with other members of your group. The questions below should help you organize your discussion. Take notes of what others are saying during the discussion.

- What did you notice in the bag of elodea exposed to light?
- What did you observe in the elodea that was kept in the dark?
- What do you think caused the changes you observed?

Answer the following questions:

1. What is produced during photosynthesis? Record at least two things that are the result of photosynthesis.

2. How do you think the investigation you conducted with elodea is related to photosynthesis?

3. How do the results of this investigation help you answer the investigation question *How do plants react to changes in the amount of sunlight they receive?*

4. What do the results of your investigation suggest will happen to water plants in a river that has very high turbidity? How might that affect other living things in the ecosystem?

LT 104

Project-Based Inquiry Science

△ **Guide**

Prepare students for an *Investigation Expo* by reminding them that they will use their recorded observations as evidence to make a formal explanation and presentation of their results after they answer the analysis questions.

☐ **Assess**

After students answer the analysis questions, they should compare answers within their group. You should hear students include the following in their discussions.

1. Photosynthesis produces oxygen and sugar (*food/glucose*).

2. The investigation with elodea is related to photosynthesis because a green plant was exposed to a source of light energy for 30 minutes and seems to be releasing the waste product, oxygen. The green

plant that was kept in the dark (*representing turbid conditions*) is not producing oxygen bubbles. (*Students will not be able to determine whether sugar has been formed in this investigation because the investigation does not test for the presence of sugar.*)

3. As green plants in water are exposed to more light, photosynthesis increases, producing more oxygen gas. In increasingly turbid conditions, less photosynthesis will take place; therefore, less oxygen in the form of bubbles of gas will be released from plants in turbid water.

4. A river with high turbidity is cloudy and blocks sunlight from getting to the plants in the water. Plants need sunlight to photosynthesize. As the amount of sunlight decreases, the amount of photosynthesis decreases. The experiment with elodea gives evidence of this. In streams of high turbidity, there is less plant life and the level of dissolved oxygen decreases. The population levels of living things that depend on dissolved oxygen will decrease as the oxygen decreases.

META NOTES

These questions scaffold students' developing understanding of the effects of turbidity on the quality of water in streams. Students make inferences needed to connect turbidity, sunlight levels, energy, and the ability of streams to support living organisms in aquatic ecosystems.

NOTES

Explain
10 min.

In an Investigation Expo, *students present their findings based on evidence and build on their knowledge base to answer the* Big Challenge.

Explain

Decide as a group on the best explanation for what happened. Remember, the best explanation is one that is best supported by the evidence. Record your group's explanation. Be prepared to share your conclusions and explanation with your class.

In an *Investigation Expo*, share with the class your analysis of your data, your answer to Question 4, and your explanation. Listen carefully as other groups share their findings and explanations.

Demonstration

Your teacher may present to your class the results of another investigation with elodea. This investigation was done while you were running yours.

You observed what happened to the elodea over 30 minutes. Your teacher also used an instrument to measure the changes observed in the two elodea samples. You collected only visual observation data.

Discuss with your class the results of your teacher's investigation. Use the following questions as a guide:

- How do the results of your teacher's investigation compare with yours?
- How do the results from your teacher's investigation help you to better understand what was happening to the elodea exposed to light?

Revise Your Explanation

As a class, revise your explanation of what caused the changes you observed. Base your revised explanation on the results of your teacher's investigation.

○ Engage

Remind students that they now have evidence and science knowledge concerning how turbidity affects living organisms, in particular, how it affects photosynthesis of aquatic plants. Eventually, they will use this information when they write recommendations for the Wamego town council about possible changes in the town's water conditions.

△ Guide

Guide students in formulating an explanation of the results of their photosynthesis/turbidity investigation. First, review the use of claims, evidence, and science knowledge.

META NOTES

Class discussions assist students in developing deeper understanding of concepts.

TEACHER TALK

❝What is a claim in an investigation? *(Listen for students to say that a claim is a statement or conclusion reached as a result of an investigation.)* Where do we get evidence in an investigation?**❞** *(Students should respond that evidence is the data collected during an experiment.)*

Finally, have them recall that *science knowledge* is information based on what experts have previously investigated and found to be consistently true.

Have each group conference with their investigation observations to complete statements for each item (*claim, evidence, and science knowledge*) and prepare an explanation for their *Investigation Expo*.

△ Guide Discussions and Presentations

Remind students that in an *Investigation Expo*, the group is to address the following points:

- The question you were trying to answer in your investigation.
- Your predictions.
- Your group's procedures and why your test was a fair comparison of the conditions.
- Your group's results and how confident you are in them.
- Your group's interpretation of the results and how confident you are in it.

Remind students that during the presentation, they should be focused on the presenters' use of claims, evidence, and science knowledge to answer the question: *How do plants react to changes in the amount of sunlight they receive?*

Model how to ask questions politely. Remind students to listen for the insertion of opinion and politely remind the presenters if opinion statements are used.

Because each group used the same materials and procedure for the investigation, choose one group to present their findings. At the end of that group's presentation, have other groups explain one at a time, ways in which their findings were similar or different. Use this information to transition into a class discussion.

Bring the discussion to a close after all questions have been asked. Have the class as a whole decide on a conclusion statement that answers the investigation question based on the students' findings.

Demonstration

5 min.

Demonstration

Your teacher may present to your class the results of another investigation with elodea. This investigation was done while you were running yours.

You observed what happened to the elodea over 30 minutes. Your teacher also used an instrument to measure the changes observed in the two elodea samples. You collected only visual observation data.

Discuss with your class the results of your teacher's investigation. Use the following questions as a guide:

- How do the results of your teacher's investigation compare with yours?
- How do the results from your teacher's investigation help you to better understand what was happening to the elodea exposed to light?

○ Engage

Following the discussion, tell students that you have a demonstration to test the same question but that you will use different equipment and procedures.

Show students the demonstration equipment (*elodea, test tubes, dissolved oxygen probe*) and explain what it will measure (*dissolved oxygen*).

△ Guide

Ask students to observe the bubbles around the stem when they become apparent.

Record the data on the board so that all students will have it available.

Ask the class to compare the test you demonstrated with the investigation that they have done. Ask students if or how they might revise their explanation based on the results from the demonstration.

Facilitate a class discussion until all are in agreement with an explanation as to why elodea reacts to changes in the amount of light. Record the class based explanation.

Revise Your Explanation

5 min.

Scientists often modify or revise their explanations when new ways of performing tests become available. New technology may also make test results more accurate.

Revise Your Explanation

As a class, revise your explanation of what caused the changes you observed. Base your revised explanation on the results of your teacher's investigation.

Tell students that now that they have observed two ways to test for how plants react to different amounts of light energy, and that they may have new evidence to use to revise their original explanation. Students can use the data that was recorded from the demonstration to modify or revise their original explanations.

Reflect

Answer the following questions. Be prepared to discuss your answers with your class.

1. How might changing the other influences on plant growth, like amount of water or nutrients, affect plants? Compare these changes to changing the amount of sunlight a plant receives.

2. High turbidity can cause changes in a plant's ability to photosynthesize. What other water-quality indicators might signal a problem for photosynthesis in plants?

3. What could happen to the other living things in the ecosystem if a plant could not complete photosynthesis? Provide an example or draw a diagram to describe what might happen.

Update the *Project Board*

As a class, update the *Project Board*. Include your knowledge from your investigation about the effects of light on plants. Include your explanations about what was happening in the *What are we learning?* column of the *Project Board*. Make sure you support your conclusions with evidence you have collected. Fill in the columns *What are we learning?* and *What is our evidence?*

What's the Point?

Animals and plants need a source of energy to survive. In this section you investigated the effect of light on plants. You discovered that plants exposed to light were able to carry out photosynthesis and produce oxygen and sugars. Without light, plants cannot produce the sugars needed to grow.

In the next *Learning Set,* you will further investigate the relationships between animals, plants, and their environment. You will continue to build your understanding of how water quality can affect the ecosystem.

Reflect

20 min.

The Reflect *questions introduce students to the complexity of aquatic ecosystems.*

◇ Evaluate

Have students answer the *Reflect* questions with their groups, then discuss the answers as a class. Responses will vary. Suggested answers are given below.

1. Increasing the amount of water or nutrients may make plants grow faster or larger. Decreasing the amount of water or nutrients will usually make plants remain small or not grow at all. Decreasing the amount of available light would restrict the plants from making their own food and therefore, affect the ability of the plants to remain alive.

2. High nitrate and phosphate levels could increase the growth of green algae on the surface, causing sunlight to be blocked from reaching plants below the surface. The decreased sunlight would restrict the plants below the surface from photosynthesizing.

3. If an aquatic plant cannot carry out photosynthesis, then the levels of oxygen in the water will decrease. Less oxygen becomes available for animals that depend on dissolved oxygen to stay alive in the water. Animals that depend on aquatic plants for food would also be affected negatively.

Update the Project Board

10 min.

Students use what they have learned in this section to add evidence to the Project Board *that will help them answer the* Big Challenge.

Update the *Project Board*

As a class, update the ro ect   oar . Include your knowledge from your investigation about the effects of light on plants. Include your explanations about what was happening in the   *hat are we learning?* column of the   ro ect   oar . Make sure you support your conclusions with evidence you have collected. Fill in the columns   *hat are we learning?* and   *hat is our evi ence?*

△ Guide

Inform students that they will update the *Project Board* with information, evidence, and new questions that result from their study in this section.

Guide students' choice of what to add to the *Project Board* with the following questions:

- What question were you trying to answer in this investigation? *(The response might be: What happens when plants are exposed to different levels of sunlight?)*

- What did you find out in the investigation that might answer this question? How do plants and sunlight interact? *(The response might be: Plants use sunlight to produce their own food through photosynthesis.)*

- What evidence do you have to support the idea that plants use sunlight for photosynthesis? *(The response might be: From science knowledge you know that during photosynthesis, plants release oxygen as a waste product. In the investigation, you saw more gas bubbles released by plants that were exposed to light than by plants that were blocked from light.)*

- How are plants and photosynthesis connected to the Unit's *Big Question: How does water quality affect the ecology of a community?* *(Responses might include: Increased turbidity can result in decreased photosynthesis in aquatic plants. Less food produced by these plants means less dissolved oxygen and probably less of different types of aquatic animals available.)*

As a result of the discussion, students should have something such as the following to add to their class *Project Board*:

- **Claim:** Plants need sunlight for photosynthesis.

- **Evidence:** The elodea in the box with the lid (*without light*) did not release as much oxygen as the elodea in the box that was exposed to light. Therefore, light is an important factor in photosynthesis.

- **Claim:** Turbidity can affect the amount of photosynthesis that takes place in aquatic plants.

- **Evidence:** Aquatic plants (*such as elodea*) that release oxygen as a result of photosynthesis decrease their level of photosynthesis as the light they are exposed to decreases. Available light for photosynthesis decreases as turbidity increases.

Assessment Options

Targeted Concepts, Skills, and Nature of Science	How do I know if students got it?
The chlorophyll in plants uses energy from sunlight to make food in a process called photosynthesis.	**ASK:** What is the relationship between chlorophyll and the process of photosynthesis? **LISTEN:** Students should say that chlorophyll in green plants traps energy from sunlight and changes it into food energy during the process of photosynthesis.
Organisms may interact in several ways as producer and consumer, predator and prey.	**ASK:** What is the difference between how plants obtain energy and how animals obtain energy? **LISTEN:** Students should respond that green plants make their own food through the process of photosynthesis. Animals must depend on consuming other organisms to obtain energy.

NOTES

Targeted Concepts, Skills, and Nature of Science	How do I know if students got it?
By following how water flows in and over land in an ecosystem, ecologists can learn how water is affected by organisms and by the land in the ecosystem.	**ASK:** How can decreased water quality affect an ecosystem? **LISTEN:** Students should say that decreased water quality causes less photosynthesis to take place, reduces the amount of dissolved oxygen available to animals that live in water and might even cause an increase in temperature. If these changes take place, then animals that depend on particular conditions will not survive.
Plant growth can affect water quality.	**ASK:** How might excessive plant growth affect water quality? **LISTEN:** Students should say that excessive plant growth can block sunlight from reaching plants below the surface.

Teacher Reflection Questions

- It can be difficult to make clear connections between the laboratory and the real world. What difficulties did students have in relating the closed shoebox to a turbid stream?

- In this section, there were many times students were asked to answer questions. What methods could be used to support students in answering questions other than class discussion or writing out answers?

- How well did the investigation results match what should have happened? What improvements could be made to the investigation to support students' learning and make it possible to obtain more accurate, more measurable results?

NOTES

....

....

....

SECTION 3.5 INTRODUCTION

3.5 Explore

Connections to Other Living Things

◀ *1 class period* *

Overview

Students look at the connections between organisms through a common breakfast of cereal and milk that they might eat. Then, they consider a more complex meal of a cheeseburger and fries. From these examples, students are introduced to models called *food chains*. They follow the path that nutrients take from organism to organism in a food chain. Then, they consider the transfer of energy in a food chain as producers *(green plants)* use light energy to make their own food *(glucose)* and consumers *(animals)* obtain some of this energy by eating other organisms. Students are introduced to the feeding classifications such as herbivores, carnivores, and omnivores and feeding relationships such as the predator-prey relationship.

*A class period is considered to be one 40 to 50 minute class.

Targeted Concepts, Skills, and Nature of Science	Performance Expectations
Living things, such as green plants, trap energy from sunlight and produce their own food. They are called producers.	Students should be able to say that plants are producers because they make their own food.
Living things that consume other organisms are called consumers.	Students should be able to say that organisms that get the energy they need to stay alive by eating other organisms are consumers.
Consumers may be classified as herbivores, carnivores, or omnivores.	Students should be able to classify consumers that eat plants as herbivores; consumers that feed exclusively on other consumers are carnivores, and consumers that obtain energy by feeding on both plants and animals are omnivores.
All living things need energy from the Sun and matter to survive. Energy gets trapped by producers and is transferred along with matter to consumers through food chains.	Students should be able to respond that all things need energy to survive, and that ultimately, the Sun is the source of all energy on Earth.

Materials	
1 per classroom	Set of dominoes
1 per class	Projection of "cereal and milk" food chain
1 per group	Poster paper
1 per group	Set of markers

Homework Options

Reflection

- **Science Content:** Choose one meal that you ate today. Draw a food from the meal and follow the ingredients back to the Sun. *(Student food choices will vary but all should include the Sun as the ultimate source of energy for all organisms on Earth.)*

- **Science Content:** Choose one meal that you ate today. List the ingredients. Classify each ingredient as a producer or consumer. Classify each consumer as an herbivore, carnivore, or omnivore. *(Student responses will vary but should classify all plant-based foods, such as vegetables and fruits, as producers. Foods of consumer origin, such as milk, beef, fish, deer, cheeses (other than soy cheeses), or eggs, should be classified as herbivores. Few, if any, students will have eaten carnivores.)*

Preparation for 3.6

- **Science Content and Science Process:** Draw a food chain you are familiar with. Describe where the organisms in your food chain live. *(Students' choices for a food chain will vary but all diagrams should use arrows showing the flow of matter (nutrients) and energy from producers to consumers (from producers to herbivores, to carnivores).)*

- **Science Content:** Describe how growth, reproduction, and death are factors in the life cycle of an organism. *(Student responses will vary.)*

SECTION 3.5 IMPLEMENTATION

3.5 Explore

Connections to Other Living Things

You looked at how small organisms and plants in an aquatic ecosystem can be affected by changes in water quality. It might seem obvious that organisms that interact with the water would be affected. The question for this *Learning Set*, however, is how water quality affects living things in an ecosystem. So, the important question to consider now is *How might the effects of water quality on a few living things affect all of the living things in the ecosystem?*

To get a sense of how connected other living things are to one another, you can look at your own interaction with living things. Earlier, you investigated about the needs of living things. Most living things require food or nutrients to survive. Consider exactly where you get your nutrients.

Procedure

1. Think about a simple breakfast of cereal and milk. You can buy cereal and milk at a grocery store. But where does it actually come from? With your class, create a diagram like this one shown on the right. Trace the parts of the breakfast back to their sources.

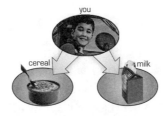

2. With your group, work together on another example. Examine a meal that some students your age enjoy, like a cheeseburger and fries.

LT 107

3.5 Explore

Connections to Other Living Things
5 min.

Students describe ways they think they are connected to other living things.

◯ Engage

Ahead of time, set up a series of dominoes on a table or box in the middle of the room so that they will fall when tapped lightly. Ask students if they have ever played with dominoes. Remind students that up to this point, they have been focusing on how water quality can affect living things. Tell them that now they will change their focus slightly because they will begin to consider how all living things in an ecosystem can be affected when water quality is changed. Tap the first domino. Ask students to use the dominoes as an analogy of what might happen to organisms in an ecosystem when water quality changes the ecosystem. Record students' ideas. *(Students' interpretations of the analogy will vary but probably will include that when something happens to change the water quality, other things in the*

*A class period is considered to be one 40 to 50 minute class.

ecosystem are affected. Water quality is signified by the standing dominoes. Changes in the quality are signified by the first domino falling. Changes in the ecosystem are signified by the subsequent falling of the other dominoes.)

TEACHER TALK

"How would you describe what just happened to the dominoes? If the dominoes were an ecosystem, how would you describe what might have happened in the ecosystem?"

△ Guide

Now guide students away from using symbols (*the dominoes*) to using experiences they have had with real organisms.

META NOTES

The domino demonstration is an analogy. An *analogy* compares an easily understood concept with a more complex concept. It enables students to begin to grasp the more complex concept.

TEACHER TALK

"Up to now, you have concentrated on what can happen to one kind of living organism when water quality changes. You saw this in the investigation about duckweed in *Learning Set 2*. In Section 3.4, you experimented with restricting light that green plants need to undergo photosynthesis. In real life, organisms live with many other kinds of organisms around them. Therefore, in real life, what will a change in water quality probably affect?" *(Students should say that all organisms in the area will be affected, not just one.)*

NOTES

....

....

....

....

....

....

....

Procedure

1. Think about a simple breakfast of cereal and milk. You can buy cereal and milk at a grocery store. But where does it actually come from? With your class, create a diagram like this one shown on the right. Trace the parts of the breakfast back to their sources.

you

cereal   milk

2. With your group, work together on another example. Examine a meal that some students your age enjoy, like a cheeseburger and fries.

LT 107

Procedure

15 min.

Students learn about the basic parts of a food chain by creating one of their own.

○ Engage

Project a transparency of the cereal and milk diagram to get students thinking about how they are connected to other living things, in this case, food sources that humans depend upon to stay alive. Explain that the graphic is a model or a concept map that shows a relationship.

△ Guide

Ask students to help you complete the diagram by following the cereal and milk back to the plants and the Sun. Your final projection should look like the one to the right.

Then, tell students to work within their groups to create a similar diagram on a poster representing the relationships within the second example—a cheeseburger and fries. Tell students that after they have completed their analyses and posters, they will present their posters to the class.

◇ Evaluate

As students work within their groups, listen for what they are saying about the origins of each food. Some students may not be aware that cheese is made from milk, which comes from cows *(consumers)* or that potatoes *(producers)* are plants. Some cheeses are made from soy products *(producers)*. Some students may know that most fries are usually cooked in oil, which comes from plant seeds or other plant parts. Sometimes fries are roasted in the oven, which reduces the amount of oil that is needed. Ground beef comes from cattle *(consumers)*, which feed on pasture grasses and hay *(producers)*. You may want to use some of this information as students present their posters.

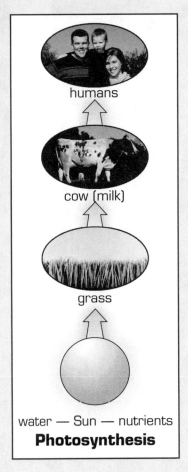

humans

cow (milk)

grass

water — Sun — nutrients
Photosynthesis

Communicate Your Ideas

Idea Briefing

10 min.

Social interactions such as an Idea Briefing *help students to articulate and refine their ideas. They expand their ideas when exposed to those of others as they share during a discussion.*

PBIS *Learning Set 3 • How Can Changes in Water Quality Affect the Living Things in an Ecosystem?*

Break apart the components of this meal. Trace each part back to its source. Use the photo to help you identify all the different parts. Record your group's analysis on a sheet of paper. You might want to use a pencil. In that way, you can erase mistakes or make changes as you break down the meal.

Communicate Your Ideas

Idea Briefing

Once your group has reached agreement, prepare a poster to share your analysis with your class. Your teacher will collect all the groups' posters and display them together. As a class, you will compare and contrast the various ideas and discuss the analyses.

Be sure to look for differences between analyses. Think about the following questions as you discuss the posters:

- Are there items that a group or groups forgot to include? What are those items?
- How did groups break down the food differently?
- What trends do you see in the way students identified the sources?
- Which organisms seem to eat only plants? Which organisms seem to eat meat?

LT 108

Project-Based Inquiry Science

Observe if students are beginning their diagrams with the Sun as the ultimate source of energy on which all organisms depend. Watch for the direction of arrows on the diagrams as well. The arrows should point in the direction from the ultimate source to each organism.

Collect posters as each group completes their product. Display them where all students can see them. Give students about five minutes to quickly become acquainted with all the posters before the *Idea Briefing* begins.

△ Guide

Draw students' attention to the four questions on page 108 in their textbooks. Remind students to listen for whether each group answers these questions as they present their posters.

△ Guide Presentations and Discussions

Remind students to listen politely while each group presents its poster and its analysis. Model the attitude that students should display as they ask questions.

Listen as students discuss and ask questions about each group's analysis. Listen for trends among the sources of the foods that students identify. Has each group been able to trace all food items back to plants and to the Sun?

If students have difficulty tracing one of the foods back to the Sun, review photosynthesis in *Lesson 3.4*.

NOTES

Food Chains Organisms in a Food Chain

10 min.

The reading is intended to help students understand the structure of a food chain, how one is written, and why it is written that way.

Food Chains
Organisms in a Food Chain

Recall that ecology is the study of how communities of plants, animals, and humans interact with each other and the physical environment. The activity you just completed points out one aspect of ecology. Organisms can be connected in a **food chain**. A food chain is a path of connected organisms where one organism relies on another as a food source. Look at a food chain that uses the cereal and milk example.

In this food chain, the grass grows because of photosynthesis. The cow eats this grass. The nutrients from the grass allow the cow to produce milk. Humans drink the milk.

You may have noticed that in both the cheeseburger example and the cereal and milk example, everything in the diagram always leads back to plants. Plants use light from the Sun during photosynthesis to make their food.

You may have also noticed that some non-plant organisms rely on other non-plant organisms as sources of food. Each of the organisms in the food chain can be classified by

- its location in the food chain, and
- the role it plays in the food chain.

Some organisms are **producers**. Producers are organisms that are capable of making their own food. Plants and some bacteria make their own food. They are producers. The grass in the food chain to the right is a producer.

food chain: a sequence that shows what eats what in an ecosystem.

producer: an organism capable of making its own food.

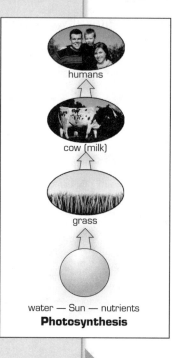

humans

cow (milk)

grass

water — Sun — nutrients
Photosynthesis

LT 109

LIVING TOGETHER

○ Engage

Draw students' attention to the diagram on page 109 in the student text that shows a path from the Sun to humans. Explain that this diagram is an expanded version of the simpler diagram on page 107 and that it is called a *food chain*. Ask, "Why is 'food chain' a reasonable name for this diagram?"

△ Guide

Have students read pages 109 and 110 to tie together concepts they have already learned about, such as photosynthesis and the roles of producers and consumers.

Ask, "What are some other predator-prey relationships that you know about?" Record their examples.

◇ Evaluate

Ask students to name examples of herbivores, omnivores, and carnivores. Record these on the board. Have students arrange them in order (*beginning with the Sun*) through all the possible levels of consumers. The organisms that the students might include are: hawks, snakes, rabbits, grass, bacteria, deer, wolves, fish, foxes, trees, seed-eating birds, and worms.

NOTES

Food Chains and Energy

5 min.

The focus of this reading is on the movement of matter and energy in an ecosystem as reflected in the interactions between living things.

Some organisms are **consumers**. Consumers cannot make their own food. They take in food by consuming other organisms. Food chains can contain several different types of consumers.

- Primary consumers eat producers.
- Secondary consumers eat primary consumers.
- Tertiary (TER-shee-air-ee) consumers eat secondary consumers.

In the food chain shown on the previous page, the cow is a primary consumer and the human is a secondary consumer.

Consumers can also be classified as herbivores, carnivores, and omnivores. A **herbivore** is a consumer that eats only plants. Examples include cattle, horses, some insects, koala bears, and elephants. A **carnivore** is a consumer that eats mainly the meat of other organisms. Examples include hawks, eagles, snakes, spiders, and sharks. An **omnivore** is a consumer that eats both plants and the meat of other organisms. Examples include pigs, bears, some rodents, humans, and foxes.

Also, carnivorous consumers (organisms that are carnivores) are known as **predators**. These organisms hunt and kill other organisms, which are called **prey**. Prey does not have to be a carnivore. Prey can be a herbivore like an insect. However, the organism that hunts and eats it is always a predator. The photo on the right shows a predator, the hawk, and its prey, the mouse.

consumer: an organism that must get its food by eating other organisms.

herbivore: an organism that obtains its food only from plants.

carnivore: an organism that obtains its food only from other animals.

omnivore: an organism that will feed on many different kinds of food, including plants and animals.

predator: an organism that hunts and kills other organisms.

prey: organisms that are hunted and killed by other organisms.

Food Chains and Energy
Your cheeseburger diagrams are not complete food chains. You might have identified items that are only parts of an organism as consumers or producers.

△ Guide

Focus students' attention on the use of arrows in the diagram on page 111 in the student text. Ask, "Why do all of the arrows point upward?" Explain that the arrows represent matter moving from producers through consumers. They also represent energy as it moves through each level of the food chain. In the case of energy, ask, "Where does this energy come from?" Listen for students to say that the energy in a food chain comes from the Sun.

For example, you might have identified "tomato" and then drawn a line to "tomato plant." Food chains connect organisms that are producers and consumers. Tomatoes do not consume tomato plants. However, your diagram helped you identify the parts of the meal and analyze their sources. Some complete food chains for the cheeseburger meal might look like the ones on the right. Notice how the arrows point *up* the food chain, toward the consumer.

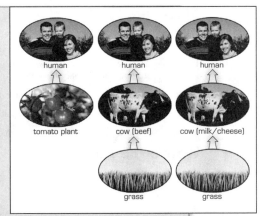

human   human   human

tomato plant   cow (beef)   cow (milk/cheese)

grass   grass

The arrows in a food chain show connections. When you examine the whole set of connections in each food chain, you can learn something very important. Remember earlier you saw that most food chains start with a plant that grows by photosynthesis. The plant takes energy (in the form of light) from the Sun. This energy is used to produce glucose (a type of sugar). The glucose is stored in the material that makes up the plant. It is stored in its stalk, its leaves, and its flowers.

In the cheeseburger food chains, the glucose is stored in the blades and seeds of the grass. The cow eats the grass. Next, it digests the grass. (Digest means to break down food so the body can use it.) The glucose is then passed on to the cow. The glucose and other nutrients are used to help the cow grow. They are stored in the tissues that make up the cow. When humans eat the hamburger, they digest the meat of the cow. By doing so, they obtain the nutrients in the meat.

The matter that makes up each member of the food chain is passed from the producer on to the highest-level consumer. This is a very important idea in understanding how living things are connected and rely on each other. The energy from the Sun that was processed by the plant is passed on to every member of the food chain as each member of the food chain is consumed. People would not have anything to eat without the Sun's energy.

Stop and Think

10 min.

A class discussion further refines ideas. If possible, try to have students make connections with their local environment.

Stop and Think

Use what you have read to answer the following questions. Be prepared to discuss your answers with the class.

1. Identify all of the producers and consumers in your diagrams. Which consumers are primary, secondary, or tertiary?

2. Identify the organisms that are herbivores, carnivores, or omnivores.

3. List all of the predators in your diagram. List all of the prey.

4. Explain how it is possible to be both a predator and prey.

What's the Point?

Using your diagrams, you were able to identify producers, consumers, carnivores, omnivores, and predators and prey. These food diagrams showed you how these organisms are connected.

Food chains are very important in understanding the health and condition of an ecosystem. The living things in an ecosystem rely on other living things as food sources. Energy from the Sun is, essentially, transferred up the food chain. This is a very critical aspect of understanding the ecology of a community. In the next section, your class will investigate changes in food chains and how these changes could affect the living things in an ecosystem.

Bears are omnivores.

LT 112

Project-Based Inquiry Science

○ Get Going

Prepare students for a class discussion by telling them to use information from their reading and the *Idea Briefing* poster discussion to answer the *Stop and Think* questions on page 112 in the student text. After they answer the questions, begin by asking the first question to initiate a class discussion.

1. Producers are the grain, tomato, and lettuce plants; consumers are the cows and humans.

2. Herbivores: cows because they feed on grain and grass; omnivores: humans because they feed on the cow and the plant matter in this particular food chain. Note that there are no carnivores in this food chain. (*If necessary, explain or accept the fact that some humans*

are vegetarians and vegans. Vegetarians may feed on some animal products (milk and eggs) but vegans abstain from all animal products.)

3. Humans might be classified by some as predatory because they feed on meat products, however, humans do not usually eat live prey.

4. Accept all reasonable responses. Hawks, wolves, or bears are predators. If someone traps and eats one of these predators raw, it then becomes prey. Cats are predators that eat live mice. Some dogs have been known to attack cats, which then makes the cat the prey.

△ Guide Discussions

Facilitate a class discussion about these four questions, reviewing the concepts of *producers, consumers, herbivore, carnivore, omnivore, predator* and *prey.* Ask students to provide context for their examples.

NOTES

....

....

....

....

....

....

....

....

....

....

....

Assessment Options

Targeted Concepts, Skills, and Nature of Science	How do I know if students got it?
Living things, such as green plants, trap energy from sunlight and produce their own food. They are called producers.	**ASK:** Explain why a green plant is classified as a producer? **LISTEN:** Students should be able to say that green plants trap the Sun's energy and produce their own food during photosynthesis.
Living things that consume other organisms are called consumers.	**ASK:** Why do herbivores, carnivores, and omnivores feed on other organisms? **LISTEN:** Students' answers should include that herbivores, carnivores, and omnivores feed on other organisms because they are incapable of producing their own food and depend on eating other organisms to get their energy.
Consumers may be classified as herbivores, carnivores, or omnivores.	**ASK:** Deer feed on leaves, bears feed on fish and berries, and bobcats feed on hares. Classify the deer, bear, and bobcat as herbivores, carnivores, or omnivores by what they eat. **LISTEN:** Students should be able to identify the deer as an herbivore, the fish as an omnivore, and the bobcat as a carnivore.
All living things need energy from the Sun and matter to survive. Energy gets trapped by producers and is transferred along with matter to consumers through food chains.	**ASK:** What moves from organism to organism through a food chain? **LISTEN:** Listen for students to say that matter and energy are moved through a food chain from organism to organism.

Teacher Reflection Questions

- Students may be familiar with predator-prey relationships. What examples of this relationship were students able to provide? How can you provide more support for learning about relationships in their own ecosystem?

- There are several discussions in this section. What steps did you take to make sure that the students stayed focused on concepts during the discussions?

- What improvements can be made to managing the poster session and the class discussion that follows it?

NOTES

3.6 Investigate

2 class periods* ▶

Modeling Changes in Food Chains

Overview

In the last section, students were introduced to the basic structure of food chains. In this section, students use a computer simulation to see how changes in one population affect other species in a food chain. Using the *NetLogo* computer simulation, all students run a computer-based population model on a community consisting of grass, mice, and coyotes. From this, they learn how changes in one population affect the other two populations in a community. Separate groups then investigate six different scenarios of changes to the grass-mice-coyote food chain to see how these changes impact the three species. Students design and present an *Investigation Expo* in which they articulate and refine their ideas as they discuss the results from their scenarios.

*A class period is considered to be one 40 to 50 minute class.

Targeted Concepts, Skills, and Nature of Science	Performance Expectations
A population consists of all members of a species within a certain area.	When asked what forms a population, students should be able to say that all members of the same species living in a particular area form a population.
Different populations living in an area form a community.	Students should be able to identify a community as being made up of populations of different species living in a given area.
The different populations that make up a community are connected through feeding relationships.	Students should be able to recognize the feeding relationships that exist among populations in a community.
Predators hunt and eat other organisms, known as prey.	Students should be able to identify organisms as predators and prey by how they feed or are used as food.
Populations in an ecosystem survive because of factors such as the size of the previous generations, availability of food, the number of surviving offspring, and certain abiotic factors.	Students should be able to recognize the effects of certain biotic and abiotic factors on the ability of some populations to survive.

Targeted Concepts, Skills, and Nature of Science	Performance Expectations
Scientists use models to simulate processes that happen too fast, too slow, on a scale that cannot be observed directly (either too small or too large), or that are too dangerous.	Students should be able to use computer models to explain phenomena they cannot control or observe in nature.
Scientists use models to help predict patterns related to various ecological phenomena.	Students should be able to use computer programs to predict patterns when given certain ecological phenomena.

Materials

For Teacher demonstration:
Computer running *NetLogo* version 3.1.2** (use Coyote/Mice/Grass model)
For students:

1 per group	Computer
1 per group	*NetLogo* version 3.1.2 (Coyote/Mice/Grass model)
1 per group	Poster paper
1 per group	Set of markers
1 per student	*Modeling Populations Predictions and Observations* page

Activity Setup and Preparation

- Run the simulations with *NetLogo* software well in advance so that you are familiar with the software, know what students should be observing, and can deal with any difficulties students might encounter. Pay particular attention to the graph window. Consider how you can project this so it can be viewed clearly.

- Load the *NetLogo* software on each student computer and run each ahead of time to make certain that it will work smoothly once students begin work on their own.

- The ideal situation is to have one computer for each student group. If this is not possible, use one computer to project the simulation interactively with the class. Information is provided in the *Implementation* segment of this section.

Homework Options

Reflection

- **Science Content:** If there were a population explosion of mice or coyotes in your population model, why might that population not increase indefinitely? *(Students' answers should say that the population could not continue to increase because a limited amount of food would not be able to support an unlimited number of mice or coyotes.)*

- **Science Content:** You are an ecologist studying the populations of a group of animals in a large lake. Write a letter to a fellow ecologist describing a food chain of animals in the lake. Identify producers and consumers in your food chain. Choose one consumer in your food chain. Describe what would happen if the numbers of that type of consumer suddenly doubled. *(Answers will vary depending on which consumer a student chooses. If the consumer that increases is a herbivore or primary consumer, then the population of producers will decrease as the consumer feeds on it. This will be followed by a decrease in the consumer as food becomes unavailable. If the student chooses a secondary consumer as its increased population, then the population of primary consumers will decrease and the producer population will increase. Eventually, the secondary consumer population will begin to decrease because its food supply is low or nonexistent.)*

Preparation for 3.7

- **Science Content:** In a community with populations of grass, mice, and coyotes, what other organisms might interact with these organisms? Draw a food web showing these relationships. *(Student choices and answers will vary but should show two or more food chains interacting with one another.)*

- **Nature of Science:** Select two land biomes or two aquatic biomes. Compare their major characteristics, such as temperature and rainfall in land biomes; depth and temperature variations in aquatic biomes. Decide how these characteristics might affect the types of organisms that live there. *(Responses will vary with the specific biomes chosen.)*

SECTION 3.6 IMPLEMENTATION

3.6 Investigate

Modeling Changes in Food Chains

You have investigated how organisms are connected in food chains. When you drink milk, you become part of a food chain that connects you, a cow, the grass, nutrients that help the grass grow, and the Sun.

You read that consumers need to get their food from other organisms. Therefore, all organisms are part of a food chain. On one side of a link in a food chain, an organism is connected to the organism it consumes. On the other side of the link, the organism is connected to the organism that consumes it.

Sunlight, grass, cows, and people form one type of a food chain.

In this section, you will be looking more closely at how the interactions in food chains work. Instead of looking at individual organisms, you will investigate how many individuals from different groups within a food chain interact. In the real world, there are many individuals of each type of organism. There are many blades of grass that a cow will eat. There are also many cows that produce milk. There are many humans who drink the milk produced by the cows.

LT 113

LIVING TOGETHER

3.6 Investigate

Modeling Changes in Food Chains

5 min.

○ Engage

Explain to students that they are going to use a computer simulation program called *NetLogo* to find out what would happen in a small community made up of three different populations if the individual populations were to change in some way.

*A class period is considered to be one 40 to 50 minute class.

Design a Population Model

10 min.

In their groups, students determine the roles of different populations in a basic food chain in preparation for using NetLogo, a computer simulation.

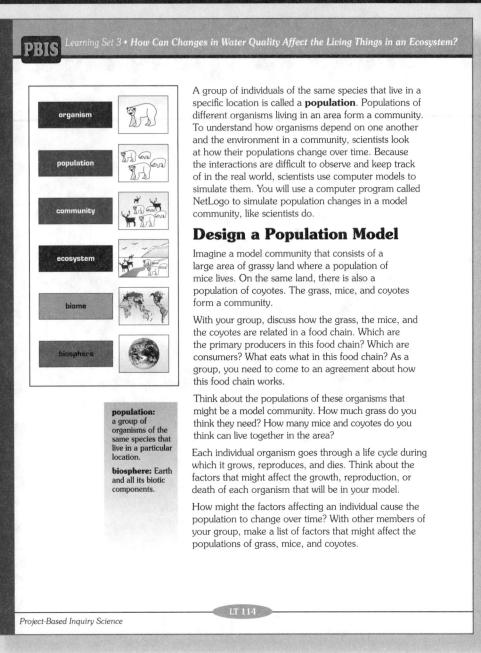

A group of individuals of the same species that live in a specific location is called a **population**. Populations of different organisms living in an area form a community. To understand how organisms depend on one another and the environment in a community, scientists look at how their populations change over time. Because the interactions are difficult to observe and keep track of in the real world, scientists use computer models to simulate them. You will use a computer program called NetLogo to simulate population changes in a model community, like scientists do.

Design a Population Model

Imagine a model community that consists of a large area of grassy land where a population of mice lives. On the same land, there is also a population of coyotes. The grass, mice, and coyotes form a community.

With your group, discuss how the grass, the mice, and the coyotes are related in a food chain. Which are the primary producers in this food chain? Which are consumers? What eats what in this food chain? As a group, you need to come to an agreement about how this food chain works.

Think about the populations of these organisms that might be a model community. How much grass do you think they need? How many mice and coyotes do you think can live together in the area?

Each individual organism goes through a life cycle during which it grows, reproduces, and dies. Think about the factors that might affect the growth, reproduction, or death of each organism that will be in your model.

How might the factors affecting an individual cause the population to change over time? With other members of your group, make a list of factors that might affect the populations of grass, mice, and coyotes.

population: a group of organisms of the same species that live in a particular location.

biosphere: Earth and all its biotic components.

○ Engage

Draw a basic food chain of grass, mice, and coyotes on the board. Have students work in their groups to think about the role of each species in the community *(producer, primary consumer, secondary consumer)* and to identify and agree upon which species would probably have more individuals in the food chain. *(Some students may think the mice will be most abundant, but the grass will have the largest population.).*

△ Guide

Ask the groups to create a list of factors that might affect the population of mice, grass, and coyotes. The two questions on page 133 will help focus students on factors that might increase or decrease the populations.

3.6 Investigate

The answers to the following questions will help you create the list:

- What might cause the amount of grass to change? List all the factors that might increase or decrease the amount of grass in your model community.

- What might cause the numbers of mice or coyotes to change? List all the factors that might increase or decrease the numbers of mice or coyotes in your model community.

Draw the food chain you imagine. Then record the factors that affect each population in your *Model Population Predictions and Observations* chart.

Model Population Predictions and Observations

Name: _____ Date: _____

1. Draw a diagram of the food chain that links the grass, mice, and coyotes in the space below.

2. List the factors that you think affect each population.

	Factors that might increase the population	Factors that might decrease the population
grass patches		
mice		
coyotes		

3. Record the data you collect when you run the model with Baseline Conditions. The first row of data has been entered for you.

	Baseline Conditions		
Generations	Number of grass patches	Number of mice	Number of coyotes
0 (start)	1412	205	70

4. Record the name of your scenario. Write down your prediction about what will happen to the number of grass patches, mice, and coyote. Then record the data you collect when you run the model.

	Baseline Conditions		
Generations	Number of grass patches	Number of mice	Number of coyotes
Prediction			
0 (start)			

How to Use NetLogo

You will be using a computer software program to investigate how organisms connected in a food chain can be affected by changes in the population of one organism. Open the program in your computer. Your teacher will tell you which model to open.

Once the model is loaded, you should see a screen like the one on the next page.

The screen is divided into three main sections:

- The *model input window* is on the left.
- The *graphics window* is on the right.
- The *plot graph window* is on the bottom.

The *model input window* is where you control the populations of grass, mice, and coyotes. You can slide the red bars on the blue slides to change the **parameters** for each organism.

parameters: the limits or boundaries (in this case, values).

How to Use NetLogo
20 min.

The students become familiar with the NetLogo program through a teacher's demonstration and suggestions for values to be used.

Provide each student with a blank copy of the *Model Population Predictions and Observations* page to record information in preparation for using the computer simulation and to keep track of their predictions.

Explain to students that you will demonstrate how to use the *NetLogo* software program before they start to use it.

○ Get Going

- **Start:** Use a computer connected to a projector to demonstrate *NetLogo*. Show students how to begin a specific example by pressing the *<setup>* button. The program is run using the *<go>* button.

- **Identify:** What each symbol in the image means and what the color coding indicates in images and graphs.

- **Demonstrate:** Show how to change the value of initial populations (*number-of-mice, number-of-coyotes*), birth rates (*max-mouse-offspring, max-coyote-offspring*), and death rate (*max-mouse-age, max-coyote-age*) for the variables by using the slide bars. Reassure students that these descriptions are also found in their text.

- **Demonstrate:** How to change the rate at which generations pass. The horizontal axis, labeled "Time" represents the number of *generations*, not minutes, hours, or days.

- **Demonstrate:** How to control the speed at which the generations pass by moving the "speed" indicator to the left to slow it down and make it easier to control.

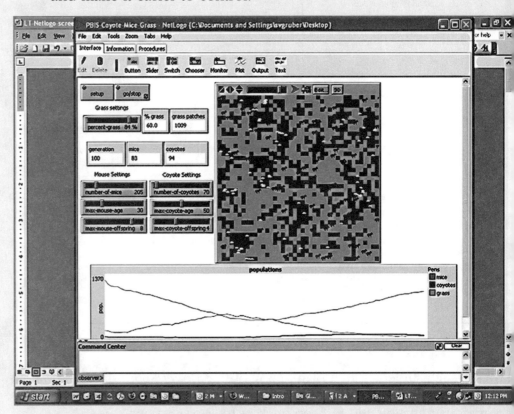

Set the factors for the grass, mice, and coyote as shown in the graphic on page 117 in the student text.

△ Guide

Emphasize watching the graph at the bottom of the screen. Support students' understanding of the graph by explaining what each line represents. Relate the key to the right of the graph so that students understand that the colors on the developing graph represent the colors of the key organisms. Ask, "What patterns do you see developing as the (*mouse population goes up and the grass population does down*)? What do you see happening to the coyote population?" *(The coyote population is difficult to see because the axis is compacted, but it does change slightly.)*

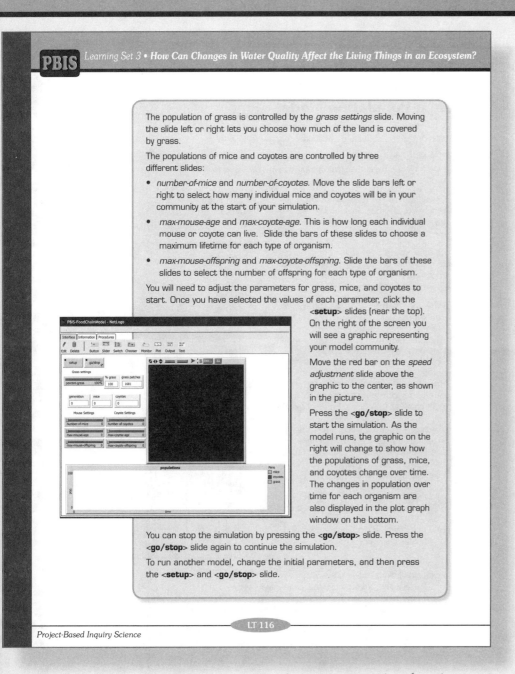

The population of grass is controlled by the *grass settings* slide. Moving the slide left or right lets you choose how much of the land is covered by grass.

The populations of mice and coyotes are controlled by three different slides:

- *number-of-mice* and *number-of-coyotes*. Move the slide bars left or right to select how many individual mice and coyotes will be in your community at the start of your simulation.

- *max-mouse-age* and *max-coyote-age*. This is how long each individual mouse or coyote can live. Slide the bars of these slides to choose a maximum lifetime for each type of organism.

- *max-mouse-offspring* and *max-coyote-offspring*. Slide the bars of these slides to select the number of offspring for each type of organism.

You will need to adjust the parameters for grass, mice, and coyotes to start. Once you have selected the values of each parameter, click the **<setup>** slides (near the top). On the right of the screen you will see a graphic representing your model community.

Move the red bar on the *speed adjustment* slide above the graphic to the center, as shown in the picture.

Press the **<go/stop>** slide to start the simulation. As the model runs, the graphic on the right will change to show how the populations of grass, mice, and coyotes change over time. The changes in population over time for each organism are also displayed in the plot graph window on the bottom.

You can stop the simulation by pressing the **<go/stop>** slide. Press the **<go/stop>** slide again to continue the simulation.

To run another model, change the initial parameters, and then press the **<setup>** and **<go/stop>** slide.

LT 116

Project-Based Inquiry Science

After the demonstration, tell students to read pages 115 and 116 in the student text again because they will better understand now that they have seen a demonstration. The reading will help prepare them to run the program on their own.

PBIS

Run a Population Model

15 min.

To learn how to use the NetLogo simulation on their own, students practice entering data. Student groups observe what happens during a simulation of a population model for realistic initial settings of grass, mice, and coyotes in a community. Groups discuss and analyze the data they obtain.

META NOTES

As much as possible, students work in their groups at a computer. If this is not possible, have students use a computer set up with a project so that they can all see the results of the changing values.

META NOTES

Before they attempt to run a simulation for a particular set of values, allow several minutes for mess-around time so that students get comfortable with how the program operates.

Run a Population Model

Now get ready to use the computer program NetLogo to simulate how the populations of grass, mice, and coyotes change.

1. Before you can use the program, you need some information about how it works. Read the section *How to Use NetLogo* to learn about the program.

2. Open the NetLogo model as instructed by your teacher. For the first simulation, all groups in your class will assume the same initial conditions for the model community. Your initial conditions are shown on this page.

3. Use the slide bars to adjust the parameters to the values on this page. These values represent a typical situation in a community of grass, mice, and coyotes.

4. Press the <**setup**> slide. Slide the *speed adjustment* bar toward the center. Observe the graphic that appears in the *graphics window*. Identify the objects in the graphic.

   a) What does each object represent?

5. Press the <**go/stop**> slide to start the simulation. Notice how the graphic on the right changes to represent changes in the populations after each generation. Let the simulation run for a while. Press the <**go/stop**> slide after about 100 generations. Look at the generation counter on the left of the screen to know when to stop. Notice any changes in the graphic at the right.

   a) Does the amount of grass look the same as when you started? How has it changed?

   b) How have the numbers of mice and coyotes changed?

6. To find the answers to the questions you made predictions about, you will run the simulation again. You will use the same initial conditions you just used. But this time, during the run, you will collect data to analyze later.

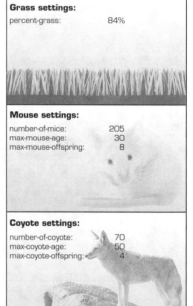

Grass settings:

| percent-grass: | 84% |

Mouse settings:

number-of-mice:	205
max-mouse-age:	30
max-mouse-offspring:	8

Coyote settings:

number-of-coyote:	70
max-coyote-age:	50
max-coyote-offspring:	4

LT 117

△ Guide and Assess

Monitor students as they begin to use the program on their own. Listen and watch for students who may not know what to do. If you see this, stop and guide them step by step, using the same figures you used in the demonstration. Make certain all students realize that they can change the individual values by sliding the bar for each component. If some students don't understand what "percent grass" means, explain that it means the part of the ground that is covered with grass plants. Show them how 100 percent would appear, and then have them move the slide to 84 percent. As they change the percent of grass, they will see this reflected on the screen image.

△ Guide

Point out to students that there are questions at specific points during the simulation practice on page 117 that they need to answer (*See question "a" at instruction 4 and questions "a" and "b" at instruction 5*). Answering these questions will help students to focus on what the ecosystem looks like as values are changed. These practice simulations will provide students with experiences they will use when they study ecosystems on page 122 in which other variables are introduced.

Below are some sample data that students might derive as they work through the simulation on pages 117 and 118.

META NOTES

In project-based science, computers and software support learning when they help teachers and students carry out investigations and create products.

Generations	Percent of ground covered by grasses	Number of mice	Number of coyotes
Initial settings: 0 generations	83.8%	205	70
20 generations	43.5%	367	45
40 generations	26.4%	514	35
60 generations	28.6%	321	59
80 generations	41.7%	144	66
100 generations	54.8%	117	75

△ Guide

After completing the simulations, tell students they will work in groups to complete the *Analyze Your Data* questions on page 118 and 119. Remind them that they will use their group results and conclusions in a class discussion that follows.

NOTES

....

....

....

....

....

Analyze Your Data

20 min.

Students analyze the data they collected and discuss it as a class to refine their analysis and ideas. Finally, students predict what they think will happen in the next 100 generations. This prepares them for learning about what populations depend upon in an ecosystem.

Record the initial conditions for the parameters in your data chart.

7. Restart the simulation by pressing first the <**setup**>, and then the <**go/stop**> slides. Stop the simulation after 20, 40, 60, 80, and 100 generations. After each stop, record in your data chart the following numbers, as they appear in the screen:

   • which *generation*—e.g., 0 (at start); 20; 40; 60; 80; 100

   • *number of grass patches*—locate this number on the screen. Record the value shown when you stop.

   • *number of mice*—locate this number on the screen. Record the value shown when you stop.

   • *number of coyotes*—locate this number on the screen. Record the value shown when you stop.

Don't worry if you cannot stop the program exactly after 20, 40, etc., generations. Try to stop as close to those numbers as possible. But remember to record in the data chart the *generation* that appears on your screen when you stop.

Analyze Your Data

Discuss the data you collected during the simulation with other members of your group.

Answer the following questions during the discussion. Record the answers. Be prepared to share your answers with the class.

1. What happened to the amount of grass during the simulation? Support your answer with evidence from the data.

2. How did the population of mice change during the simulation? Use examples from the data you collected when answering the question.

3. How did the population of coyotes change?

LT 118

In preparation for answering the *Analyze Your Data* questions, remind students that they must present evidence when answering a question. Tell students they have 10 minutes to answer the questions in their groups. Then, they will take about 10 minutes to discuss the answers as a class.

As they answer analysis questions, tell students to describe what happened to each population as compared to the original values or populations.

1. See the table on the next page. During the simulation, the percent of ground covered by grasses declined during the first 60 generations as the mice fed on it to the point where the ground was a little less than 30 percent grass. Then, more grass began to grow until about half the ground was covered by grass by the 100th generation.

Initial settings: 0 generations	83.8%
20 generations	43.5%
40 generations	26.4%
60 generations	28.6%
80 generations	41.7%
100 generations	54.8%

2. See the table below. The population of mice almost doubled in the first 20 generations (*from 205 to 367 mice*). By the 40th generation, the population had continued to grow to about 2.5 times the size of the original population (*205 × 2.5 = 512*). Then, the population started to decline. By the 60th generation, the total population of mice was down to about $1\frac{2}{3}$ the size of the original value (*205 × 1.66 = 340*). By the 80th generation, it was down to about $\frac{2}{3}$rd the original value (*205 × 0.66 = 135*) and was just a little lower (*117*) at the 100th generation.

Generations	Number of mice
Initial settings: 0 generations	205
20 generations	367
40 generations	572
60 generations	340
80 generations	135
100 generations	117

3. See the table. Overall, the coyote population decreased through the 40th generation then began to increase and by the 100th generation, surpassing its original population. Between the start of the study and the 20th generation, the population of coyotes decreased to about $\frac{2}{3}$rd (*70 × 0.66 = 46*) and to one-half its original value by the 40th generation (*70 × 0.5 = 35*). It increased to $\frac{4}{5}$th its original number (*70 × 0.8 = 56*) by the 60th generation; to $\frac{9}{10}$th its original population by the 80th generation (*70 × 0.9 = 63*) and to $1\frac{1}{10}$th its original number by the 100th generation (*70 × 1.1= 77*).

Generations	Number of coyotes
Initial settings: 0 generations	70
20 generations	46
40 generations	35
60 generations	56
80 generations	63
100 generations	77

4. When the mouse population increased, the grass population was decreasing, probably because there were more mice to eat the grass. As the coyote population started to increase, there was a noticeable decrease in the numbers of mice. This is probably because there were a larger number of coyotes feeding on mice.

5. From the graph and the table data shown previously-through the 40th generation-the numbers of mice increased while the percentage of grass cover decreased, and the numbers of coyotes decreased. Then, the number of coyotes began to increase and the mice population began to decrease. This pattern continued through the 100th generation. During this time, the grass population began to recover and increased as there were fewer mice to feed on the grass.

6. Student answers will vary. Students should be able to use their results to say that they observed an increase in mice while the grass decreased. Later, the population of mice started to decrease. At the same time, the grass population started to increase. This probably occurred because the coyotes had many mice to feed on. As fewer mice were feeding on the grass, the grass population could recover and increase.

7. Answers will vary. Overall, the coyote population appears to have changed the least. During the whole time, there were probably enough mice to keep the coyote population alive.

8. Students' predictions will vary. Accept all reasonable responses but require students to back their reasoning with evidence based on their results. Students should figure out that if the mice population continues to decrease, then, in time, the coyote population will decrease because there are fewer mice to eat.

4. Compare the change in the population of coyotes to that of the grass and mice in your data. Did the coyotes change more, less, or about the same as the grass and mice? Use evidence from your data to answer the question.

5. What relationships did you find among the populations of the different organisms? If you have a computer program available, you may want to create a graph of your results to see the relationships. The graph will help you to communicate your ideas better when you discuss the results with your class.

6. When did you observe periods during which one population is increasing while another is decreasing? Why do you think this might happen? Why do you think this happened?

7. Which population changed the least during the simulation? Why do you think a population might change only very little?

8. What do you think might happen to the populations in your model community in the future 100 generations? Describe what you think might happen using information from the data, and what you know about how the grass, mice, and coyotes are connected.

LT 119

LIVING TOGETHER

△ Guide

Have students run the NetLogo™ software again, but this time, let it go for 200 generations. If you run NetLogo™ for another 100 generations, this is what you will find: The mice population reaches its lowest level around 140 generations. Then, the coyote population begins to decrease, reaches its lowest values (*of around 7 coyotes*) around the 160th generation. As this occurs, the mice population begins to increase again. Help students recognize the cyclical nature in this.

Population Changes in an Ecosystem

10 min.

The reading helps students focus on factors in organisms that affect population size.

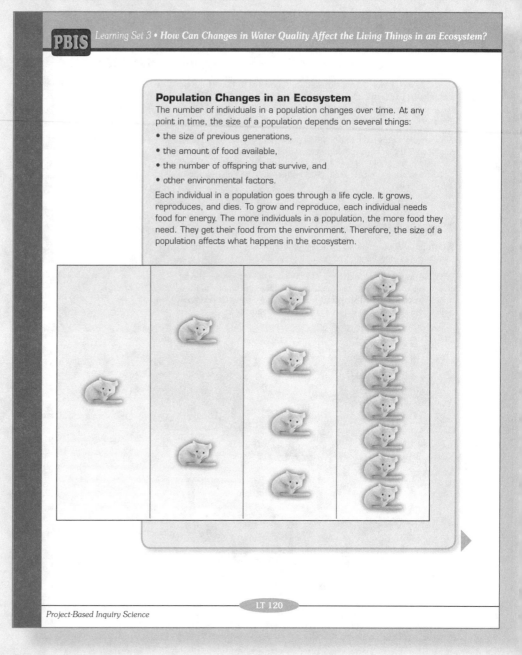

Population Changes in an Ecosystem

The number of individuals in a population changes over time. At any point in time, the size of a population depends on several things:

• the size of previous generations,

• the amount of food available,

• the number of offspring that survive, and

• other environmental factors.

Each individual in a population goes through a life cycle. It grows, reproduces, and dies. To grow and reproduce, each individual needs food for energy. The more individuals in a population, the more food they need. They get their food from the environment. Therefore, the size of a population affects what happens in the ecosystem.

LT 120

Project-Based Inquiry Science

◯ Get Going

Have students read the science information on pages 120 and 121 in the student text to help place the results of their simulation in a realistic context and prepare them for a class discussion. Draw students' attention to the four criteria in a population (*size of the populations, amount of food, survival of offspring, and environmental factors*). The reading reinforces why animals consume food-to obtain energy-and the influence the amount of food has on a population's ability to reproduce.

Math Connection

If environmental conditions are right, and food is abundant, populations can grow very fast. For instance, a population of the bacterium *Escherichia coli* (*E. coli*) can double in size every 20 minutes. A population of 100 individuals of *E. coli* can grow to over 50,000 individuals in just three hours.

In your computer simulation, mice get their energy from eating grass. The more mice, the more grass they will eat. As the population of mice increases, the amount of grass decreases. The mice have a direct effect on the amount of grass. However, with less grass, fewer mice can get enough food to survive. Therefore, the population of mice starts to decrease after awhile. The population of mice has a direct effect on the grass. The grass has a direct effect on the mice.

Suppose every mouse in your model gave birth to two mice. The number of mice would grow very quickly. The rate of reproduction has a large effect on how a population changes over time. When there is lots of food, more individuals are able to reproduce. More young become adults. This will increase the population over time.

This did not happen in your simulation. In fact, it rarely happens in the real world. This is because not all individuals reproduce. Some die of natural causes before they reproduce. Some are consumed by predators before they have offspring. When food is scarce, some die of starvation before they reproduce.

△ Guide Discussions

For the class discussion, have students focus on how changes in one factor in an environment can affect population levels of organisms. Remind groups to use evidence from their simulation in their answers.

To save time, select three groups to describe the separate changes in the mice population, the grass population, and the coyote population, and give reasons based on their evidence. If students cannot give a reason for the outcome, redirect the question so that other students can provide a reason.

Remind students of how to ask questions during a presentation. Tell students that in their questions, they can help focus on the person who will answer the question. Model the way students are to ask questions of each other, such as, "Ask the presenting groups a question that makes the group use data from their investigation, such as, 'At about what generation did you see the mice population go down and the coyote population increase?' "

After each group presents, ask students to agree on a summary as to what conditions caused population decreases and population increases. Answers will vary but students should include that as certain foods became scarce, the organism feeding on that food would decrease. As certain foods became more available, populations of organisms feeding on that food source would begin to recover and their numbers would increase.

NOTES

3.6

Run a Population Model When the Ecosystem Changes

equilibrium: a condition of a system in which all influences cancel one another. The result is an unchanging, or balanced, system.

You have used NetLogo to model changes in the populations of grass, mice, and coyotes in a healthy ecosystem. When an ecosystem is healthy, population sizes change temporarily. They can increase or decrease over a few generations. But they tend to recover and become stable over time. Over many generations, the numbers of grass patches, mice, and coyotes reach **equilibrium**. These conditions represent a balanced ecosystem.

You are now ready to use NetLogo to simulate how populations in a community change when the conditions in the ecosystem change.

Procedure

1. Each group will model one of the six scenarios on the next pages. Read your assigned scenario. As you are reading, pay attention to the section describing the conditions that have developed. Look at how those changes correspond to parameter settings in your model.

2. Predict what will happen in your simulation based on what you learned about how this ecosystem works in your previous simulation. Record what will happen to the populations of grass, mice, and coyotes when conditions change as described in your scenario. Report your prediction in your data chart.

3. Select the initial parameters for the grass, mice, and coyotes.

4. Make sure the speed adjustment bar is set at mid-range. Press the <**setup**> and <**go/stop**> slides to start the simulation.

5. Stop the simulation after 20, 40, 60, 80, and 100 generations as you did earlier. Collect data for the numbers of grass patches, mice, and coyotes each time you stop. Record this data in your data chart, together with each generation number.

A wood mouse eating fallen sunflower seeds from a bird feeder in a garden.

Run a Population Model When the Ecosystem Changes
5 min.

Students have been observing the reactions of a balanced environment or equilibrium. Students now investigate changes that occur in a community when conditions in the ecosystem change.

Procedure
15 min.

Students run a new simulation with some variables changed.

○ Engage

Now that students are comfortable with running the software on the balanced ecosystem, explain that each group will run the program again using a slight variation in the parameters.

Tell students that there will be six variations and that these are described as six different scenarios on pages 123 and 124 of their textbook. Assign the six scenarios among six groups.

Establish a connection between the work students are about to do and the concepts they have been learning, namely that populations in a community are connected in some way through food chains in the community.

"Before you begin work on the scenarios, let's think about what is taking place. Look back at pages 113 and 114 and in the practice simulations. What did you learn about organisms there? *(Food chains are made up of populations of organisms that interact with one another in a community.)* These were fairly simple relationships that you saw in the first one hundred generations. As you went into the second hundred generations, you could see trends develop. The way the populations increased or decreased might have become predictable. They were *stable* or *balanced,* or in *equilibrium.* Nothing happened to change the patterns of relationships in the food chains. If you went on for another hundred generations, you would probably become pretty good at predicting what would happen because you saw how the mice and grass and coyotes acted before.

As you begin the new simulations, you will continue to work with populations of the same kinds of organisms, but in each scenario, something will happen to upset that balance. In each case, some kind of restriction will be placed on the food chain. Your job will be to think about how the food chain is being affected *(lack of water, disease, or pesticides)* and predict what will happen to the populations in the food chain. Then, you will run the simulation and see what happens to the population.**"**

Remind students that they will be responsible for:

- the setup of their scenario,
- making a prediction, and
- collecting data.

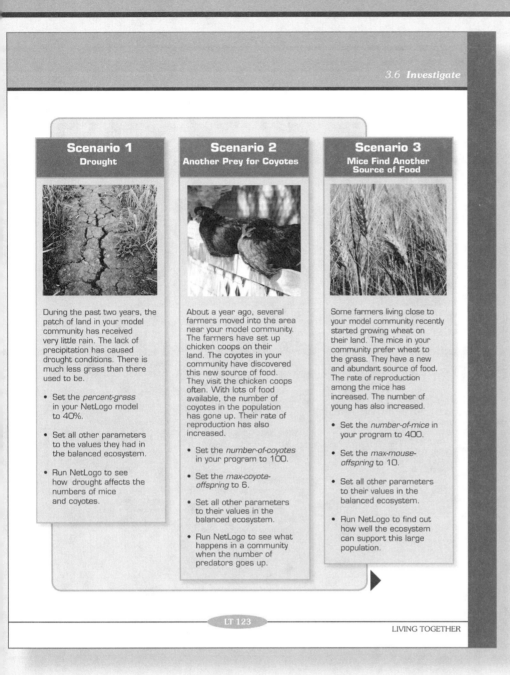

Scenario 1
Drought

During the past two years, the patch of land in your model community has received very little rain. The lack of precipitation has caused drought conditions. There is much less grass than there used to be.

- Set the *percent-grass* in your NetLogo model to 40%.

- Set all other parameters to the values they had in the balanced ecosystem.

- Run NetLogo to see how drought affects the numbers of mice and coyotes.

Scenario 2
Another Prey for Coyotes

About a year ago, several farmers moved into the area near your model community. The farmers have set up chicken coops on their land. The coyotes in your community have discovered this new source of food. They visit the chicken coops often. With lots of food available, the number of coyotes in the population has gone up. Their rate of reproduction has also increased.

- Set the *number-of-coyotes* in your program to 100.

- Set the *max-coyote-offspring* to 6.

- Set all other parameters to their values in the balanced ecosystem.

- Run NetLogo to see what happens in a community when the number of predators goes up.

Scenario 3
Mice Find Another Source of Food

Some farmers living close to your model community recently started growing wheat on their land. The mice in your community prefer wheat to the grass. They have a new and abundant source of food. The rate of reproduction among the mice has increased. The number of young has also increased.

- Set the *number-of-mice* in your program to 400.

- Set the *max-mouse-offspring* to 10.

- Set all other parameters to their values in the balanced ecosystem.

- Run NetLogo to find out how well the ecosystem can support this large population.

LT 123

LIVING TOGETHER

Explain to students that they will use the evidence they collect during the scenario simulations to prepare for the *Investigation Expo* after the simulations have been run.

◯ Get Going

Before students set up their parameters, have each group read their scenario. Ask students to make certain of the factors in their scenario and to discuss this in their group. Based on their discussion, remind them to record a prediction of what they think will happen because of the change in the parameters.

META NOTES

Ideally, these simulations can be run on six different computers simultaneously. If that is not possible, select three scenarios and run each one, projecting the results. Two student groups can be responsible for the data of one of these projected scenarios for the *Investigation Expo.*

Scenario 4
A Season of Abundant Rains

For about one year now, the patch of land where your community lives has received lots of rain. As a result, there is a lush carpet of grass covering the entire area.

- Set the *percent-grass* in your program to 100%.

- Set all other parameters to their values in the balanced ecosystem.

- Run NetLogo to see what is the effect of this abundant source of food for the mice and coyotes in the community.

Scenario 5
A Coyote-Virus Epidemic

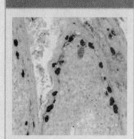

Over the past several months, a virus has infected the population of coyotes in your model community. First, some coyotes got the virus from drinking contaminated water. The infection then spread quickly among the population. Many adult coyotes have died. The number of offspring has also dropped dramatically.

- Set the *number-of-coyotes* in your program to 8.

- Set the *max-coyote-offspring* to 2.

- Set all other parameters to their values in the balanced ecosystem.

- Run NetLogo to see how a reduction in the number of coyotes affects the mouse population. Observe how the number grass patches changes too.

Scenario 6
Pesticides in the Environment

Large areas of land near your community are cleared to grow crops. To prevent bugs from destroying the crops, the fields are sprayed with **pesticides**. Unfortunately, these chemicals affect the young mice in your community. Many of them die before reaching maturity and reproducing.

- Set the *max-mouse-offspring* to 2.

- Set all other parameters to their values in the balanced ecosystem.

- Run NetLogo to see how a reduction in primary consumers affects the predator population and the number of grass patches.

LT 124

△ Guide and Assess

Monitor students as they set up their scenarios. Are they running the correct scenario? If students appear unsure of what to do, stop and suggest some ways to help them organize and focus.

- What is my scenario about? Do I understand what is described? (*What is a "drought?" What is a chicken coop used for? What does "drop dramatically" mean?*)

- What is the new parameter?

- What are the "other parameters" from the balanced ecosystem for my scenario?

- What will my prediction be?

366

Analyze Your Data

1. Compare the results of your scenario simulation with your prediction of what would happen. Was the prediction accurate? Why or why not?

2. Compare the results of your scenario simulation with the simulation of the balanced ecosystem you ran earlier. What changes in the populations have occurred?

3. What was the effect of the change in conditions in your simulation on the populations of organisms in your community? Do your best to explain why.

pesticide: a chemical used to prevent bugs from destroying crops.

Communicate Your Results

Investigation Expo

Use the *Analyze Your Data* questions as a way to discuss with your group the results of your simulation.

Create a poster describing the conditions you were modeling in your simulation. Make the description as detailed as you possibly can. Include all the parameters you selected to run your simulation. Illustrate how the populations of mice, grass, and coyotes changed. Indicate on your illustration the relationships between the populations. Also include an explanation of how the changes in one population caused changes in the others.

During the *Investigation Expo,* you are going to explain how your model community worked. You need to include enough details in your presentation so your class will understand how the changes in one population, or conditions in the environment, affected the other populations. Answer all of the following questions in your presentation:

1. How did the populations of grass, mice, and coyotes change in your simulation?

2. Was the amount of grass after 100 generations more than, less than, or about the same as at the beginning?

3. Did the numbers of mice and coyotes increase, decrease, or stay the same after 100 generations?

4. What changed in your new scenario compared to the balanced ecosystem?

5. Do you think the conditions of your simulation improved or worsened the health of the balanced ecosystem?

LT 125

Analyze Your Data

10 min.

Students analyze their results and prepare for the Investigation Expo.

△ Guide

Tell students that they have 10 minutes to discuss and formulate answers to the *Analyze Your Data* questions for their scenario. Explain that after that, they will create a poster that reflects their scenario outcome. Tell them that the poster and the answers will be used in an *Investigation Expo* where all scenarios will be presented, and students will have the opportunity to compare their results with others.

Sample Data

Scenario 1: Drought

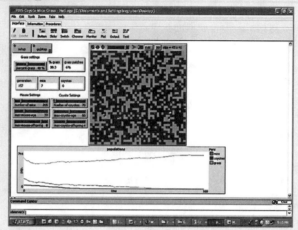

1. Answers (to *Analyze question 1*) will vary with the scenario depending on how the prediction aligns with the results of the simulation. Remind students that their reasons are to be supported by evidence from their simulation data.

2. Answers will vary but should include that in the balanced simulation, there was (apparently) sufficient rain to maintain a grass population so that the mice could feed and live and the coyotes had enough mice to live on. In the drought simulation, the population of mice gradually died down to very few and the coyote population died down. The grass population initially died back somewhat, but in time, with fewer mice to feed on it, the grass population was increasing again by the 100th generation.

3. The number of mice has decreased dramatically because there wasn't enough grass to support them. The coyotes died out because there weren't enough mice to feed on. According to the graph, the grass appears to be increasing toward the 100th generation.

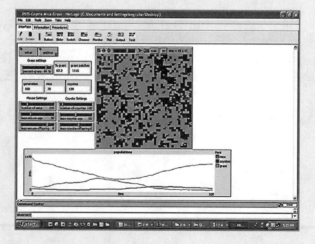

Scenario 2: Another Prey for Coyotes

1. Answers (to *Analyze question 1*) will vary with the scenario depending on how the prediction aligns with the results of the simulation. Remind students that their reasons are to be supported by evidence from their simulation data.

2. The coyote population has increased; the mice population increased for awhile and then decreased as the coyote population started to rise. The grass population decreased while the mice population increased, but after the mice declined, the grass increased.

3. The coyote population increased due to additional food (*chickens*), and the mice population declined as the coyote population rose because there were more coyotes to feed on them; the grass population increased because there were fewer mice feeding on it.

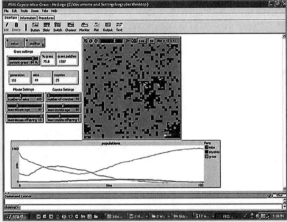

Scenario 3: Mice Find Another Source of Food

1. Answers (to *Analyze question 1*) will vary with the scenario depending on how the prediction aligns with the results of the simulation. Remind students that their reasons are to be supported by evidence from their simulation data.

2. The mice population by 100 generations is less than ¼ of what it was initially because the coyotes more than doubled in population. By this same point, there are fewer coyotes overall because the mice population has declined. The grasses are flourishing because there are so few mice to feed on it.

3. Initially, there were many mice and the coyote had plenty to eat. The coyote population increased to almost three times its initial value by the 60th generation. This caused a large drop in the mice population. The decrease in the mice population caused a decrease in the coyote population. By the end of 100 generations, both mice and coyote populations were below where they started. With very little to feed on the grass, the grass population increased.

Scenario 4: A Season of Abundant Rain

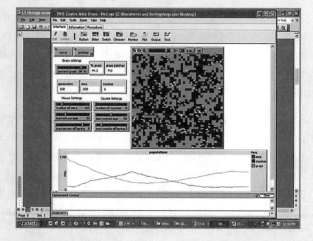

1. Answers (to *Analyze question 1*) will vary with the scenario depending on how the prediction aligns with the results of the simulation. Remind students that their reasons are to be supported by evidence from their simulation data.

2. The mouse population increased and the grass population decreased to about the 30th generation. Then the coyote population began to increase and the mouse population declined. As the mice declined, the grass population was able to increase again.

3. Abundant rainfall caused the grass to increase and provided much food for the mice. The mice population increased to about the 34th generation. At about that time, because there were so many mice, the coyote population began to increase. The coyote population more than tripled by the 65th generation. This resulted in a steady decrease in mice until, around the 90th generation, the mice population reached a low of less than half its original population. By then, the coyote population was decreasing due to fewer mice being available to feed on. As the number of mice declined, the grass had a chance to recover.

Scenario 5: A Coyote-Virus Epidemic

1. Answers (to *Analyze question 1*) will vary with the scenario depending on how the prediction aligns with the results of the simulation. Remind students that their reasons are to be supported by evidence from their simulation data.

2. The mice population seemed to cycle with the grass population, increasing when there was plenty of grass and decreasing when the grass reached a lower percent of coverage. The coyote population remained constantly low for the entire 100 generations.

3. The mice population cycled with the grass population, probably because there were so few coyotes to feed on the mice. It was as if the coyote population did not exist. The coyote population hardly changed, only becoming slightly lower because so few offspring survived. The coyote population was so low that it had no impact on the mice population.

Scenario 6: Pesticides in the Environment

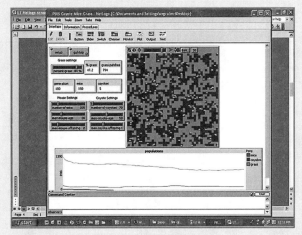

1. Answers (to *Analyze question 1*) will vary with the scenario depending on how the prediction aligns with the results of the simulation. Remind students that their reasons are to be supported by evidence from their simulation data.

2. The mice population steadily decreases. The coyote population also decreases. The grass population stays at a high level.

3. Because the mice population is kept so low by early die-off, it has little effect on the grass, which seems to grow unhindered. Because the mice population is kept low, there is little for coyotes to feed on. Therefore, the coyote population is also kept at a low level.

Communicate Your Results
Investigation Expo

20 min.

Posters and class discussions help students articulate and refine their ideas about how communities can change as they share the results of various scenarios.

on _____ organisms in yo_____ your best
to exp_____

Communicate Your Results

Investigation Expo

Use the *Analyze Your Data* questions as a way to discuss with your group the results of your simulation.

Create a poster describing the conditions you were modeling in your simulation. Make the description as detailed as you possibly can. Include all the parameters you selected to run your simulation. Illustrate how the populations of mice, grass, and coyotes changed. Indicate on your illustration the relationships between the populations. Also include an explanation of how the changes in one population caused changes in the others.

During the *Investigation Expo,* you are going to explain how your model community worked. You need to include enough details in your presentation so your class will understand how the changes in one population, or conditions in the environment, affected the other populations. Answer all of the following questions in your presentation:

1. How did the populations of grass, mice, and coyotes change in your simulation?
2. Was the amount of grass after 100 generations more than, less than, or about the same as at the beginning?
3. Did the numbers of mice and coyotes increase, decrease, or stay the same after 100 generations?
4. What changed in your new scenario compared to the balanced ecosystem?
5. Do you think the conditions of your simulation improved or worsened the health of the balanced ecosystem?

LT 125

LIVING TOGETHER

○ Engage

For the *Investigation Expo,* remind students that each group will need the following:

- data from the ecosystem in equilibrium,
- data and conclusions from the ecosystem described in the scenario,
- their group poster.

△ Guide

Explain that to get the most benefit from the presentations, there are some criteria for the presentation.

Posters should include:

- the *conditions* that were modeled,
- all *parameters* clearly spelled out,
- the *prediction* made about what the group thought would happen
- an *illustration* of how the populations changed (*students might use the graph from the run*)

- an *explanation* of what the illustration means, and
- how their results supported or did not support their prediction.

Quickly review what an *explanation* is based upon: claim, evidence, and science knowledge.

△ Guide Discussions and Presentations

For the immediate task of discussing the individual scenarios, focus students on identifying their prediction, representing their data accurately, and presenting the groups' outcome clearly on their poster.

Collect finished posters and arrange them in the room so that students will be able to compare the results of one scenario with that of another. Give students five minutes to peruse posters from other groups.

Remind students to listen carefully and ask questions of the presenting students if there is something that they do not understand. Remind them to direct their question to the presenter and not to you (*the teacher*). If necessary, model this behavior for the class as you have previously.

NOTES

As you listen to the investigation presentations of the other groups, observe how the populations change for each scenario. Compare the relationships among the populations in the different scenarios. Are the relationships the same? Did any of the scenarios result in the disappearance of some organisms? How do you think that might affect the populations of the other organisms in the community?

What's the Point?

Populations of organisms vary over time because of interactions with other organisms in the food chain and because of changes in the environment. Given enough time to adjust, a community in an ecosystem reaches a balance where the populations of organisms fluctuate up and down a little bit but remain relatively stable.

This balance can be upset when conditions in the environment cause changes in one population. Because the organisms interact with one another and their environment, all other populations are affected as well.

Sometimes the changes introduced in the environment result in the disappearance of an organism. In the next section, you will continue to investigate the relationships among the organisms in a community. You will also examine the relationship between organisms and their environments.

Populations of animals in a healthy ecosystem change temporarily but will reach equilibrium over time.

Each group should be prepared with answers to questions 1 through 5 on page 125. Students who are listening can ask one of these questions if the presenting group neglects to include it in their presentation.

Conclude the discussion by telling students that during the next section, (*Section 3.7*) they will continue to explore the relationships between organisms in ecosystems. They will expand the number of types of organisms that are important in these relationships.

Assessment Options

Targeted Concepts, Skills, and Nature of Science	How do I know if students got it?
A population consists of all members of a species within a certain area.	**ASK:** How many populations of birds are there in a flock of 114 house finches, 28 white-breasted nuthatches, 38 dark-eyed juncos, and three red-tailed hawks? **LISTEN:** Students should be able to say that there are four populations in the flock. Students' answers should include that all members of a population belong to the same species.
Different populations living in an area form a community.	**ASK:** At a small pond, there are four populations of birds, one population of frogs, twelve populations of insects, eight populations of trees, six populations of grasses and water plants, and two populations of fish. Why is this pond a community? **LISTEN:** Students should be able to say that the pond is a community because it is made up of different populations that live in the area.
The different populations that make up a community are connected through feeding relationships.	**ASK:** How are the pond community members described interconnected? **LISTEN:** Students' answers should be able to say that the pond-community members are interconnected by feeding relationships and form several food chains.
Predators hunt and eat other organisms, known as prey.	**ASK:** In the grass-mice-coyote food chain, which organism is a predator and which is prey? **LISTEN:** Students should be able to identify the coyote as a predator and the mice as prey.

Targeted Concepts, Skills, and Nature of Science	How do I know if students got it?
Populations in an ecosystem survive because of factors such as the size of the previous generations, availability of food, the number of surviving offspring, and certain abiotic factors.	**ASK:** In an ecosystem of grass and mice, what do you predict will happen to the mice population if there is twice the normal amount of rainfall one year? **LISTEN:** Students will probably predict that the mice population will increase because there will be abundant grass to feed on.
Scientists use models to simulate processes that happen too fast, too slow, on a scale that cannot be observed directly (either too small or too large), or that are too dangerous.	**ASK:** Why would a scientist use a simulation to understand food chain relationships in a hard to reach mountain area? **LISTEN:** Students' answers will vary but might include that the area is hard to reach. This might invite ideas that the scientist could not spend much time there, that it is too expensive to visit, that he or she collects information on one visit and then constructs different models with that information in the laboratory.
Scientists use models to help predict patterns related to various ecological phenomena.	**ASK:** Why would a scientist model the effects of different amounts of rainfall or food availability or numbers of offspring? **LISTEN:** Answers will vary. Reasonable responses might include that predicting based on models might help scientists to prevent future problems in that ecosystem.

NOTES

....

....

....

....

....

Teacher Reflection Questions

- Setting up computer simulations can be difficult. How did students handle this section? How well did they understand the scenarios they investigated and the graphs that were generated? What can be done the next time to improve their understanding of the outcomes?

- How accurately did students compare their predictions with their results? Were they able to develop claims and support them with evidence from their simulations? How can they be better supported in connecting evidence with their claims?

- What improvements can be made to managing the use of computers in the classroom, especially when a group is working on one computer?

NOTES

....

....

....

....

....

....

....

....

....

....

....

....

SECTION 3.7 INTRODUCTION

3.7 Explore

1 class period* ▶

Connections between Living Things in an Ecosystem

Overview

Using Food Web Cards, pairs of students construct food chains and find that they are not always one-dimensional. Working with other pairs, students combine food chains that they have constructed to create food webs. In this way, students learn that *food webs* are multidimensional food chains connected to each other through interacting organisms. Groups present their findings to the class in an *Idea Briefing* and observe that other food webs can be created using the same ecosystem cards.

Students also learn that energy is transferred via food webs through the different levels of an ecosystem, and discover the role of decomposers in an ecosystem. In closing, students learn that the world's biomes are large aquatic or terrestrial ecosystems, distinguished from each other by differences in temperatures and the availability of water. They close by updating the class *Project Board* and prepare to address the *Big Challenge* presented to them at the start of the Unit.

*A class period is considered to be one 40 to 50 minute class.

Targeted Concepts, Skills, and Nature of Science	Performance Expectations
All living things need energy from the Sun and matter to survive. Energy gets trapped by producers and is transferred along with matter to consumers through food chains.	Students should be able to say that all that energy is ultimately supplied by the Sun and is transferred from organism to organism through food chains and food webs.
Food webs show a complex network of interlocking food chains and the direction of energy flow through an ecosystem.	Students should recognize food webs as complex networks through which energy flows in an ecosystem.
Earth is divided into large ecosystems called biomes, which vary by climate and in which specific organisms survive because they are adapted to those conditions.	Students should be able to say that biomes are large ecosystems with organisms adapted to those particular conditions.

Materials

1 per class	Set color copied laminated Food Web Cards
1 per student	Blank U.S. biomes map
1 per class	Map of U.S. biomes
1 per group	Blue erasable transparency marker
1 per group	Green erasable transparency marker
1 per class	Masking tape
6 per group	Popsicle sticks with a prominent arrow pointing to one end of the stick
1 per group	Poster paper
1 per group	Set of markers
1 per student	*Food Chain Records* page
1 per group	Blank transparency map of the U.S.
1 per class	Class *Project Board*

Activity Setup and Preparations

- Create a food web using the Food Web Cards for the watershed ecosystem before teaching the class. This will prepare you for the various combinations students might devise for the activity of this *Learning Set*.

Homework Options

Reflection

- **Science Content:** Explain why decomposers and detritivores are important in an ecosystem. (*Answers will vary but students should be able to explain that decomposers are important because they break down and return nutrients to the soil where they can be used again. Detritivores are important because they obtain energy from feeding on dead organisms.*)

- **Science Content:** In what ways are food webs like a recycling facility? (*Answers will vary but should include that energy and materials are moved or transferred from one organism to another* *in a food web. Along the way, organisms die and their remains are fed on by detritivores and finally broken down completely by decomposers such as bacteria and fungi. These organisms return raw materials to the soil where they can be used again.)*

Preparation for Answer the Big Question/ the Big Challenge

- **Science Content:** Think about the description of the ecology around the town of Wamego. Select one or more terrestrial biomes into which Wamego would fit. Justify your biome choice with evidence from your study of the area. *(Answers will vary. Most students will probably choose the grassland biome because of the extensive farming (nearly 95 percent of the residents are employed by Wamego's farming businesses) that was described in the beginning of the Unit.)*

- **Nature of Science:** How do you think the challenges facing Wamego will be settled? *(Answers will vary. The challenge will probably be a compromise between scientific evidence and the necessity of meeting the economic needs of the community.)*

NOTES

....

....

....

....

....

....

....

....

....

....

SECTION 3.7 IMPLEMENTATION

3.7 Explore

Connections between Living Things in an Ecosystem

You investigated the effects of sunlight on the growth of plants and how plants use energy from the Sun during photosynthesis to create food. The energy is used to make sugars. The plants use the sugars to function and grow. Plants are producers. They are the foundation of the food chain. Without the Sun, plants cannot grow. In your model food chains, you saw how populations can change when plant growth changes.

ithin this quiet country scene are many organisms that e en on a healthy aquatic ecosystem.

Many organisms live in a watershed along with plants. You will work with your group to explore how organisms in a watershed ecosystem rely on one another. Your teacher will provide you with a set of cards. Each card has a picture of a single organism. The card also contains information about the organism. You will use these cards to explore possible connections among these living things.

Procedure: Simple Connections

1. With a partner, examine the Food Web Cards. Read and discuss all the information on each card. The cards will help you think of possible connections between the organisms pictured. If you have any questions about the information, ask your teacher for more explanation.

2. Assemble a food chain using the cards. Place the cards on a flat surface (table or desk). Make connections between the cards using $1\frac{1}{2}$ in. x 2 in. sticky notes on which you have drawn arrows. Draw one arrow on each sticky note. Make sure each sticky note has the arrow pointing to the right. Once you have built your idea of the food chain, record your food chain on a *Foo Chain Recor s* page, like the one shown on the next page. Circle the organisms in the chain, and make sure the arrows point in the direction of the energy flow.

Materials
- **Food Web Cards**
- **sticky notes**

LT 127

LIVING TOGETHER

3.7 Explore

Connections between Living Things in an Ecosystem
5 min.

The reading reviews the central concepts about the importance of the Sun and photosynthesis to food chains.

○ **Engage**

Have students review basic concepts about why producers are the basis of food chains. Tell them that reviewing this central factor will help them to build food chains that illustrate the connections between organisms that share resources in a watershed.

---

*A class period is considered to be one 40 to 50 minute class.

Procedure: Simple Connections

10 min.

Students work in groups to construct food chains using Ecosystem Flash Cards with watershed organisms.

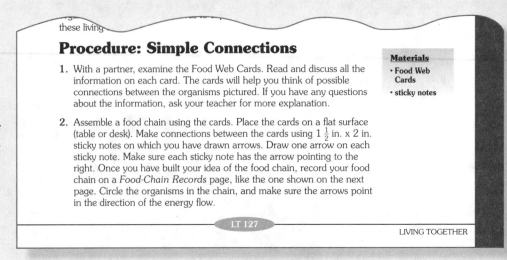

these living

Procedure: Simple Connections

1. With a partner, examine the Food Web Cards. Read and discuss all the information on each card. The cards will help you think of possible connections between the organisms pictured. If you have any questions about the information, ask your teacher for more explanation.

2. Assemble a food chain using the cards. Place the cards on a flat surface (table or desk). Make connections between the cards using $1\frac{1}{2}$ in. x 2 in. sticky notes on which you have drawn arrows. Draw one arrow on each sticky note. Make sure each sticky note has the arrow pointing to the right. Once you have built your idea of the food chain, record your food chain on a *Food-Chain Records* page, like the one shown on the next page. Circle the organisms in the chain, and make sure the arrows point in the direction of the energy flow.

Materials
- Food Web Cards
- sticky notes

LT 127

LIVING TOGETHER

△ Guide

Explain that students will works in groups. Initially each group will break into two pairs. Each pair will construct a food chain from *Ecosystem Flash Cards* and flat popsicle sticks that you provide. Then, the pairs will join in their group to work together to combine their food chains. Do not tell them at this point, but they will find that the separate food chains will combine to form a more complex series of feeding relationships, namely, a *food web*.

Distribute the *Ecosystem Flash Cards* and popsicle sticks. Provide each student with a *Food-Chain Records* page.

⬡ Get Going

Tell students that they have three minutes to work with a partner to read the information on the back of the cards and assemble a simple food chain. Remind students to record their food chain on a Food-*Chain Records* page.

Signal the end of three minutes and have each pair join with the rest of their group.

Students should continue to follow the directions in their textbook, making sure they record the food chain constructed by the second group. Allow students about 7 minutes to complete this second phase.

3. Meet with the other half of your group. Compare your cards and food chains. Record their food chain on your *Food-Chain Records* pages in the space indicated.

4. As a group, combine your two food chains into one connected food chain. Be careful. There is a catch: *If both food chains have the same organism, your new configuration can have that organism listed only once.*

   For example, if you both have a card with a rabbit on it, you can use only one of the rabbit cards. As before, place the cards on a flat surface (table or desk). Make connections between the cards using the sticky notes. (Hint: It is possible to place more than one note to and from an organism.)

5. Once you have built this larger food chain, record this third food chain on your *Food-Chain Records* pages in the space indicated. Circle the organisms in the chain, and make sure the arrows are pointing in the direction of the energy flow.

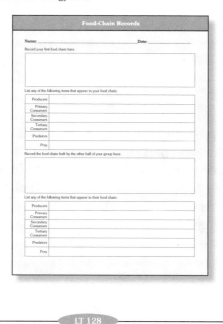

△ Guide and Assess

Circulate among the groups, especially as they begin to discover that their food chains may overlap in some way. Listen for problems that might occur at Procedure 4, when students learn that they can use an organism only once. If necessary, draw their attention to the *Hint* about being able to use more than one arrow stick with a single organism. If necessary, stop the entire class and demonstrate how one organism can be part of two food chains using an example based on experiences students are familiar with. Draw arrows as you demonstrate the example. Make certain that you draw two arrows at one point coming from or going to one of the organisms.

TEACHER TALK

"Think about a food chain that you are part of, probably every day."

Food Chain 1: Sun ⟶ oat or wheat plant ⟶ student

Food Chain 2: grass ⟶ cow

TEACHER TALK

"Each day, you probably eat cereal or bread that is made from oats or wheat. That makes you part of that food chain. You also probably drink milk, eat butter, or some other milk-based product *(pudding)*. This means that you *(and the Sun)* are part of both food chains and will have two arrows going to or coming from that entry."

3.7 Explore

The Missing Link in the Food Chain: Decomposers

Consider what happens to the matter and energy after the final consumer. There is another type of role in the food chain in addition to producers and consumers. The third role is the **decomposer**. Decomposers break down the tissues of dead plants and animals. Bacteria and fungi are examples of decomposers that break down animal tissue. There are also animals known as **detritivores**. These include buzzards, flies, earthworms, and cockroaches. They feed on dead plants and animals. In turn, other decomposers break down their bodies when they die. The role of the decomposer is to get rid of all the waste and tissues of dead plants and animals.

In the end, decomposers leave behind nutrients taken from the plant or animal tissue and leave them in the surrounding soil and water. These nutrients are then processed by plants during photosynthesis. The whole food chain begins again. Once again, energy from the Sun is passed on to other organisms.

The fungus growing on the dead tree trunk helps it decompose by breaking down the tissues of the dead tree.

decomposer: an organism that breaks down the wastes and remains of other organisms.

detritivores: organisms that feed on dead plants and animals.

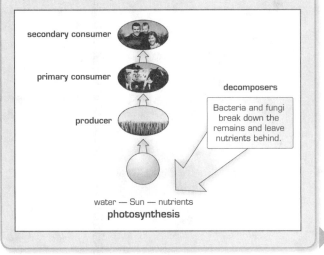

secondary consumer

primary consumer

decomposers

producer

Bacteria and fungi break down the remains and leave nutrients behind.

water — Sun — nutrients
photosynthesis

LIVING TOGETHER

The Missing Link in the Food Chain: Decomposers
5 min.

Students learn the role and importance of decomposers in food chains.

META NOTES

Every bit of organic matter on Earth is processed by some type of living organism. Detritus is made up of organic non-living leftovers, found in all kinds of ecosystems. It is the source of food and energy for organisms living in food chains deep in fresh and saltwater habitats where sunlight cannot reach. It is also composed of the leaves and other organic debris found in streams and on forest floors. Decomposers are nature's recyclers. They obtain food and energy and release wastes that can be used again and again in other food chains.

○ Engage

Have students read *The Missing Link in the Food Chain: Decomposers* section about the role of decomposers in the transfer of matter and energy.

Food Chains Connect to Form Food Webs

5 min.

Students read about food webs and are prepared for constructing a food web.

food web: a series of interlocking food chains. They show the transfer of energy through the different levels in an ecosystem.

Food Chains Connect to Form Food Webs

You may have noticed that your food chains look different. Up until now, your food chains have always connected one organism to another in a straight line. One organism follows the next in a single line. To connect the food chains in the activity you just did, it was impossible to keep everything in a straight line.

Food chains show how at least three organisms are connected. Suppose one of the organisms in that food chain consumes an organism not listed in the food chain. In that case, the food chain does not tell the entire story. The picture of how organisms interact with one another in an ecosystem is not complete.

Most animals are part of more than one food chain. They eat more than one kind of food to stay alive. Thus, food chains cross each other. These interconnected food chains form a **food web**. Food webs expand the food chain concept from a single line of organisms into a network of interactions.

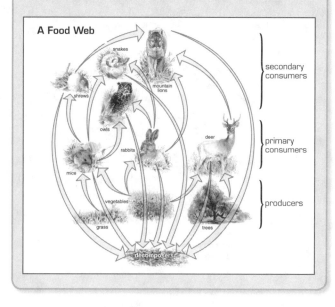

A Food Web

snakes
shrews
mountain lions
owls
deer
rabbits
mice
vegetables
grass
trees
decomposers

secondary consumers

primary consumers

producers

Project-Based Inquiry Science

Food chain illustrations are, in essence, concept maps in which connecting terms have been replaced by arrows. The arrows signify the direction in which matter or energy is moving— from the organism being eaten, to the organism that is eating it. The head of the arrow is always pointed to the organism that benefits from the relationship.

○ Engage

Project an image of the food web shown in the student text. Emphasize that a food web is a way to show how organisms are connected in nature.

◇ Evaluate

Use the projection to have students identify at least two individual food chains. To clarify that they understand all of the parts ask, "Where does each food chain begin?" *(With a producer.)* "What producer seems to be part of three different food chains?" *(The vegetables.)* "What do the arrows signify?" *(Transfer of matter and energy to the next level.)* "Where does each food chain end?" *(With decomposers.)*

Procedure: Complex Connections

1. Your teacher will partner your group with another group.
2. Begin by recording the food chain of each on another set of *Food-Chain Records* pages. Using each group's cards and arrows, build one food web that combines the two small food webs on your table or desk. Once again, you may not use a card more than once from your set of cards — one organism, one card. To make longer connections (arrows), place the sticky notes to reach across your surface.
3. Once your group has settled on a food web, record your group food web on your new *Food-Chain Records* pages in the space indicated. Then obtain two blank sheets of poster paper from your teacher. Each group will use these two sheets of paper to present their food web to the rest of the class.
4. On the first sheet, place your Food Web Cards on the sheet (as you did on the table or desk). Securely tape them to the sheet. Instead of using sticky notes, use a marker to draw connections between organisms. Remember that the arrows should point toward the consumers in the direction of the energy flow.
5. On the second sheet of poster paper, identify
   - organisms that are producers
   - which organisms are primary consumers, which are secondary consumers, and which are tertiary consumers
   - the herbivores, the carnivores, and the omnivores in your chain
   - the organisms your group omitted from the food chain. Be sure to group these cards together and tape them to the second sheet of poster paper.

Be sure to write large enough so others will be able to read your posters during your presentation.

Communicate Your Ideas

Idea Briefing

Each group will have five minutes to share its food webs with the other groups in the class. Your teacher will lead you in a discussion of the various food webs the groups constructed. You will discuss the producers and consumers you have identified and any possible organisms your group omitted from the food web.

While listening to the presentations, be sure to look for differences in food webs. Think about the following questions as you discuss the posters:

Procedure: Complex Connections

15 min.

Pairs of groups organize food webs and prepare posters to present in an Idea Briefing.

△ Guide

Read through the procedure with students to make sure they understand what they will do. Provide each student with a new *Food-Chain Records* page, as they will be creating new, more complex food chain relationships in the form of a food web.

Explain to students that they have about 10 minutes to compile a food web from the food chains they already put together in their previous groups. Remind them that they can use an organism only once. As students begin to assemble their webs, they may need to be reminded or shown that they can use more than one arrow going to or coming from a single organism.

Provide each group with two pieces of poster paper after they have spent about 5 minutes with their cards. Refer students to their text for directions on how to prepare their posters for the *Idea Briefing*.

Near the end of 10 minutes, see how groups are progressing. If most groups are not done in that time, give the class additional time.

Communicate Your Ideas
Idea Briefing
15 min.

Students present their watershed ecosystem food webs. Students discuss the role of each organism and why some organisms could not be included in the food webs.

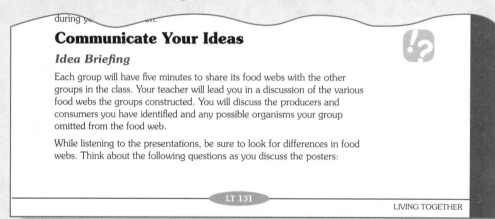

during y...

Communicate Your Ideas

Idea Briefing

Each group will have five minutes to share its food webs with the other groups in the class. Your teacher will lead you in a discussion of the various food webs the groups constructed. You will discuss the producers and consumers you have identified and any possible organisms your group omitted from the food web.

While listening to the presentations, be sure to look for differences in food webs. Think about the following questions as you discuss the posters:

LT 131

LIVING TOGETHER

○ Engage

Collect and display all of the posters in such a way that students can see them as separate ideas.

△ Guide Presentations and Discussions

Tell students to think about criteria and constraints for the *Idea Briefing* by reading the three bulleted questions on pages 131 and 132 before groups present their posters. Remind them to prepare their presentations with reasons for why organisms are shown in the roles they are (*as producers or consumers*).

If time is limited, select two groups to present their posters.

Facilitate the *Idea Briefing*. After each group presents its poster, initiate a discussion based on the three questions. Remind students to provide reasons for their answers.

- Are there items that a group or groups forgot to include?
- How did groups organize the organisms differently?
- What similarities and differences do you see among groups in the list of organisms that could not be incorporated into the food web?

Reflect

Work with your small group to answer the following questions. Be prepared to share your answers with the class.

1. What patterns or trends did you notice across each of the food webs?

2. What similarities exist between all of the food webs?

3. What do you notice about the numbers of producers and tertiary consumers compared to the numbers of primary and secondary consumers? What does this comparison say about the energy required for tertiary consumers to survive?

4. List the possible connections you see between your group's food web and some of the other food webs presented.

5. Do you think there is a way to combine all these food webs into one single food web? Why or why not?

Update the *Project Board*

Recall that the question for this *Learning Set* is *How can changes in water quality affect the living things in an ecosystem?* Discuss with your class your knowledge of the biotic parts of an ecosystem. Record on the *ro ect oar* what you discovered about the relationships among these organisms.

What's the Point?

Food chains are important in understanding the condition of an ecosystem. Often, many different organisms live in any ecosystem. Most are part of more than one food chain. Therefore, a food chain alone does not provide a complete picture. You cannot see all of the relationships among the organisms in a community just by looking at a single food chain.

To see all of the relationships among organisms, you need to look at all the feeding relationships among the organisms. In most ecosystems, they form a complex network. Ecologists call this network a food web. A food web connects all the food chains in an ecosystem. A food web provides you with a big picture that can help you understand how the effect of water quality on one organism can affect an entire ecosystem.

LT 132

ro ect ase nquiry Science

Reflect
10 min.

Students use the food webs that were presented and make inferences about food webs.

◯ Get Going

Have students answer the five *Reflect* questions in their groups. Select two or three questions to discuss as a class if time becomes short. Students answers will vary with the food webs that they have constructed, but possible answers are:

1. The patterns or trends I noticed across the food webs were... that the energy and matter moved from producer to consumer to decomposer; that there were fewer consumers as the food web continued; that everything always ended with decomposers.

2. The similarities that exist are... that each food web started with producers and ended with decomposers; that there are more producers in a food web than there are tertiary consumers.

3. The number of producers is more or greater than the number of primary consumers, and the number of tertiary consumers is less or fewer than the number of secondary consumers. Tertiary consumers need more energy to survive than do producers.

4. Student responses will vary with the food webs presented.

5. Accept all reasonable responses. Yes, probably because all of the organisms used were watershed organisms and all interact with at least one other organism in that ecosystem.

Update the Project Board
10 min.

Students complete the last update of the Project Board *in preparation for answering the Unit challenge.*

> ### Update the *Project Board*
> Recall that the question for this *Learning Set* is *How can changes in water quality affect the living things in an ecosystem?* Discuss with your class your knowledge of the biotic parts of an ecosystem. Record on the *ro ect oar* what you discovered about the relationships among these organisms.
>
>
> **What's the Point?**

Remind students that in the third column, they record what they learned (*claims*) and in the fourth column, they record results (*evidence*) supporting what they have learned.

Sample entries might include:

- **Claim:** Ecosystems can be described by 1) food webs that depict how particular organisms interact, and 2) that energy flows from producers to tertiary level consumers.

  Evidence: Food webs are constructed from multiple, overlapping food chains.

- **Claim:** Changes in the environment can cause a population to severely decrease in number or remove it from an ecosystem.

  Evidence: In the *NetLogo* simulation, some situations such as drought, disease, or pesticides, caused the environment to become unbalanced. This was reflected in the reaction of the coyote or mice populations.

◇ Evaluate

Student participation in adding items to the *Project Board* can be used as an informal check of the student's understanding of the targeted ideas.

Have students work in their groups to begin to formulate an answer to the *Learning Set* question: *How can changes in water quality affect the living things in an ecosystem?* Students' answers should make use of information and observations collected during this *Learning Set.* Inform them that thinking through an answer to this question now will help prepare them to answer the *Big Challenge* question.

More to Learn

Earth's Biomes

In this Unit, you investigated the living and nonliving parts of a river ecosystem. There are many different types of ecosystems on Earth. To better understand the natural world, scientists have classified ecosystems into a small number of different types, called biomes.

Rivers, streams and most lakes are considered freshwater because they do not have salt in them. The freshwater ecosystem of rivers is one example of a biome. A biome is a community of plants and animals best **adapted to** an area's natural environment and climate. In the freshwater environment you have been studying, the plants and animals are well adapted to living there. You saw that in *Learning Set* as you studied the different types of animals and plants that live in this biome.

Another word for a water environment is aquatic. **Aquatic biomes** are biomes that depend on water. Aquatic biomes can be freshwater like the river you studied, or they can depend on salt water. Freshwater and saltwater biomes are the most important biomes on Earth. Water is necessary to all life. All life on land depends on water for survival. Many species live in water for all or part of their lives.

adapted to:
suited for living in a given environment.

aquatic biomes: biomes that depend on water.

LT 133

LIVING TOGETHER

More to Learn: *Earth's Biomes*

10 min.

Students use information to acquaint themselves with large ecosystems called biomes and investigate biomes that make up the United States.

○ Engage

Project a transparency of Earth's biomes or project a variety of biomes to introduce the idea that Earth is made up of a variety of different ecosystems.

△ Guide

Prior to doing the investigation, assign students to read for homework the information about the terrestrial and aquatic biomes of the world in the *More to Learn* section. Students may seek out other references or online resources for information about biomes.

META NOTES

Reference material can be in the form of descriptive information in reputable references or data from investigations that have been published in reliable and recognized sources.

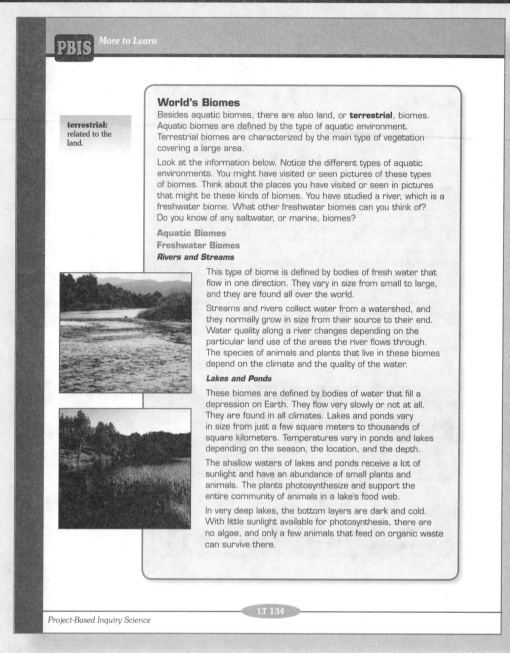

PBIS *More to Learn*

World's Biomes

Besides aquatic biomes, there are also land, or **terrestrial**, biomes. Aquatic biomes are defined by the type of aquatic environment. Terrestrial biomes are characterized by the main type of vegetation covering a large area.

Look at the information below. Notice the different types of aquatic environments. You might have visited or seen pictures of these types of biomes. Think about the places you have visited or seen in pictures that might be these kinds of biomes. You have studied a river, which is a freshwater biome. What other freshwater biomes can you think of? Do you know of any saltwater, or marine, biomes?

terrestrial: related to the land.

Aquatic Biomes
Freshwater Biomes
Rivers and Streams

This type of biome is defined by bodies of fresh water that flow in one direction. They vary in size from small to large, and they are found all over the world.

Streams and rivers collect water from a watershed, and they normally grow in size from their source to their end. Water quality along a river changes depending on the particular land use of the areas the river flows through. The species of animals and plants that live in these biomes depend on the climate and the quality of the water.

Lakes and Ponds

These biomes are defined by bodies of water that fill a depression on Earth. They flow very slowly or not at all. They are found in all climates. Lakes and ponds vary in size from just a few square meters to thousands of square kilometers. Temperatures vary in ponds and lakes depending on the season, the location, and the depth.

The shallow waters of lakes and ponds receive a lot of sunlight and have an abundance of small plants and animals. The plants photosynthesize and support the entire community of animals in a lake's food web.

In very deep lakes, the bottom layers are dark and cold. With little sunlight available for photosynthesis, there are no algae, and only a few animals that feed on organic waste can survive there.

LT 134

Project-Based Inquiry Science

Divide the class into groups and assign each group one or more of the biomes described in the student text. Give students ten minutes to study their biomes in preparation for the investigation to come.

More to Learn

Wetlands

Wetlands are areas that are under water for at least part of the year. They are found in all climates. The vegetation in wetlands can survive flooding. Because the water and vegetation can offer shelter and protection, wetlands support a large diversity of plants and animals. Plants found in wetlands include both grasses and trees. In wetlands close to ocean coasts, the water often contains a lot of salt. Only animals and plants that tolerate high levels of salt can survive in this environment.

Marine Biomes

Shorelines

This type of habitat occurs where oceans and land meet. There are many types of shorelines depending on the geographical features. Sandy shores, rocky shores, and salt marshes are some examples of shorelines. Coastal areas are subject to tides, the periodic rising and lowering of the sea level. The
rising tides push salt water inland, and the water covers parts of the coastline exposed during low tides. Because of tides and the wave motion, animals and plants living along shorelines can tolerate high levels of salt and survive periodic submerging and exposure.

Temperate Oceans

Oceans are the largest habitats on Earth. Oceans cover three-quarters of the surface of our planet and are home to a large variety of animals and plants. Oceans can be divided into three zones, based on the amount of light received.

The sunlit zone, closest to the surface, extends about 100 meters down. A large variety of microscopic floating algae live there and support the entire ocean food web. Many animals and plants also live in this area.

The second layer is the twilight zone. Very little light penetrates here, and therefore there are no plants. This layer is home to animals that can live without a lot of light.

Deeper down is the dark zone. This area is completely dark and very cold. This deep in the ocean, there is a lot of pressure from the water above. Very few organisms can live in these conditions.

Tropical Oceans

Tropical oceans are near the Equator. They receive direct sunlight all year long and are very warm.

Tropical oceans are home to coral reefs. Coral reefs are structures built by a community of several thousand tiny organisms living together.

Terrestrial Biomes

Tundra

At the North and South Poles, the weather is very harsh. The tundra biome is characteristic of the polar climate. Temperatures are very cold year-round, and the soil is permanently frozen. There is very little **precipitation**, and the growing season is very short. There are no trees on the tundra. The vegetation consists of small plants called lichens and mosses. All animals and plants living in the tundra can live in very cold conditions.

Temperate Forests

Many of the trees in temperate forests have large leaves that capture a lot of sunlight for photosynthesis. These trees shed their leaves in the fall, at the end of the growing season, and become **dormant** during winter. They regrow their leaves the following spring to restart photosynthesis. This way, trees adapt to the changing seasons.

Grasslands

Grasslands are big open spaces covered by grasses. There are very few trees and shrubs. Grasslands are found all over the world where average temperatures are mild and precipitation is moderate. They often occur between forests and deserts. The amount of rain precipitation determines the difference between an area being a grassland, a desert, or a forest. With a lot of precipitation, a grassland will become a forest. With less precipitation, it will become a desert.

Grasslands have rich soils suitable for agriculture. Often, large areas of grassland are used to grow crops.

Taiga (coniferous forests)

The land of the taiga is covered by vast forests. Winters are very cold, but the ground is not permanently frozen. Precipitation is also higher than in the tundra. These conditions allow trees to grow. The trees in the taiga are evergreen; they retain their leaves year-round. The leaves are needle-like, and they are protected from cold by a waxy coating. Animals that inhabit the taiga can live in very cold weather.

Rainforests

There are two types of rainforests: tropical and temperate. Both types have lush vegetation and are very wet. Rainforests receive a lot of rain. The difference between tropical and temperate rainforests is in their average annual temperatures: tropical rainforests are warm, and temperate rainforests are cool.

Rainforests have an amazing variety of plants and animal species and are the most diverse habitats on land. Trees are very tall and form a green cover with their crowns, called the canopy. The canopy filters the light from the Sun. The vegetation below is adapted to live in shaded conditions.

Deserts

Deserts are areas that receive little precipitation and experience extreme variations in temperature. Many deserts are very hot during the day and very cold during the night. Despite the extreme conditions of temperature and humidity, deserts are home to many different species of plants and animals.

Plants that live in this environment look very different from other plants. Because of the dry conditions, they have developed special adaptations to collect and store water. Shrubs are the dominant form of vegetation. Desert plants typically have very thick stems to store water and small or spiny leaves. They also have large root systems to collect the infrequent rainwater.

precipitation: water that falls to the ground as rain, snow, hail, or sleet.

dormant: temporarily not active.

LT 137

LIVING TOGETHER

Investigate: United States Biomes

10 min.

Students use information from their reading to determine that the United States is made up of a variety of biomes.

Investigate

United States Biomes

So far you have read about the main types of different world biomes and the climates where they are found. It is now time for you to work with your group to predict the types of biomes that might be found in the United States.

Use what you read about different biomes to prepare a list of the types of biomes you think might be part of the U.S. These questions can help you identify U.S. biomes.

- What types of aquatic biomes do you think there are? List all the aquatic biomes you think can be found in the U.S. Make sure to include both freshwater and marine biomes in your list.

- What types of terrestrial biomes do you think there are? Prepare a list of all terrestrial biomes in the U.S. Identify at least one place in the United States where you think a particular biome might exist and a reason for your thinking.

Communicate Your Ideas

Idea Briefing

On the blank transparency map of the U.S. provided by your teacher, draw the areas where you think each biome might occur. Use a blue marker to color aquatic biomes and a green marker to color terrestrial biomes. Label each biome. Be prepared to share your map with your class. Make sure you have a reason for including each biome and where it might occur.

The maps may not all be the same. As each group presents, look at where they placed their biomes and listen carefully to their reasoning. Discuss the evidence and reasoning each group is using, and try to agree, as a class, on where U.S. biomes are located.

○ Engage

Explain that students will be using the information on biomes to predict what kinds of biomes there are in the U.S. To help with this prediction, draw attention to the science information about adaptations. Explain what an adaptation is and how it can be used as evidence.

"To help you make decisions about the kinds of biomes there are in the United States, it might be helpful to think about the kinds of plants and animals that are found in different parts of the country. Each lives where it lives because it has traits or characteristics for a specific set of conditions- the temperature or amount of water, or the kinds of food that are available.

An *adaptation* is a characteristic that an organism *(whether it is a plant, animal, bacterium, or fungus)* has inherited. An organism's adaptations enable it to survive the particular conditions in which it lives. As you think about the kinds of biomes there are in the United States, think about some of the animals that live in each area. Ask yourself if a polar bear has the traits for living in Florida. Thinking like this will give you some of the evidence you need to explain your choice of certain biomes."

⬡ Get Going

Have students work in their groups to predict the kinds of aquatic and terrestrial biomes they think they will find in the United States. Have them record their predictions.

After writing their predictions, have students prepare lists of aquatic and of terrestrial U.S. biomes. Tell students that they will use this information in an *Idea Briefing* to follow.

Their lists should include:

- Aquatic Biomes in the U.S.
    - Freshwater biomes: rivers and streams, lakes and ponds, wetlands
    - Marine biomes: shorelines, temperate oceans
- Terrestrial Biomes in the U.S.
    - Tundra, temperate forests, grasslands, taiga, rainforest, deserts

Communicate Your Ideas
Idea Briefing

10 min.

Students estimate the location of different kinds of biomes in the United States and present evidence for their choices in a class discussion.

Communicate Your Ideas

Idea Briefing

On the blank transparency map of the U.S. provided by your teacher, draw the areas where you think each biome might occur. Use a blue marker to color aquatic biomes and a green marker to color terrestrial biomes. Label each biome. Be prepared to share your map with your class. Make sure you have a reason for including each biome and where it might occur.

The maps may not all be the same. As each group presents, look at where they placed their biomes and listen carefully to their reasoning. Discuss the evidence and reasoning each group is using, and try to agree, as a class, on where U.S. biomes are located.

○ Engage

Inform students that they will mark a blank map of the United States with the biome choices that they listed previously. Provide each group with a blank map and tell them they have five minutes to mark their maps.

☐ Assess

While groups are working on the map, remind students that each person should have some input. Take note of issues that students might struggle with as they work together. *(Someone may disagree with how the map is drawn, neatly or not so neatly; someone may be concerned that his or her choice is not covered sufficiently.)*

△ Guide Presentations

Remind groups to have their evidence ready when they go to present their map.

Ask for one group to volunteer to present their map first. Remind other students to listen quietly and respond politely if they need to have the reason or evidence for the placement of a particular biome explained.

NOTES

....

....

....

....

....

....

Examine the map of U.S. biomes that your teacher will make available. How close were your predictions to the biomes scientists have identified? Was your class more accurate at identifying terrestrial biomes or aquatic biomes? Which were easier to identify? Discuss reasons for your answers.

adaptation: a special trait that allows an animal to survive in its environment.

natural selection: the competition for survival where organisms with more desirable traits survive and reproduce more than organisms with less desirable traits. Over time, desirable traits build up in a species and unfavorable ones disappear.

evolution: the process of change in a species over time.

What are Adaptations?

Adaptations are special traits that help an organism survive in a given environment. They are characteristics or behaviors that are inherited by a plant or an animal. They are passed on from parent to offspring. Adaptations may be physical traits. Giraffes have long necks. Their long necks let them reach leaves of trees that are too high for other grazing animals to reach. Cacti have spiny "leaves." The spines help cacti conserve water in hot and dry climates. Water from plants with large leaves can escape through the leaves.

Adaptations may also be behaviors. Some animals, such as birds, migrate south when temperatures get cold. Other animals, such as bears, escape cold temperatures by hibernating.

In the competition for survival, organisms that have favorable adaptations have a greater chance of living longer and reproducing. Over many years, these desirable traits build up in a species. Unfavorable ones disappear. This process is called **natural selection**. The result is **evolution**. Evolution can be a long, slow process.

As each group finishes its presentation, post the map where everyone can see it so that it might contribute to the discussion when the next group does their presentation.

◇ Evaluate

Once all maps have been presented, distribute or project a map of biomes in the United States for students to check their predictions against.

Assessment Options

Targeted Concepts, Skills, and Nature of Science	How do I know if students got it?
All living things need energy from the Sun and matter to survive. Energy gets trapped by producers and is transferred along with matter to consumers through food chains.	**ASK:** Explain how each kind of organism gets its energy in a food chain that consists of plants, rabbits, and coyotes. **LISTEN:** Students should respond that plants get energy from the Sun through photosynthesis; the rabbit gets its energy by feeding on the red clover; coyotes get their energy by eating or preying on rabbits.
Food webs show a complex network of interlocking food chains and the direction of energy flow through an ecosystem.	**ASK:** Explain why, in a food web, several arrows might point to a single organism? What do the arrows represent and why would there be more than one arrow pointing toward a single organism such as a mountain lion? **LISTEN:** Students' responses should include that the arrows represent the flow of energy and matter from one organism to another. There can be more than one arrow pointing to a single organism because the one organism might get its energy from several different organisms.
Earth is divided into large ecosystems called *biomes*, which vary by climate and in which specific organisms survive because they are adapted to those conditions.	**ASK:** Select a biome and using information from references, describe how adaptations of its plants and animals reflect the climate in that biome. **LISTEN:** Students' responses should include the temperature and rainfall in the biome, representative plants and animals that live there and the ways they are adapted to survive in that climate.

Teacher Reflection Questions

- What difficulties did students have in constructing a food web once two groups joined their ideas?

- What more can be done to make certain that students understand that matter and energy are being transferred through a food web?

- What improvements can be made to the *Project Board* for this section to prepare students to use it in answering the *Big Challenge?*

NOTES

ANSWER THE BIG QUESTION INTRODUCTION

Answer the Big Question

Address the Big Challenge

1 class period * ▶

*A class period is considered to be one 40 to 50 minute class.

Overview

The class refocuses on the question: *How does water quality affect the ecology of a community?* They read again about the challenge faced by Wamego as the town determines how to save itself economically without destroying the quality of its land and water. The students work in groups, and assume the role of four individuals, all of whom have interests in Wamego. In a *Solution Showcase*, these representatives present recommendations to the town council for and against allowing FabCo, a fictional company, to set up a new manufacturing facility along the river that borders the town. The recommendations of individuals against allowing FabCo to set up along the river are based upon evidence derived from investigations into how various factors affect water and land quality. Each group makes its presentation using a poster, software program, or a skit.

Targeted Concepts, Skills, and Nature of Science	Performance Expectations
Water is an essential substance in an ecosystem. By following how water flows in and over the land, and how it affects and is affected by the organisms and substances in its path, ecologists can determine how water quality will change.	Students should be able to present evidence to the Wamego town council that would explain what will or might happen to the town's water and land resources if a new manufacturing facility is built along the river.

Materials	
1 per class	Pair of maps of the "current" and "proposed" changes to the town
1 per class	Class *Project Board*
1 per group	Computer with presentation software program
2 per group	*Create Your Explanation* pages

Activity Setup and Preparations

Arrange the room into four working areas, one for each representative group. Make certain that a computer is available for software program presentations if necessary.

Homework Options

Reflection

- **Nature of Science:** Summarize the stand each of the four groups represents for the town. *(William Waters wants what is best for the ecology of the town. He understands the conditions that trout require to stay alive and also understands how water quality can change when the land changes. He knows from investigative evidence that the water quality will change when the land is disturbed with new building. Sara Song represents FabCo and thinks she has a plan to accomplish new development without disturbing the ecology of the area. Ramone Ramirez is a farmer who practices careful management of his land and understands that water quality can change if the land is disturbed by building new homes as well as a new factory. Asha Adu lives in another town downstream from Wamego. She understands that if the Wamego infrastructure fails, then water flowing in her direction will also affect her town.)*

- **Science Content:** What are the five criteria each group discussed? *(Water-quality issues, sources of pollution that FabCo could cause, the effect of changing water quality on organisms in the local food web, ways to reduce harmful effects, and water-quality tests that should be recommended.)*

NOTES

....

....

....

....

◀ *1 class period**

Learning Set 3

Answer the Big Question: Address the Big Challenge

Recall the Big Challenge

5 min.

Students read again the story of Wamego and the problems it faces ecologically and economically, and prepare to represent different interest groups in Wamego.

Answer the Big Question

Address the Big Challenge

You began this Unit by reading about a small town that needed your help. The town was faced with an important decision. You were asked to help them understand the possible results of different decisions they might make. As you worked through the Unit, you kept in mind the *Big Question How does water quality affect the ecology of a community?* and the advice you need to give the community. You've recorded claims and recommendations in the last column of the *Project Board*.

You are now ready to complete this Unit. Read about the town's situation one more time. Over the next few days, you will answer the *Big Question* as you address the *Big Challenge* the town is facing.

Recall the Big Challenge

Wamego Needs Help!

Wamego (hwah-MEE-goh) is a small town with a population of about 1800. It is on the banks of the Crystal River. This town has always been a farming community. Most of the farmers grow corn and soybeans. These are the best crops to grow in this area. Nearly 95% of the residents are employed by Wamego's farming businesses. The local economy depends on farming. The other businesses in town all depend on the farmers and their employees (workers). These businesses include a grocery store, gas stations, a movie theater, and several restaurants.

The Crystal River is also important to Wamego. The river is a source of water for the crops. The river is also known as a good trout-fishing river. Trout need

LT 140

○ **Engage**

In preparation for presenting solutions to the problems faced by the town of Wamego, have students read again about the details of the town, its history, its interests, and its current economic concerns. Initially, have students read through to page 143, and stop just before *Plan Your Answer*.

---

*A class period is considered to be one 40 to 50 minute class.

very clean, cold water to thrive. Crystal River suits their needs. Every summer, Wamego has a Trout Festival. Many people who enjoy fishing travel to the area. The festival celebrates trout fishing and preservation. The festival also educates people about what trout need to thrive. The goal of the education effort is to keep the number of trout at a healthy level. In that way, people can enjoy fishing there for many years to come. This festival is fun for many residents and tourists. It is also another income source for the residents of Wamego.

Lately, the farming business has been poor. Crop prices have dropped. The farmers are not making very much money. There is not enough to pay their workers or to support themselves. Some of the farmers have gone bankrupt. As a result, Wamego has lost 15% of its population during the last five years.

The town council is very concerned. They know farming will always be a part of life in Wamego. But they worry about the town losing too many people. They do not want to get so small that there will be very few businesses and residents in Wamego.

FabCo Wants to Move In

A mid-sized manufacturing company called FabCo has contacted the town council. FabCo manufactures cloth. The cloth is sold to companies that make clothes. FabCo is looking for a new location to build their company headquarters and manufacturing plant. FabCo is very interested in relocating to Wamego for several reasons.

LT 141

LIVING TOGETHER

It might be helpful if, as they read, students summarize each aspect of the problem and record these summaries where everyone can see the same version.

- Wamego has a fairly large river and a train line running through town. This, along with roads, would provide transportation routes for their products.

- The cost of living in the town is low. Their employees would like that.

- The river provides a natural resource (water). Water is important to the production of their cloth.

If FabCo is allowed to move to Wamego, the town could benefit as well. It would mean the following benefits:

- About 15,000 new residents would relocate to Wamego. This would require the building of many new homes, roads, and parks. A new school would need to be built. New businesses offering services to the company and the new residents would be needed. This means more buildings, parking lots, and roads would appear in Wamego.

- FabCo would offer many new jobs to Wamego's residents.

- The town would have money from taxes collected from FabCo and the new residents. This extra money could be used to improve life in Wamego in many ways, including a new hospital.

- The town would not have to depend on farming alone.

Sounds Great! So, What's the Problem?

Many of the residents, including some town council members, are concerned. They worry that FabCo's presence could mean problems for their community. Currently, the land is used for agriculture. If FabCo comes to town, the use of the land will change. The land will be needed for residential, commercial, and industrial purposes. Some people, including the organizers of the Trout Festival, wonder if this will change the river and the wildlife of Wamego.

Wamego residents are not the only ones concerned. Ten miles downstream is the town of St. George. It is also located along the Crystal River. St. George is an even smaller town than Wamego. It is a resort town. People travel from all over to vacation in St. George, using the river for recreation. There is fishing, swimming, boating, hiking, and camping in the area. There are several

LT 142

hotels and bed & breakfasts that provide accommodations for tourists. The Crystal River's water quality is very important to St. George's economy and residents. The residents of St. George are worried that the changes in Wamego might affect their lives.

You have investigated the effects of changing land uses on water quality and the effect this can have on living things. You have learned how interconnected the different parts of an ecosystem are. To end this Unit, your group will create a presentation to answer the *Big Question* for the town of Wamego.

Plan Your Answer

The town council knows that FabCo can bring money into the area. However, the council wants to know what to expect or what might happen if changes take place. The council is asking four people for their ideas.

- William Waters — organizer of the annual Wamego Trout Festival, fisher
- Sara Song — FabCo executive, native of Wamego
- Ramone Ramirez — farmer
- Asha Adu — resort owner, town of St. George

Your teacher will assign you to a new group. Each member of your group will be from one of the land-use groups you worked in during the Unit. Your group will represent one of these four people. You will prepare a presentation for the

△ Guide

After students have finished summarizing the *Big Challenge* about Wamego, have them read to page 146, stopping just before *Communicate Your Ideas*. In this reading, students will be introduced to four individuals (*William Waters, Sara Song, Ramone Ramirez, and Ashu Adu*), each of whom has a significant interest in the future of Wamego. Explain how the class will be used to represent each of these people.

Plan Your Answer
20 min.

Students are introduced to and take on the roles of four individuals interested in a secure future for Wamego.

"In this reading, you have been introduced to four people. It is obvious that each of these people have a deep interest in Wamego, but each for a different reason. These four will each talk to the Wamego town council and present their concerns to the council. They will talk about the problems and the benefits of having a new business move to town.

Each of you will get to know the viewpoints of one of these people very thoroughly. The class will be divided into four groups and each group will play the role one of these individuals. The members of each group will work together, using the results of their investigations and information from the *Project Board* to represent the person's point of view.

When you give your presentation to the town council, it will be in the form of a *Solution Showcase*. You may demonstrate your points using a poster, a PowerPoint presentation, or as a skit."

Pull out the *Project Board* where everyone can have access to it.

Call attention to the proposed *before* and *after* maps on page 144. It might be helpful to make enlargements of these maps so that each presenter has access to them during the *Solution Showcase*.

Suggest that students assemble all the evidence and notes they may have from the Unit activities and discussions.

⬡ Get Going

Once students have the *Big Challenge,* divide the class into four groups, making sure that each group has at least one person from each of the different land use investigations (*residential, commercial, industrial, and agricultural*). Suggest that students review the four types of land use. Assign one of the four people to each group. Give students about 5 minutes to become acquainted with the individual they represent.

△ Guide and Assess

Circulate around the room as students settle in to learning about their individual. Listen carefully for signs of confusion. If you become aware of some general confusion, stop the class and guide students. Students should be aware that they are taking on the role of a specific viewpoint, but that each member of each group brings some expertise in that they each represent a different land use.

Listen also for how clearly each group understands the person it represents. If a group does not seem to understand their person's point of view, stop and ask questions to get students focused on their person and his or her side of the story. You might ask, "What does the person do for a living?" "What does he or she understand about water quality?" "How does the evidence you have found in the investigations support this person's side of the story?"

PBIS

town council. Your presentation will focus on how FabCo's arrival could possibly change water quality from the point of view of the person you represent. You will need to think about why water quality is important to your assigned person. You will have to apply what you know about water quality and ecology to represent one of these individuals in your presentation.

Below is more information about how Wamego would change if FabCo arrived. On the next few pages is information on each of the individuals you are representing and their concerns. Use this information as you put together your argument.

Below: a current map of the area

Bottom: map showing proposed changes

• The 15,000 new residents would need homes and apartment buildings to be built.

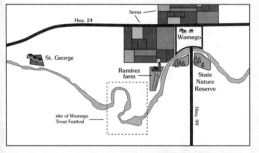

• New businesses and commercial areas would need to be developed to meet the needs of the population (restaurants, dry cleaners, day care centers, grocery stores, hospital, recreational facilities).

• New roads would be needed for the increased traffic to homes and businesses.

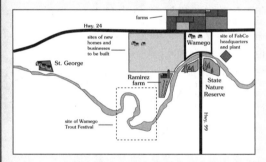

• All new storm drains for the new neighborhoods, businesses, parking lots, and streets would eventually drain into the Crystal River.

• Textile (cloth) manufacturing can often produce runoff (water) that can be acidic or basic and high in temperature.

Remind students that it would be easy to become very emotional when presenting reasons why they want or do not want the new factory, but that when they present to the town council, they need to present their positions with claims and with evidence. To accomplish this, they need to make use of data recorded throughout the Unit, the *Project Board*, and maps. If necessary, explain bias and tell students to avoid biased statements during their presentations.

- The size of Wamego's water-treatment plant could not handle all the new residents' needs. It would require updating. Until the upgrade is complete, the sewer system could have periodic overflows with raw sewage spilling into storm drains.

William Waters — organizer of the annual Wamego Trout Festival, fisher

William has lived in Wamego his entire life. His family has been fishing the Crystal River for over 100 years. He works hard to protect the health of the Crystal River and the trout population in the river.

The Wamego Trout Festival location is downstream from the new plant location. William is worried about what might get into the stream to affect the trout population. He knows that trout are very sensitive to changes in temperature, acidity, and dissolved oxygen in their water. Also, William is very concerned about all of the development exposing a lot of soil. He wonders what effect this condition would have on the quality of the stream and the plants in the stream.

Sara Song — FabCo executive, native of Wamego

Sara works for FabCo. Her current role for the company is to help relocate their headquarters and plant. Sara was born and raised in Wamego. She has not lived in Wamego for fifteen years, but she has been back to visit family often. Wamego has a very special place in Sara's heart. It was Sara's idea to move FabCo to Wamego. She was familiar with the town, and she knew it would be a good match for the company. Sara also thinks that Wamego will benefit from this change without having to sacrifice its farming culture or its environment. Sara wants to deal with problems that could occur from

- the waste the plant's manufacturing process produces, and
- erosion at the site of the plant or near the new homes and businesses.

She wants the town council to know she is aware of these issues, how they might be caused, and what problems they could cause. She will present this information along with some suggestions of how to prevent problems. Her goal is to ensure that both the town and the company benefit from this partnership.

Check to see if students need poster paper, a computer for a software program presentation, or other small props for role playing.

Tell students how much time they have left for planning their presentation.

Ramone Ramirez — farmer

Ramone owns and operates one of the largest soybean farms in the state. His farm is one of the few financially successful farms in the area. Ramone is very environmentally aware. He has been careful to ensure that his farm does not harm the Crystal River. Ramone has also successfully avoided the use of pesticides to keep herbivore insects from consuming his crops. Ramone uses predator insects to control the herbivore insect population.

Ramone irrigates (waters) his crops with water drawn from the Crystal River. The proposed site for new home and business development is next to Ramone's farm. The storm drains from these developments would enter the river upstream from his farm. He is very concerned about how changes in the condition of the water will affect his crops. Also, he wonders what changing the landscape will do to the predator insects he relies upon so much.

Asha Adu — resort owner, the town of St. George

Asha runs a Bed & Breakfast in St. George, the town ten miles downstream from Wamego. Asha was born in Africa, but when she was still a baby, her family moved to St. George. Her parents opened the Bed & Breakfast when she was very young. Now she runs the business. Asha has spent almost her whole life growing up and living on the banks of the Crystal River. Asha is very concerned that the Crystal River will become polluted. She does not want to see the beauty of the place she has known and loved her whole life destroyed. She also does not want to have her tourism business affected by these changes.

Asha's brother works for the town of Wamego. He makes repairs and maintains the sewer system. Asha is aware of the condition and limits of the sewer system. She wonders what effect this could have on the water. Asha's tourists come to St. George because the river is so clear and clean. Asha wonders what changes could happen to the clarity and cleanliness of the river.

Communicate Your Ideas

Solution Showcase

The goal for a *Solution Showcase* is to have everyone better understand how a particular group approached their question or challenge. In this case, you get the opportunity to see the variety of ideas from different points of view. You will also have the chance to see the common ecological concepts important to everyone in the Wamego area. The *Solution Showcase* provides an opportunity for groups to share what their knowledge and how they have applied it to Wamego's problem.

LT 146

Project-Based Inquiry Science

Communicate Your Ideas

15 min.

Students prepare their presentations in final form and appear before the Wamego town council.

△ Guide Presentations and Discussions

Introduce students to the idea of a *Solution Showcase*, explaining that a showcase should include a title, the history of the project under study, and the final recommendations the group has concluded should be made.

Emphasize that each group should always provide reasons for their conclusions and the evidence that backs up their reasons. Inform students that to be convincing to a group such as a town council, their presentations should all meet certain criteria and constraints. Draw students' attention to the five items for discussion listed on page 147. These items can be used as criteria.

Explain that while one group is presenting, the remaining three groups will listen as if they are members of the town council. Audience members can make certain that each criterion is discussed. If they do not hear answers to these items, then they should ask questions of the presenters.

Each group will present the argument of the person it represents and explain that person's ideas. Each group will discuss and formulate an argument based on its point of view and will present that argument in a *Solution Showcase*.

Each presentation should include several items to effectively communicate its individual's concerns and ideas. Each member of your group will be from a different one of the four land-use groups you have worked in during the Unit. This will allow you each to serve as an expert for each of the land uses. Each group member will have to remember and apply the information from this Unit.

Each presentation should focus on the following items:

- The water-quality issues your individual would be concerned about.
- Sources of pollution your individual thinks FabCo would cause.
- The problems your individual is concerned about. Consider the effect of these concerns on the organisms and food web for the area. Be very specific. The organisms living in this area are the same ones you examined in your food webs. Be sure to fully explain how the ecology problems you highlight would affect this food web.
- Ideas your individual could suggest to reduce the harmful effects they are concerned about.
- The water-quality tests your individual would recommend if the changes took place. Be sure to describe the reason your individual thinks each test is needed.

You may want to use *Create Your Explanation* pages to help you think through and justify your claims and recommendations. You will probably need to refer to data from earlier investigations. The explanations and recommendations you and others made during the Unit might help you. Be sure to refer to the *Project Board* as a resource.

○ Get Going

Let students know how much time they have for their *Solution Showcases*. Then, have each group present their *Solution Showcase*.

◇ Evaluate

As each group presents, listen for how clearly students represent the character they have been assigned. Listen for answers to the criteria statements, and watch for how thoroughly students make use of evidence that they developed during the Unit.

After all the presentations have been made, have the class volunteer what they thought Wamego should do at the beginning of the Unit and what they now think the town should do. Will Wamego let FabCo build their plant?

PBIS

Your group will select one of three formats for your presentation.

Poster	You can construct a poster or posters. You will share the poster and its elements with the class. Your poster is not the presentation. Your poster is only a prop for you to refer to during your presentation. Your presentation should involve all members of your group.
Presentation software	You can construct a presentation software presentation. Slides would have graphics and text that accompany your presentation. Avoid simply reading the slide to the class. Present your ideas using the slides as an aid. Your presentation should involve all members of your group.
Skit	You can create a skit. Your skit must have a written script that contains facts and concerns similar to what other groups will provide in their posters or presentation software presentations. Your skit should involve all members of your group.

Make sure to present the reasons you made the decisions you did. Your teacher will tell you how long you have to present. You will need to present your ideas quickly and clearly.

Assessment Options

Targeted Concepts, Skills, and Nature of Science	How do I know if students got it?
Water is an essential substance in an ecosystem. By following how water flows in and over the land, and how it affects and is affected by the organisms and substances in its path, ecologists can determine how water quality will change.	**ASK:** How might a town's water quality be affected if the land use is changed by developing a factory and building new homes and roads? **LISTEN:** Students should be able to present evidence from investigations that would explain what will or might happen to a town's water and land resources if a new manufacturing facility is built along the river.

Teacher Reflection Questions

- Which type of presentation made the best use of data from the investigations during the course of the Unit?
- How easy was it for students to avoid bias in their presentations?
- How could you evaluate the amount of time realistically needed for a *Solution Showcase*?

NOTES

Living Together Blackline Masters

* Number indicates Learning Set.section.sequence within section

Name: _____ **Date:** _____

Question

What question are you investigating and answering with this experiment?

Prediction

What do you think the answer is and why do you think that?

Variable Identification

- Which variable will you be changing in your experiment?
- What conditions and procedures will you control in your experiment?
- What will you measure as evidence of the variable's effect on growth?

Procedure and Data

Write detailed instructions for how to conduct the experiment. You need to include the following:

- how you set up the duckweed samples,
- how you measure changes,
- how you record data.

Name: _____ **Date:** _____

Use the following pages to record the plant growth in the four jars over a period of 6-10 days.

Observations may include the number of fronds, color of fronds, size of fronds, number of living plants, number of dead or dying plants, length of roots, or anything else your class decides is important.

Date: _____

Observations

Applying Indicators to My Land Use

Name: ______________________ **Date:** ______________

What water-quality indicator do you want to test?

What feature or aspect in the photos suggests you should test this?

What do you predict would be the outcome of the test? Provide a specific number or value you expect to see from the test.

What facts or useful information do you have from your research and investigations that suggest this is a good test and an important one for good water quality? Provide at least two.

What process or condition that occurs in the land use could cause possible problems with this indicator?

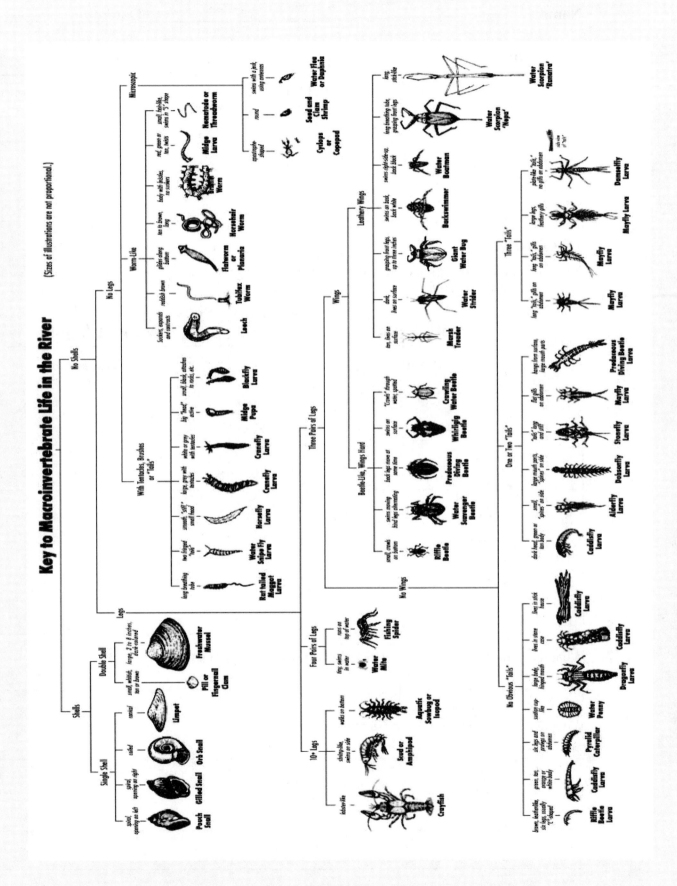

Name: _____ Date_____

Prediction

Describe what you think will be the effect of keep elodea out of the light. Also, what difference do you think you will see between the two plants at the end of the investigation?

Observations

Record details of how the plants look and what you see in the bags. Draw a sketch of the elodea to help you describe what you see.

With Light	Without Light
Before	Before
After	After

What did you notice in the bag of elodea that was exposed to light?

_____

What did you observe in the elodea that was kept in the dark?

_____

What do you think caused the changes you observed?

_____

_____

Name: _____ **Date:** _____

1. Draw a diagram of the food chain that links the grass, mice, and coyotes in the space below.

[ ]

2. List the factors that you think affect each population.

	Factors that might increase the population	Factors that might decrease the population
grass patches		
mice		
coyotes		

3. Record the data you collect when you run the model with Baseline Conditions. The first row of data has been entered for you.

Baseline Conditions			
Generations	**Number of grass patches**	**Number of mice**	**Number of coyotes**
0 (start)	1412	205	70

4. Record the name of your scenario. Write down your prediction about what will happen to the number of grass patches, mice, and coyote. Then record the data you collect when you run the model.

Baseline Conditions			
Generations	**Number of grass patches**	**Number of mice**	**Number of coyotes**
Prediction			
0 (start)			

Name: _____ **Date:** _____

Record your first food chain here.

| |
| |
| |

List any of the following items that appear in your food chain:

Producers	
Primary Consumers	
Secondary Consumers	
Tertiary Consumers	
Predators	
Prey	

Record the food chain built by the other half of your group here.

| |
| |
| |

List any of the following items that appear in their food chain:

Producers	
Primary Consumers	
Secondary Consumers	
Tertiary Consumers	
Predators	
Prey	